Magnetic Materials Based Biosensors

Magnetic Materials Based Biosensors

Special Issue Editor

Galina V. Kurlyandskaya

MDPI • Basel • Beijing • Wuhan • Barcelona • Belgrade

Special Issue Editor
Galina V. Kurlyandskaya
Universidad del Pais Vasco—Euskal Herriko Unibertsitatea
Spain

Editorial Office
MDPI
St. Alban-Anlage 66
Basel, Switzerland

This is a reprint of articles from the Special Issue published online in the open access journal *Sensors* (ISSN 1424-8220) from 2017 to 2018 (available at: http://www.mdpi.com/journal/sensors/special_issues/MagneticMaterials)

For citation purposes, cite each article independently as indicated on the article page online and as indicated below:

LastName, A.A.; LastName, B.B.; LastName, C.C. Article Title. *Journal Name* **Year**, *Article Number, Page Range.*

ISBN 978-3-03897-254-9 (Pbk)
ISBN 978-3-03897-255-6 (PDF)

Cover image courtesy of Galina V. Kurlyandskaya (in collaboration with SGIKER services of UPV-EHU).

Contents

About the Special Issue Editor

Galina V. Kurlyandskaya Prof. Dr. graduated from Ural State University A.M.Gorky, Ekaterinburg, Russia in 1983. She started her research work in 1983 at the Institute of Metal Physics. She obtained her PhD in physics of magnetic phenomena in 1990 and Doctor of Science degree in 2007 from Ural State Univesrity A.M. Gorky. Prof. Kurlyandskaya received an advanced training at the Institute of Applied Magnetism University of Complutense, University of Oviedo, University of the Basque Country-Euskal Herriko Unibertsitatea, University of Dusseldorf Heinrich Heine, ENS Cashan, University of Maryland, Ural State University A.M. Gorky, Ural Federal University B.N. Yeltsin (Laboratory of Magnetic Sensors), Brazilian Research Center of Physics (CBPF) and University of Santa Maria. Her main research areas are fabrication, magnetic and transport properties of nanostructured magnetic materials, magnetic domain structure, magnetoabsorption, magnetic sensors and magnetic biosensors.

Preface to "Magnetic Materials Based Biosensors"

This book contains peer-reviewed contributions from Special Issue "Magnetic Materials Based Biosensors" of Sensors MDPI submitted to journal in time period from August 2017 till July 2018. Book contains 13 research works representing international multidisciplinary teams from Austria, China, Germany, Greece, Iran, Russia, Serbia, Spain, Taiwan and United States of America. It can be useful for PHD students and researches working in the field of magnetic nanomaterials and biomedical applications.

Selective and quantitative detection of biocomponents is widely requested in biomedical applications, clinical diagnostics, environmental monitoring, toxicology, regenerative medicine, drug delivery, pharmacology and other fields. Creation of a new generation of biosensors with such parameters as very high sensitivity, low power consumption, stability of operation parameters,quick response, resistance to aggressive medium, small size, low price has been the subject of special attention for last two decades. Therefore, multidisciplinary area of magnetic biosensing have been extensively developed aiming to create compact analytical devices for non-expensive and low time consuming analysis provided at the point of care by non-skilled personnel. Biological samples exhibit very low magnetic background, and thus highly sensitive measurements of magnetic labels or magnetic nanoparticles enriched units can be performed without further processing. A magnetic biosensor is a compact analytical device in which magnetic transducer converts a magnetic field variation into a change of frequency, current, voltage, etc. Different types of magnetic effects are capable of creating magnetic biosensors with extra high sensitivity: inductive effect, magnetoresistance (anisotropic magnetoresistance, giant magnetoresistance, spin-valves), Hall effect, ferromagnetic resonance, giant magnetoimpedance. Magnetic biosensors use either label free or magnetic label detection principle. For magnetic label detection, the magnetic field sensor estimates the integral disturbance of the external magnetic field using the stray fields of superparamagnetic particles present in the test solution. Electrochemical impedance spectroscopy, magnetoelastic resonance of magnetoimpedance spectroscopy can be employed for label free detection.

Although, existing devices allow a quantified evaluation of small changes in the magnetic susceptibility, in the living organisms, or in magnetic field values created by the extracellular electric currents, there is a need to enhance both sensitivity and specificity and improve their design up to miniaturized analytical systems. Development of magnetic nanomaterials with properties of mimicking properties of the natural tissue is very important step. This book describes interesting examples of magnetic materials based biosensors, including the synthesis of model materials for biosensor development, new engineering solutions and theoretical contributions on the magnetic biosensor sensitivity. I would like to thank Authors, Reviewers and Editorial members of MDPI for special efforts to insure the high standard of this book.

Galina V. Kurlyandskaya

Special Issue Editor

Review

Galfenol Thin Films and Nanowires

Bethanie J. H. Stadler [1,2,*], Madhukar Reddy [2], Rajneeta Basantkumar [1], Patrick McGary [1], Eliot Estrine [1], Xiaobo Huang [1], Sang Yeob Sung [1], Liwen Tan [2], Jia Zou [1], Mazin Maqableh [1], Daniel Shore [2], Thomas Gage [2], Joseph Um [1], Matthew Hein [1] and Anirudh Sharma [1]

[1] Electrical and Computer Engineering, University of Minnesota, Minneapolis, MN 55455, USA; Rajneeta.Basantkumar@hotmail.com (R.B.); pmcgary@bju.edu (P.M.); eliotestrine@gmail.com (E.E.); Xiaobo.Huang@wdc.com (X.H.); neoesp@gmail.com (S.Y.S.); jiazou11@gmail.com (J.Z.); mazenakos@gmail.com (M.M.); umxxx023@umn.edu (J.U.); Heinx055@umn.edu (M.H.); shar0340@umn.edu (A.S.)
[2] Chemical Engineering and Materials Science, University of Minnesota, Minneapolis, MN 55455, USA; skmadhukar@gmail.com (M.R.); liwen.tan@seagate.com (L.T.); shore033@umn.edu (D.S.); tom.e.gage@gmail.com (T.G.)
* Correspondence: stadler@umn.edu; Tel.: +1-612-626-1628

Received: 4 July 2018; Accepted: 26 July 2018; Published: 12 August 2018

Abstract: Galfenol ($Fe_{1-x}Ga_x$, $10 < x < 40$) may be the only smart material that can be made by electrochemical deposition which enables thick film and nanowire structures. This article reviews the deposition, characterization, and applications of Galfenol thin films and nanowires. Galfenol films have been made by sputter deposition as well as by electrochemical deposition, which can be difficult due to the insolubility of gallium. However, a stable process has been developed, using citrate complexing, a rotating disk electrode, Cu seed layers, and pulsed deposition. Galfenol thin films and nanowires have been characterized for crystal structures and magnetostriction both by our group and by collaborators. Films and nanowires have been shown to be largely polycrystalline, with magnetostrictions that are on the same order of magnitude as textured bulk Galfenol. Electrodeposited Galfenol films were made with epitaxial texture on GaAs. Galfenol nanowires have been made by electrodeposition into anodic aluminum oxide templates using similar parameters defined for films. Segmented nanowires of Galfenol/Cu have been made to provide engineered magnetic properties. Applications of Galfenol and other magnetic nanowires include microfluidic sensors, magnetic separation, cellular radio-frequency identification (RFID) tags, magnetic resonance imaging (MRI) contrast, and hyperthermia.

Keywords: Galfenol; magnetic nanowires; electrochemical deposition

---

1. Introduction

In this paper, we will review 15 years of magnetic nanowire work at the University of Minnesota. In the early days, these magnetic nanowires were proposed as hair-like sensors [1], similar to the cilia found in many biological species. The material of choice was Galfenol ($Fe_{1-x}Ga_x$, $10 < x < 40$) [2] which was a relatively new magnetostrictive alloy, rivaling Terfenol-D (Tb-Fe-Dy) with magnetostriction (430 ppm) of the same order of magnitude, yet with much more useful (ductile) mechanical properties. Where Terfenol-D is difficult to machine even coarsely (e.g., for lamination to avoid eddy currents), Galfenol can be finely machined (e.g., with small threads for suspension mounting) [3]. In fact, Galfenol *can* be used in suspension (tension), especially when stress annealing is used to produce internal compressive stresses, but brittle Terfenol-D requires the application of a compressive strain for most applications, such as sonar transducers [4,5]. This simplification of the transducer design is significant, through which entirely new designs become possible. The ductility of Galfenol also played an important role in nanowire sensors. Although Terfenol in bulk or film forms could have higher saturation magnetostrictions (up to 1600 ppm, depending on the processing conditions), it is difficult to image a process by which nanowires of Terfenol could be made without breaking. Galfenol, however, can be made into wire-like unimorphs by rolling followed by cutting [6], and even into nanowires by electrodeposition into nanoporous templates. Unlike Terfenol and other rare earth-containing alloys, a fairly wide processing window has been defined to enable the electrodeposition of Galfenol.

The first part of this review will discuss the electrochemical processing parameters and the resulting properties of Galfenol films. Next, Galfenol nanowire synthesis and properties will be discussed. Finally, several biomedical applications of nanowires will be introduced.

2. Galfenol Films and Properties

Galfenol was first studied in bulk single-crystal and polycrystalline forms at NWSC, Carderock Division in Maryland and Ames Lab in Iowa (Figure 1) [2]. The magnetostriction is largest in the (100) direction, and it has a bimodal dependence on Ga composition with 400 ppm at 19% Ga and 430 ppm at 25% Ga, where both require optimized quenching [7]. To produce thin films of Galfenol, sputtering was attempted by Basantkumar at the University of Minnesota [8] using sputtering targets provided by Etrema, Inc. in Iowa. The magnetostriction of these films, measured using a capacitance bridge technique, also appeared to have a bimodal dependence on Ga composition, although shifted from bulk values due to the state of strain in the films with a maximum value nearing 150 ppm, Figure 1 [7,8]. This means that the ideal composition shifted from 19% to 15% Ga for the lower peak, but 25% Ga also remains a good composition even with strain. Although not as large as bulk magnetostriction values, 150 ppm is sufficient for new microelectromechanical system (MEMS) cantilever sensors made possible by the formation of the film.

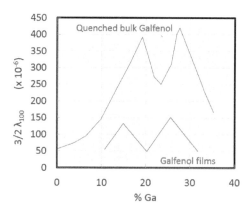

Figure 1. Magnetostriction vs. composition trends in bulk (blue, [7]) and sputtered films (red, [8]).

Thicker films (~microns) are preferred for many applications but are difficult to achieve with vacuum fabrication, so electrodeposition was explored by our group and others [9–11]. The processing window for Galfenol spanned a wide range of current densities and aqueous electrolyte compositions (Table 1). In our group, this was accomplished using a combinatorial technique in a Hull cell, which is a trapezoidal electrochemical cell where the current density (j) during deposition varies as a function of total current (I) and distance along the working electrode (x), according to Reference [12]:

$$j(x) = I \,(51.04 - 52.42 \times \log x) \text{ for } 0.635 < x < 8.255 \text{ cm.} \quad (1)$$

Table 1. Electrolytes used in combinatorial study.

ID	Fe^{2+}:Ga^{3+} Ion Ratio	$[Fe_2SO_4]$ (M)	$[Ga_2(SO_4)_3]$ (M)	Na-Cit (M)
A1	1:5	0.01	0.025	0.05
B1	1:5	0.02	0.05	0.1
B2	1:2.5	0.04	0.05	0.1
C2	1:2.5	0.04	0.05	0.1
C3	1:2.5	0.04	0.05	0.15
C4	1:2.5	0.04	0.05	0.2

It was difficult to establish a direct trend between Ga^{3+} content in the electrolyte and %Ga in the resulting films. However, through the addition of sodium citrate to the electrolyte, Ga^{3+} formed a complex with citrate (Cit) anions, and a trend was established for the dependence of %Ga in the films. The important relationship is the ratio of the Fe^{2+} concentration $[Fe^{2+}]$ vs. the sum of $[Ga^{3+}]$ *and* [Cit] in an electrolyte, as shown in the "phase diagram" of Figure 2. The deposited films were segregated into regions defined as Fe metal, Galfenol, oxides, and Ga-rich metal. Results are shown only for electrolytes that produced a Galfenol region. As expected, the deposition rates increased as a function of current density, Figure 3, until they reached a saturation value. A novel energy dispersive spectroscopy technique was used to quantify the composition and thickness of films, and these results were verified by Rutherford backscattering (RBS).

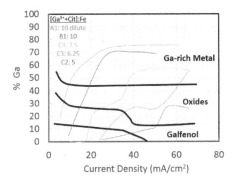

Figure 2. Electrochemical phase diagram showing the regions of ion ratios and current densities that produce Galfenol or other phases. Electrolyte ion concentrations are shown in the legend [11].

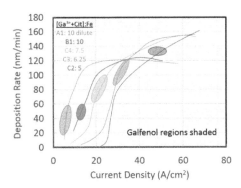

Figure 3. Deposition rates for films from electrolytes that produced Galfenol (shaded regions). Electrolyte ion concentrations are shown in the legend [11].

Non-combinatorial studies further increased the understanding of Gafenol electrodeposition. First, Reddy [13] found that it was best to make electrolytes by adding sodium citrate, sodium sulfate, iron sulfate, and gallium sulfate, in this order. With similar concentrations to prior studies, Reddy then used a rotating disk electrode to control the diffusion boundary layer in the electrolyte to achieve steady-state electrochemistry due to parallel processes of mass transfer and kinetics. The mass transfer-limited current densities increased with the square root of the electrode rotation (ω), owing to a thinning boundary layer. The kinetic proportionality constant was dependent on the number of electrons taking part in the overall reaction, the Faraday constant, the bulk concentration of the cations, the ion diffusivity in the electrolyte, and the kinematic viscosity of the electrolyte. This balance of processes is sometimes called the Koutecky-Levich relationship. Reddy used fixed potentials to obtain similar current densities as prior combinatorial studies, thereby determining the kinetic-limited current density at each voltage using a Tafel plot [14]. A mechanism was proposed based on Reference [15] in which an absorbed monovalent $[Fe(I)]_{ads}$ intermediate is the rate-determining first step, Equation (2). This intermediate then either reduces to a Fe (solid) deposit, or catalyzes Ga (solid) via an adsorbed $[Ga(III) - Fe(II)]_{ads}$ intermediate (Equations (3)–(5)). Since Fe^{2+} and Ga^{3+} have different mass-transport rates, the composition can be controlled simply by varying the rotation rate (ω), as shown in Figure 4.

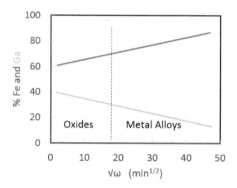

Figure 4. Due to differences in mass transport rates, the composition of Galfenol films can be tailored simply using the rotation rate (ω) of a rotation disk electrode [14].

$$Fe(II) + e^- \rightarrow [Fe(I)]_{ads} \tag{2}$$

$$[Fe(I)]_{ads} + e^- \rightarrow Fe(s) \tag{3}$$

$$[Fe(I)]_{ads} + Ga(III) \rightarrow [Ga(III) - Fe(I)]_{ads} \tag{4}$$

$$[Ga(III) - Fe(I)]_{ads} + 3e^- \rightarrow Ga(s) + [Fe(I)]_{ads} \tag{5}$$

Our next goal was to orient the Galfenol films to have the maximum magnetostriction, namely (100) texture, using n-GaAs as a substrate for epitaxial growth [13]. Reddy used both brass substrates (similar to Hull cell and rotating disk studies) and epi-ready n-GaAs substrates (2° miscut, 2×10^{17} cm^{-3} Si-doping). A slower growth rate was used than in the previous study to enable better epitaxy; as a result, some nuclei had time to grow with low-energy (110) surfaces on both substrates. Therefore, on the brass substrate, the texture parallel to the surface was (110) and (211) as before (Figure 5a,b). On GaAs, however, epitaxy in the (100) planes was also achieved, as seen in the areal X-ray diffraction pattern in Figure 5c. Rocking curves verified these results [13].

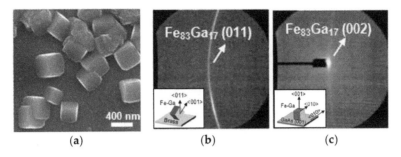

Figure 5. Galfenol grown with slow deposition rates onto (**a**) brass substrates, resulting in (**b**) primarily cube on edge texturing. (**c**) Although some grains were observed with low-energy (110) orientations, a majority of the grains were epitaxially oriented on GaAs. Reproduced from [13], with the permission of AIP Publishing.

The capacitance bridge measurements of sputtered Galfenol magnetostriction required thin, well-characterized, insulating substrates, such as glass coverslips, but electrodeposited Galfenol films require conductive substrates, mostly metals. Estrine showed that sputter-deposited Cr/Cu adhesion layers were effective contacts for the deposition of Galfenol films (170 nm) onto glass coverslips [16]. Magnetostriction was measured as high as 140 ppm for these 17–21% Ga films. This was the first time that electrodeposited Galfenol was confirmed to be magnetostrictive. The next year, Estrine measured the values at a variety of compositions [17] and found that the values agreed with what would be expected for polycrystalline Galfenol (Figure 6).

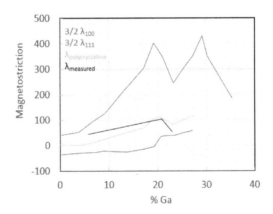

Figure 6. Measured values for the magnetostriction of Galfenol crystallographic orientations and the calculated value of polycrystalline Galfenol, which compares well to measured values [17].

A summary for the study of electrodeposition for Galfenol is that films can be engineered to have composition, crystal structure, and magnetostriction sufficient for microelectromechanical system (MEMS) devices. In fact, a recent visitor to our lab, Vargas, demonstrated that Galfenol films are also biocompatible by internalizing Galfenol-coated discs into cells [18]. Additionally, increasing interest in bio applications, such as the initially proposed cilia sensors, led this work toward the electrochemical synthesis of nanowires, which is presented next.

3. Galfenol Nanowires and Properties

To make nanowires, an electrically conductive film (e.g., Ti/Cu or W/Cu) is sputter-deposited onto one side of a nanoporous anodic aluminum oxide (AAO) template (Figure 7) [19]. The coated side is insulated with a polymer coating so that when the AAO is placed inside an electrolyte such as those discussed above, the cations are reduced to metal only at the bottom of the columnar nanopores. This reduced metal deposit grows inside the template, conforming to the shape of the nanopores and resulting in nanowires.

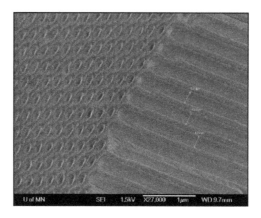

Figure 7. Anodic aluminum oxide template [19].

In initial nanowire studies, the Fe:Ga composition varied significantly along the length of the nanowires (Figure 8) [9]. This was believed to be due to a decreasing distance between the counter electrode and the working electrode, which was the growing ends of the nanowires. A [Cit] complexing agent was discovered that helped stabilize the Ga in solution, and therefore mitigated the composition variations [11]. However, highly variable lengths were often seen in Galfenol nanowire arrays to a degree that was orders of magnitude more severe than that of other metal nanowires. Modeling showed that hemi-spherical diffusion boundary layers caused a few of the nanopores to 'steal' cations from the nanopores around them [20], resulting in the strange bi-modal lengths seen by SEM in Figure 9a. Reddy applied the rotating disk electrode that was used in his steady-state film electrochemistry to keep the diffusion boundary layer planar (not hemi-spherical). This technique, together with Cu seed layer growth prior to Galfenol growth and pulsed electrodeposition, led to nanowire lengths with only 3% standard deviation, as shown in Figure 9b. It is important to mention that this severe inhomogeneity in nanowire length was unique to Galfenol and did not occur in any other metal alloys deposited in our group to date.

Once reproducible Galfenol nanowires were available, three new application-related questions revealed themselves. Can nanowires be ordered with controllable spacings? Can nanowires be integrated with useful substrates, such as Si? What can be done about shape anisotropy so that magnetostriction can be utilized?

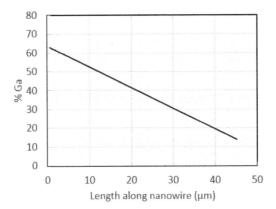

Figure 8. Initial trend of compostion change along nanowire lengths, later mitigated using [Cit] complexing [9].

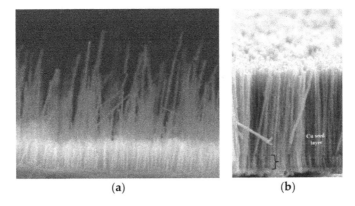

(a)   (b)

Figure 9. (a) A second problem unique to Galfenol was extremely inhomogeneous lengths, where many short nanowires (~4 µm) were seen at the growth electrode while others grew extremely fast, even to the top of the template. (b) Uniform lengths (8 µm ± 3%) were achieved using a rotating disk electrode, a Cu seed layer, and pulsed deposition [20].

First, controllable nanowire size/spacing can be made by imprinting precursor Al before it is anodized to produce the nanoporous oxide template, Figure 10. Al is a relatively soft material, so nanostamps can be made of many other materials [21]. Tan [22] used e-beam defined Si_3N_4 on Si, Figure 10a, and found that rounded pillars can imprint Al well enough (20 nm dimples) to enable nanopores to align with the imprint upon anodization, as shown in Figure 10b [21,23]. These pillars were more manufacturable than the pointed stamps that were assumed to be required by the literature at the time [24,25]. Figure 10b shows nanopores that formed at a boundary of stamped and unstamped Al after subsequent oxidization via anodization. Large areas can be difficult to obtain if a stamp is made by e-beam lithography, especially if very small pillars/spacings are desired. For example, in 2009, a 4 square inch nanostamp of hexagonally spaced (30 nm) pillars would take 23 years of ebeam writing! Today, e-beams can translate between written areas faster, but our current design is more novel and faster than pillars. Specifically, line stamps are imprinted into the Al precursor twice with 60 degrees between imprints to produce directed self-assembly of hexagonally close-packed nanopores when the Al is oxidized [26]. This technique can be used for areas on the order of square inches with little time or cost, as depicted in Figure 10c,d [26].

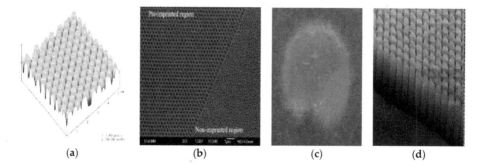

(a)   (b)   (c)   (d)

Figure 10. (a) Si_3N_4 nanostamp to imprint Al films and foils so that (b) nanopores aligned in the subsequent oxide formed by anodization of the Al. (c) Line stamps of Ni could be made large enough to use for large area (1 cm in diameter) alignment, shown diffracting blue light. (d) Magnification of nanopores (1 μm long) produced by line stamps [23,26].

The second applications-related question was whether electroplated nanowires could be integrated with convenient platforms, such as Si, for use in electronic, spintronic, photonic, MEMS/NEMS (micro/nano electromechanical systems), microfluidics, or other systems. Most early nanowire AAO papers involved four steps: anodizing Al foils, removing remnant Al, etching the oxide off of the bottom of pores, and finally growing nanowires. Zhou [23] found that Al films could be evaporated onto Ti/Cu-coated Si for subsequent anodization, and that these films could be nanoimprinted to enable ordered nanowire arrays to be integrated with Si. If the anodization is carried out until the Al is fully consumed, the barrier can be removed electrochemically to enable nanowire growth into the pores using the original Ti-Cu undercoating. Zhou also etched into the Si to form divots for subsequent ordered nanoporous Si. Huang [27] and then Maqableh [28] used this anodization to integrate 10 nm diameter spintronic nanowires inside AAO on Si. Interestingly, the sidewalls of the AAO nanopores were so smooth that the Cu nanowires grown inside of them had almost bulk resistivity, despite having diameters 4× smaller than the mean free path of the electrons [28].

Third, to actively use magnetostriction, the effect must dominate over shape anisotropy, which is strongly oriented along the nanowire axes due to the very high aspect ratios (length/width = 10–200 nm/1–10 μm). By definition, magnetostriction is a change in dimension due to an applied magnetic field. The reverse phenomenon can be useful as well; namely, a magnetization change can be measured when the material is strained. In single-component magnetic nanowires, the magnetization cannot rotate in response to an applied force or vibration due to the shape anisotropy ($2\pi M_s$, where M_s is the saturation magnetization). To get around this constraint, segmented nanowires were made, typically FeGa/Cu [20]. Hysteresis loops taken at varied angles reveal the reversal mechanism of the magnetization in these nanowires. For example, the 100 nm diameter samples depicted in Figure 11a [20] mostly reverse by vortices, characterized by a minimum coercivity (H_c) when the applied field is parallel to the nanowire axes and a maximum coercivity when the applied field is perpendicular. The severe shape anisotropy of the single component nanowires (curve "a") caused a deviation from this behavior, such that the magnetization coherently rotated after perpendicular saturation, leading to a lower H_c at that angle. Sample "b" exhibited classic behavior. Importantly, sample "c" had an isotropic magnetic signature since the Galfenol segments were isotropic in shape, and they were separated from each other by long Cu segments which minimized dipolar interactions. This isotropic sample is most likely to exhibit a change in magnetization if strained or a change in dimension if magnetized.

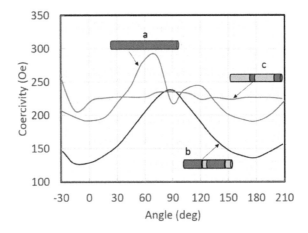

Figure 11. The coercivity from magnetic hysteresis loops for (**a**) pure Galfenol nanowires and segmented Galfenol/Cu nanowires with (**b**) 50 nm of Cu in Galfenol nanowires or (**c**) 50 nm Galfenol in Cu nanowires (diameter = 100 nm; long segments 400 and 500 nm, respectively).

During most of our Galfenol nanowire studies, we collaborated with Prof. Alison Flatau's group at the University of Maryland. Downey used a nanomanipulator to measure the Young's modulus of Galfenol nanowires to be about 58 GPa by tensile testing [29]. These values are promising in that ductile behavior is needed for many of the bio and sensing applications where these nanowires will essentially be used as nanoelectromechanical systems (NEMS). Park used magnetic force microscopy (MFM) [30] to show that the GaFe segments in GaFe/Cu nanowires exhibited vortices at remanence, but the moment was fully parallel to the nanowire when fields larger than 300 Oe were applied in this direction. Park was also able to measure the hysteresis loops [31] and magnetostriction [32] of single nanowires using atomic force microscopy (AFM) techniques. Finally, Park used in Galfenol/Cu segmented nanowires to demonstrate that they could be used as pressure sensors, where pressure was applied to strain the nanowires. This induced a magnetization change in the nanowires that was detected with giant magnetoresistive (GMR) film sensors below the nanowire array [33].

Newer collaborations include those at National Institute of Standards and Technology (NIST), Gaithersburg, where Grutter et al. used polarization-analyzed small angle neutron scattering (PASANS) to observe complex three-dimensional ordering between Galfenol segments in FeGa/Cu nanowires [34]. Combined dipolar interwire and intersegment interactions led to ordering that was ferromagnetic between nanowires and antiferromagnetic between segments parallel to the nanowires. Ponce et al. from the University of Texas, San Antonio used holographic TEM to also observe vortex and antiferromagnetic ordering in thin layers (1–2 nanowires thick) [35].

4. Bio Applications: Flow Sensors, Purification, Barcodes, RFID Nanotags, and MRI Contrast

Our first biosensor used cobalt (Co). nanowires and a magnetoresistive sensor at the bottom of a microfluidic channel. Hein et al. detected flows in microfluidic channels from Diagnostic Biosensors that varied from 0.5 mL/min to 6 mL/min with a signal-to-noise ratio (SNR) of 44 using only 140 μW of power and no amplification, Figure 12 [36]. The magnetoresistive sensors were sensitive to "in-plane" moments that were provided by the nanowires as they bent in the flow, as shown. Vibration sensors were also tested using these nanowires.

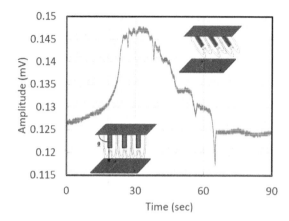

Figure 12. Signal from magnetoresistive sensors at the bottom of a microfluidic channel showing a direct proportionality to flow due to a rapid increase in pressure then a slow release with three sudden drops in pressure and a final backwash from complete pressure release [36].

Sharma et al. incubated Au-tipped Ni nanowires with osteosarcoma for our first look at internalization via endosomes [37], which was later confirmed with a more rigorous study that involved PEG and peptide (RGD) coatings [38]. These nanowires were found to have low cytotoxicity, first by acridine orange and propidium iodide dual staining [37] and later by IL-1β, TNF-α, and MTS assays. It is also worth mentioning that although these Au-tipped nanowires showed very low cytotoxicity, the Au was deposited from a cyanide-based electrolyte. A thiosulfate-sulfite electrolyte was developed for Au deposition by Estrine and Tabokovic et al. [39] as an alternative to cyanide. Interestingly, the integrin overexpression in osteosarcoma led these cells to self-dispersal of RGD-coated nanowires by attachment and motility, followed by internalization and propagation via cell splitting and proliferation. Cells with internalized nanowires could be purified from their assays via magnetic separation [40].

Results such as the ones shown in Figure 11 indicated that it could be possible to barcode cells using nanowires with distinct magnetic signatures. For example, samples a, b, and c in Figure 11 have fairly distinct signatures. By varying diameter as well as segmentation, even more barcodes are possible since coercivity vs. angle curves are highly dependent on magnetization reversal mechanisms. Reddy found that 35 nm FeGa nanowires appear to have diameters that are too small to support vortices, leading to new magnetic signatures [40].

Although coercivity vs. angle appears to contain distinguishing characteristics for different nanowires, normal means of measuring the hysteresis loops, e.g., vibrating sample magnetometry, cannot easily distinguish mixtures of barcode nanowires. Sharma demonstrated the use of first-order reversal curves to demultiplex mixtures [41]. In recent collaborations with Prof. Rhonda Franklin, we are exploring options to use high frequency (e.g., radio frequency) identification (RFID) methods to distinguish mixtures, an area in which Zhou and Um have recently presented promising results [42,43]. Conveniently, this research has cross-fertilizing potential, and Zhou designed the first coplanar waveguide circulator [44].

Many other bio applications are under study by our group and others because magnetic nanowires have potential in magnetic resonance imaging (MRI) contrast [45], targeted cell separation, and hyperthermia. Magnetic nanowires have high saturation magnetizations and high shape anisotropy compared to superparamagnetic iron oxide nanoparticles (SPIONs), which have become ubiquitous in bio-magnetic studies. Besides their magnetic advantages, these nanowires have larger surface areas for attaching biomolecules. New literature is sure to make this review article dated upon publication, thanks to an exciting frontier ahead.

5. Conclusions

Thin films and nanowires of Galfenol have been made and studied in Minnesota for several years. Beginning with sputter deposition to verify thin film properties, our group moved on to define a phase diagram for the electrochemical deposition of this important magnetostrictive alloy. To keep Gallium available to deposit, a citrate complexing agent was identified. To control the composition, a rotating disk electrode (RDE) was used to engineer the diffusion boundary layer. This same RDE was instrumental in mitigating a problem of severely inhomogeneous lengths that was unique to Galfenol nanowire growth. Cu seed layers and pulsing can be added to RDE if absolute length uniformity is required for a given application. The properties of Galfenol films have been extensively studied, showing that they are polycrystalline with the expected magnetostriction for this cubic alloy and that they can be epitaxially deposited on GaAs. Galfenol nanowires were made by electrodeposition into anodic aluminum oxide (AAO) templates. These templates were made with long range order for the careful placement of nanowires in arrays, both as free-standing samples and as integrated components on Si. Segmentation of Galfenol nanowires with Cu enabled controlled magnetic properties, including isotropy magnetic shapes. Several bio applications of magnetic nanowires have been published, including microfluidic sensors and MRI contrast agents, and others are on the immediate horizon, including magnetic separation, cellular RFID tags, and hyperthermia.

Author Contributions: Conceptualization and Supervision, B.J.H.S.; Methodology, Validation, and Formal Analysis, M.R., R.B., P.M, E.E., X.H., S.Y.S, L.T., J.Z., M.M., D.S., T.G., J.U., M.H., A.S.; Writing-Original Draft Preparation, Writing-Review & Editing, B.J.H.S.

Funding: Funding for reviewed work is acknowledged in each cited article, and no additional external funding was received for this manuscript.

Acknowledgments: This manuscript reviews published conducted done at the University of Minnesota. Please see the cited works for appropriate acknowledgements.

Conflicts of Interest: The authors declare no conflict of interest.

References

1. McGary, P.D.; Tan, L.; Zou, J.; Stadler, B.J.; Downey, P.R.; Flatau, A.B. Magnetic nanowires for acoustic sensors. *J. Appl. Phys.* **2006**, *99*, 08B310. [CrossRef]
2. Clark, A.E.; Restorff, J.B.; Wun-Fogle, M.; Lograsso, T.A.; Schlagel, D.L. Magnetostrictive properties of body-centered cubic Fe-Ga and Fe-Ga-Al alloys. *IEEE Trans. Magn.* **2000**, *36*, 3238–3240. [CrossRef]
3. Brooks, M.; Summers, E.; Restorff, J.B.; Wun-Fogle, M. Behavior of magnetic field—Annealed Galfenol steel. *J. Appl. Phys.* **2012**, *111*, 07A907. [CrossRef]
4. Restorff, J.B.; Wun-Fogle, M.; Clark, A.E.; Hathaway, K.B. Induced magnetic anisotropy in stress-annealed Galfenol alloys. *IEEE Trans. Magn.* **2006**, *42*, 3087–3089. [CrossRef]
5. Atulasimha, J.; Flatau, A.B. A review of magnetostrictive iron—Gallium alloys. *Smart Mater. Struct.* **2011**, *20*, 043001. [CrossRef]
6. Mudivarthi, C.; Datta, S.; Atulasimha, J.; Flatau, A.B. A bidirectionally coupled magnetoelastic model and its validation using a Galfenol unimorph sensor. *Smart Mater. Struct.* **2008**, *17*, 035005. [CrossRef]
7. Petculescu, G.; Hathaway, K.B.; Lograsso, T.A.; Wun-Fogle, M.; Clark, A.E. Magnetic field dependence of galfenol elastic properties. *J. Appl. Phys.* **2005**, *97*, 10M315. [CrossRef]
8. Basantkumar, R.R.; Stadler, B.H.; Robbins, W.P.; Summers, E.M. Integration of thin-film galfenol with MEMS cantilevers for magnetic actuation. *IEEE Trans. Magn.* **2006**, *42*, 3102–3104. [CrossRef]
9. McGary, P.D.; Stadler, B.J.H. Electrochemical deposition of $Fe_{1-x}Ga_x$ nanowire arrays. *J. Appl. Phys.* **2005**, *97*, 10R503. [CrossRef]

10. Lupu, N.; Chiriac, H.; Pascariu, P. Electrochemical deposition of Fe Ga/Ni Fe magnetic multilayered films and nanowire arrays. *J. Appl. Phys.* **2008**, *103*, 07B511. [CrossRef]

11. McGary, P.D.; Reddy, K.S.; Haugstad, G.D.; Stadler, B.J. Combinatorial electrodeposition of magnetostrictive $Fe_{1-x}Ga_x$. *J. Electrochem. Soc.* **2010**, *157*, D656–D665. [CrossRef]

12. Hull, R.O. Apparatus and Process for the Study of Plating Solutions. U.S. Patent 2,149,344, 7 March 1939.

13. Reddy, K.S.; Maqableh, M.M.; Stadler, B.J. Epitaxial $Fe_{1-x}Ga_x$/GaAs Structures via Electrochemistry for Spintronic Applications. *J. Appl. Phys.* **2012**, *111*, 07E502. [CrossRef]

14. Reddy, K.S.; Estrine, E.C.; Lim, D.H.; Smyrl, W.H.; Stadler, B.J. Controlled Electrochemical Deposition of Magnetostrictive $Fe_{1-x}Ga_x$ Alloys. *Electrochem. Commun.* **2012**, *18*, 127–130. [CrossRef]

15. Podlaha, E.J.; Landolt, D. Induced codeposition III. Molybdenum alloys with nickel, cobalt, and iron. *J. Electrochem. Soc.* **1997**, *144*, 1672–1680. [CrossRef]

16. Estrine, E.C.; Hein, M.; Robbins, W.P.; Stadler, B.J. Composition and Crystallinity in Electrochemically Deposited Magnetostrictive Galfenol (FeGa). *J. Appl. Phys.* **2014**, *115*, 17A918. [CrossRef]

17. Estrine, E.C.; Robbins, W.P.; Maqableh, M.M.; Stadler, B.J. Electrodeposition and characterization of magnetostrictive Galfenol (FeGa) thin films for use in MEMS. *J. Appl. Phys.* **2013**, *113*, 17A937. [CrossRef]

18. Vargas-Estevez, C.; Blanquer, A.; Dulal, P.; del Real, R.P.; Duch, M.; Ibáñez, E.; Barrios, L.; Murillo, G.; Torras, N.; Nogués, C.; et al. Study of Galfenol direct cytotoxicity and remote microactuation in cells. *Biomaterials* **2017**, *139*, 67–74. [CrossRef] [PubMed]

19. Huang, X. Constricted Current Perpendicular to Plane (CPP) Magnetic Sensor via Electroplating. Ph.D. Thesis, University of Minnesota, Minneapolis, MN, USA, 2011.

20. Reddy, S.M.; Park, J.J.; Na, S.M.; Maqableh, M.M.; Flatau, A.B.; Stadler, B.J. Electrochemical Synthesis of Magnetostrictive Fe-Ga/Cu Multilayered Nanowire Arrays with Tailored Magnetic Response. *Adv. Funct. Mater.* **2011**, *21*, 4677–4683. [CrossRef]

21. Maqablah, M.; Tan, L.; Huang, X.; Cobian, R.; Norby, G.; Victora, R.H.; Stadler, B.J.H. CPP GMR through Nanowires (Invited). *IEEE Trans. Magn.* **2012**, *48*, 1–7.

22. Tan, L. Templated Synthesis of Magnetic Nanowires by Electrochemical Deposition. Ph.D. Thesis, University of Minnesota, Minneapolis, MN, USA, 2009.

23. Zou, J.; Qi, X.; Tan, L.; Stadler, B.J.H. Nanoporous Silicon with Long-Range-Order using Imprinted Anodic Alumina Etch Masks. *Appl. Phys. Lett.* **2006**, *89*, 093106. [CrossRef]

24. Masuda, H.; Yamada, H.; Satoh, M.; Asoh, H.; Nakao, M.; Tamamura, T. Highly ordered nanochannel-array architecture in anodic alumina. *Appl. Phys. Lett.* **1997**, *71*, 2770–2772. [CrossRef]

25. Choi, J.; Nielsch, K.; Reiche, M.; Wehrspohn, R.B.; Gösele, U. Fabrication of monodomain alumina pore arrays with an interpore distance smaller than the lattice constant of the imprint stamp. *J. Vac. Sci. Technol. B* **2003**, *21*, 763–766. [CrossRef]

26. Sung, S.; Maqablah, M.; Huang, X.; Reddy, K.S.M.; Victora, R.H.; Stadler, B.J.H. Metallic 10 nm Diameter Nanowire Magnetic Sensors and Large-Scale Ordered Arrays. *IEEE Trans. Magn.* **2014**, *50*, 1–5. [CrossRef]

27. Huang, X.; Tan, L.; Cho, H.; Stadler, B.J.H. Magnetoresistance and Spin Transfer Torque in Electrodeposited Co/Cu Multilayered Nanowire Arrays with Small Diameters. *J. Appl. Phys.* **2009**, *103*, 07B504. [CrossRef]

28. Maqableh, M.M.; Huang, X.; Sung, S.Y.; Reddy, K.S.; Norby, G.; Victora, R.H.; Stadler, B.J. Low Resistivity 10 nm Diameter Magnetic Sensors. *Nano Lett.* **2012**, *12*, 4102–4109. [CrossRef] [PubMed]

29. Downey, P.R.; Flatau, A.B.; McGary, P.D.; Stadler, B.J.H. Effect of magnetic field on the mechanical properties of magnetostrictive iron-gallium nanowires. *J. Appl. Phys.* **2008**, *103*, 07D305. [CrossRef]

30. Park, J.J.; Reddy, M.; Mudivarthi, C.; Downey, P.R.; Stadler, B.J.H.; Flatau, A.B. Characterization of the magnetic properties of multilayer magnetostrictive Iron-Gallium nanowires. *J. Appl. Phys.* **2010**, *107*, 09A954. [CrossRef]

31. Park, J.J.; Reddy, M.; Stadler, B.J.H.; Flatau, A.B. Hysteresis measurement of individual multilayered Fe-Ga/Cu nanowires using magnetic force microscopy. *J. Appl. Phys.* **2013**, *113*, 17A331. [CrossRef]

32. Park, J.J.; Estrine, E.C.; Stadler, B.J.H.; Flatau, A.B. Technique for measurement of magnetostriction in an individual nanowire using atomic force microscopy. *J. Appl. Phys.* **2014**, *115*, 17A919. [CrossRef]

33. Park, J.J.; Reddy, S.M.; Stadler, B.J.H.; Flatau, A.B. Magnetostrictive Fe-Ga/Cu Nanowires Array with GMR Sensor for Sensing Applied Pressure. *IEEE Sens. J.* **2017**, *17*, 2015–2020. [CrossRef]

34. Grutter, A.J.; Krycka, K.L.; Tartakovskaya, E.V.; Borchers, J.A.; Reddy, K.S.; Ortega, E.; Ponce, A.; Stadler, B.J. Frustrated Long-Range Magnetic Domain Structure in High-Density Nanowire Arrays. *ACS Nano* **2017**, *11*, 8311–8319. [CrossRef] [PubMed]

35. Ortega, E.; Reddy, S.M.; Betancourt, I.; Roughani, S.; Stadler, B.J.H.; Ponce, A. Magnetic ordering in 45 nm-diameter multisegmented FeGa/Cu nanowires: Single nanowires and arrays. *J. Mater. Chem. C* **2017**, *5*, 7546–7552. [CrossRef]

36. Hein, M.; Maqableh, M.; Delahunt, M.; Tondra, M.; Flatau, A.; Shield, C.; Stadler, B. Fabrication of BioInspired Inorganic Nanocilia Sensors. *IEEE Trans. Magn.* **2013**, *49*, 191–194. [CrossRef]

37. Sharma, A.; Zhu, Y.; Thor, S.; Zhou, F.; Stadler, B.; Hubel, A. Magnetic Barcode Nanowires for Osteosarcoma Cell Control, Detection, and Separation. *IEEE Trans. Magn.* **2013**, *49*, 453–456. [CrossRef]

38. Sharma, A.; Orlowski, G.M.; Zhu, Y.; Shore, D.; Kim, S.Y.; DiVito, M.; Hubel, A.; Stadler, B.J.H. Inducing cells to self-disperse non-toxic Ni nanowires via integrin-mediated responses. *Nanotechnology* **2015**, *26*, 135102. [CrossRef] [PubMed]

39. Estrine, E.; Riemer, S.; Venkatosamy, V.; Stadler, B.; Tabakovic, I. Mechanism and Stability Study of Gold Electrodeposition from Thiosulfate-Sulfite Solution. *J. Electrochem. Soc.* **2014**, *161*, D687–D696. [CrossRef]

40. Reddy, K.S.M.; Park, J.J.; Maqableh, M.M.; Flatau, A.B.; Stadler, B.J.H. Magnetization Reversal Mechanisms in 35-nm Diameter $Fe_{1-x}Ga_x$/Cu Multilayered Nanowires. *J. Appl. Phys.* **2012**, *111*, 07A920. [CrossRef]

41. Sharma, A.; DiVito, M.D.; Shore, D.E.; Block, A.D.; Pollock, K.; Solheid, P.; Feinberg, J.M.; Modiano, J.; Lam, C.H.; Hubel, A.; et al. Alignment of collagen matrices using magnetic nanowires and magnetic barcodereadout using first order reversal curves (FORC). *J. Magn. Magn. Mater.* **2018**, *459*, 176–181. [CrossRef]

42. Zhou, W.; Um, J.; Zhang, Y.; Nelson, A.; Stadler, B.; Franklin, R. Ferromagnetic Resonance Characterization of Magnetic Nanowires for Biolabel Applications. In Proceedings of the IMBioC of the 2018 IEEE MTT-S International Microwave Symposium, Philadelphia, PA, USA, 10–15 June 2018.

43. Um, J.; Zhou, W.; Franklin, R.; Stadler, B. Detection of Nanowires for RFID Biolabels using Ferromagnetic Resonance. In Proceedings of the Spring 2018 Materials Research Society Meeting, Phoenix, AZ, USA, 2–6 April 2018.

44. Zhou, W.; Um, J.; Stadler, B.; Franklin, R. Design of self-biased coplanar circulator with ferromagnetic nanowires. In Proceedings of the Radio and Wireless Symposium (RWS), Anaheim, CA, USA, 15–18 January 2018; pp. 240–242.

45. Shore, D.; Pailloux, S.L.; Zhang, J.; Gage, T.; Flannigan, D.J.; Garwood, M.; Pierre, V.C.; Stadler, B.J. Electrodeposited Fe and Fe-Au Nanowires as MRI Contrast Agents. *Chem. Commun.* **2016**, *52*, 12634–12637. [CrossRef] [PubMed]

Article

Magnetorelaxometry in the Presence of a DC Bias Field of Ferromagnetic Nanoparticles Bearing a Viscoelastic Corona

Victor Rusakov [†] and Yuriy Raikher [†,*]

Institute of Continuous Media Mechanics, Russian Academy of Sciences, Ural Branch, Perm 614013, Russia; vvr@icmm.ru
* Correspondence: raikher@icmm.ru; Tel.: +7-342-237-8323
† These authors contributed equally to this work.

Received: 16 April 2018; Accepted: 18 May 2018; Published: 22 May 2018

Abstract: With allowance for orientational Brownian motion, the magnetorelaxometry (MRX) signal, i.e., the decay of magnetization generated by an ensemble of ferromagnet nanoparticles, each of which bears a macromolecular corona (a loose layer of polymer gel) is studied. The rheology of corona is modelled by the Jeffreys scheme. The latter, although comprising only three phenomenological parameters, enables one to describe a wide spectrum of viscoelastic media: from linearly viscous liquids to weakly-fluent gels. The "transverse" configuration of MRX is considered where the system is subjected to a DC (constant bias) field, whereas the probing field is applied perpendicularly to the bias one. The analysis shows that the rate of magnetization decay strongly depends on the state of corona and slows down with enhancement of the corona elasticity. In addition, for the case of "transverse" MRX, we consider the integral time, i.e., the characteristic that is applicable to relaxation processes with an arbitrary number of decay modes. Expressions for the dependence of the integral time on the corona elasticity parameter and temperature are derived.

Keywords: magnetic nanoparticles; viscoelasticity; magnetorelaxometry

---

1. Introduction: Magnetorelaxometry in Linear Approximation

Magnetic nanoparticles are successfully used in diverse bioengineering and medical applications both as themselves (e.g., as contrasting agents in Magnetic Resonance Imaging (MRI)) or as essential components of physicochemical complexes (magnetic polymerosomes [1,2], microferrogels [3]). These techniques span from object-oriented drug delivery [4] to active thermal action on malignant cells (magneto-inductive hyperthermia [5]) or forced penetration through cell membranes [6].

Nowadays, one of the most developed applications of magnetic nanoparticles is their introduction as sensors for diagnosing the state and content of complex media both of non-organic and biological origin. In the context of conventional microrheology approach, it is assumed that the particles do not chemically react with the medium under study and, thus, serve as the means of specific "nondestructive testing". On the contrary, in biochemical analysis, the nanoparticles are functionalized, and the main interest is focused on the degree of reaction between the molecules of tested solution with the markers grafted to the particles or with their bare but chemically active surface. The macromolecular coating that a particle acquires as a result of this adsorption is often termed as a protein corona [7].

A unique advantage of magnetic nanosensors is that, with the aid of an applied magnetic field, one is able to remotely excite their motion and analyze the generated response to that.

The chemical processes resulting in formation of the corona on the particle surface affect its dynamic behavior. Those changes could be sensed by either measuring the magnetic spectrum of the system (if the probing is done with an AC field) or—in pulse regime—by registering the signal of

the magnetization decay. The latter experimental technique is known as magnetorelaxometry (MRX).
In many cases, MRX as a laboratory test is more preferable in comparison with magnetic spectroscopy,
in particular, due to its much easier technical implementation [8–10].

Consider a ferromagnet nanoparticle that, due to its smallness, is single-domain and, thus, bears
a magnetic moment $\vec{\mu}$ of constant magnitude. This particle floats in a water solution of macromolecules
some of which, possessing affinity to the particle surface, adsorb on it, so that the particle environment
transforms from a Newtonian liquid into a viscoelastic polymer gel.

For simplicity of the following considerations, we assume that vector $\vec{\mu}$ is "frozen" in the particle
body and treat the nanoparticle as a miniature permanent magnet. This means neglecting the
superparamagnetic effect: thermal fluctuations of $\vec{\mu}$ inside the particle. Such an approach implies that
the Néel (internal) relaxation time of $\vec{\mu}$ is much greater than any other reference time of the problem.
Certainly, this approximation is not universally valid. However, as it follows from a number of works
on the subject, see, for example, reviews [11,12] for general considerations and papers [13,14] for
realistic examples. The particles of 15–20 nm in diameter made of a moderately magnetically hard
substance (e.g., cobalt ferrite) matches this requirement fairly well in the frequency range below 1 MHz,
which is most relevant for MRX.

For the above-described magnetically hard nanoparticle suspended in a fluid medium, the only
fluctuational process affecting vector $\vec{\mu}$ is its rotary Brownian motion together with the particle. Due to
that, in the absence of field, the directions of magnetic moments are distributed at random, and the net
magnetization of the ensemble is zero. An MRX measurement begins with application of a uniform
magnetic field that orients the particle magnetic moments, thus inducing a non-zero equilibrium
magnetization of the system. Then, the magnitude (or direction) of the field is abruptly changed,
and the signal generated by the magnetization evolving to a new equilibrium is registered for further
analysis. By that, MRX not only accomplishes its general purpose—to evaluate the amount of particles
(that is proportional to the signal intensity)—but delivers information on the details of the particle
rotary motion.

If the exerted field is turned off completely, the magnetization of the particle ensemble decays
freely down to zero, and this process is governed solely by the rotary Brownian diffusion. There are
no external factors. If the field is switched between two finite values—from $\vec{H}$ to $\vec{H} + \vec{H}_1$—the same
occurs to the system magnetization. Under those conditions, the MRX signal becomes a function
of both values ($\vec{H}$ and $\vec{H}_1$), by variation of which one can extract more information from the same
experiment than in the case where the field is just turned off. A convenient way to study this relaxation
is to change the bias field in small ($\sim H_1 \ll H$) steps to be able to use the linear response theory. In that
case, the obtained MRX signal depends parametrically only on that value H of the field, around which
the variation is done. The "linear" MRX of that kind is the subject of the present work.

2. Model

We consider a nanoparticle that, as a result of formation of a corona, dwells in a viscoelastic
environment, i.e., a medium with retarded response. As a rheological model for the latter, the Jeffreys
scheme (see [15], for example) is used because, unlike the plain Maxwell one, it is robust when applied
to Brownian motion and is free of artifacts [16–18]. The viscoelastic properties of the Jeffreys model
are fully rendered by three parameters (see the scheme outlined in Figure 1). A single-element chain
(narrow damper) associated with the solvent there—a Newtonian fluid with the viscosity coefficient
η_N—is set in parallel with a two-element Maxwell chain associated with the corona. The latter is
assumed to possess both elasticity with modulus G (the spring) and intrinsic viscosity η_M (wide
damper). Evidently, in a medium with a pronounced visoelasticty, the Jeffreys viscosity coefficients are
substantially different: the intrinsic viscosity of the macromolecular gel is much greater than that of
the low-molecular solvent, $\eta_M \gg \eta_N$.

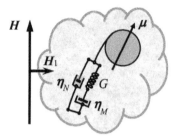

Figure 1. Sketch of a magnetic particle bearing a corona of Jeffreys medium. The system is set under constant field $\vec{H}$, and the existing equilibrium is perturbed by the field $\vec{H}_1$ switched on/off stepwise or alternating harmonically.

To justify application of the Jeffreys scheme to the rotary motion of the particles (Figure 1), we remind readers that this phenomenology works rather well for the translational Brownian motion in semi-dilute polymer solutions. In such models [19,20], it is assumed that each particle is surrounded by a depletion layer where concentration of macromolecules is much lower than that in the bulk. Inside this thin layer, the viscosity is effectively small (η_N), and the diffusion process is fast (short times, small distances). However, when larger displacements (of the order of the particle size or greater) are considered, the particle experiences its environment as a medium with a high viscosity η_M. Similar behavior is inherent to biofilms [21], which, at short time intervals, respond as low viscous fluids but at longer scale react as weakly-fluent gels.

Following this line, we infer that a scheme with two rather different viscosity coefficients should be appropriate for the rotary diffusion of the particles furnished with a corona. Indeed, a loose macromolecular coating, just slightly changing the total mass of the system, at the same time strongly affects its rotary friction. Moreover, the presence of corona cannot be accounted for by just a simple renormalization of the particle hydrodynamic diameter since the corona brings in the particle dynamics a substantial retardation component. That is, if the particle turns over large angles, this motion entrains the whole corona. On the other hand, the corona is virtually insensitive to small angle displacements because such a motion concerns only the molecular fragments in close vicinity of the particle surface, and, due to that, the contribution of corona to resistance and retardation is insignificant. Therefore, at short time intervals, the particle may be considered as floating in a Newtonian fluid with low viscosity η_N.

For nanoparticles embedded in any viscous environment, the effect of inertia is negligible, and the process of rotary relaxation is always monotonic (overdamped regime). Given that, we chose as a main indicator of the state of corona the relaxation time of its elastic stress: $\tau_M = \eta_M/G$. In the theoretical consideration below, we analyze how the presence and magnitude of τ_M is reflected in the MRX spectra of the particle ensemble subjected to a bias DC field.

As in the adopted model, the magnetic moment $\vec{\mu}$ is "frozen" in the particle, and we take unit vector $\vec{e} = \vec{\mu}/\mu$ as a marker of the particle orientation. The equations of rotary motion for a Brownian particle in the inertialess limit take the form [17,18]:

$$\dot{\vec{e}} = \left(\vec{\Omega} \times \vec{e}\right), \quad \vec{\Omega} = \frac{1}{\zeta_N}\left[-\hat{\vec{L}}U + \vec{Q} + \vec{y}_N(t)\right], \quad \hat{\vec{L}} = \left(\vec{e} \times \frac{\partial}{\partial \vec{e}}\right), \tag{1}$$

$$\left(1 + \tau_M \frac{\partial}{\partial t}\right)\vec{Q} = -\zeta_M \vec{\Omega} + \vec{y}_M(t).$$

Here, $\vec{\Omega}$ is angular velocity of vector $\vec{e}$, and U is the orientation-dependent part of the particle energy, while $\hat{\vec{L}}$ is the operator of infinitesimal rotation with respect to $\vec{e}$. Vector $\vec{Q}$ in Equation (1) has the meaning of a torque acting on the particle on the part of the corona. The response coefficients of the viscoelastic medium are defined in a standard way [22] as

$$\zeta_\alpha = 6\eta_\alpha V, \quad K = 6GV, \quad \alpha = N, M, \tag{2}$$

where V is the particle volume (we assume a sphere), and the subscript indicates either Newtonian ($\alpha = N$) or Maxwell ($\alpha = M$) viscosity (see Figure 1). With Notations (2), the stress relaxation time may be equivalently written as $\tau_M = \zeta_M/K$. We remark that there are no universal expressions to replace Formulas (2) provided the particles are non-spherical (anisometric). However, quite reliable estimates could be obtained if to approximate a particle with an ellipsoid of revolution (spheroid). For that shape, the effect of non-sphericity is rendered by a formfactor $\mathcal{F}$ that is to be inserted in Formulas (2) alongside η_α and G, respectively. The dependence of $\mathcal{F}$ as a function of the particle aspect ratio is known (see [23]) for an example.

Correlators of the random forces that model thermal noise in the system are expressed with the aid of the fluctuation–dissipation theorem:

$$\langle y_{i\alpha}(t)\, y_{j\beta}(t+\tau)\rangle = 2T\zeta_\alpha \delta_{\alpha\beta}\delta_{ij}\delta(\tau), \quad \alpha = N, M; \tag{3}$$

note that, hereafter, we scale temperature in energy units.

The kinetic equation for the distribution function $W(\vec{e}, \vec{Q}, t)$ that corresponds to the set of stochastic Equation (1) is obtained via standard procedure (see [18,24,25]) for example:

$$2\tau_D \frac{\partial}{\partial t} W = \left[(1+q^{-1})\beta T \frac{\partial}{\partial \vec{Q}} - \hat{\vec{L}} \right] \left(\frac{1}{T}\vec{Q} + \beta T \frac{\partial}{\partial \vec{Q}} \right) W - \left(\beta T \frac{\partial}{\partial \vec{Q}} - \hat{\vec{L}} \right) \left[\frac{1}{T}\left(\hat{\vec{L}}U\right) + \hat{\vec{L}} \right] W. \tag{4}$$

Here, the following notations are used:

$$\tau_D = \zeta_N/2T = 3\eta_N V/T \tag{5}$$

for the Debye time of orientational diffusion of a spherical particle in a fluid of viscosity, and

$$q = \zeta_M/\zeta_N = \eta_M/\eta_N, \quad \beta = K/T \tag{6}$$

for non-dimensional rheological parameters. The first one defines "maxwellity" of the Jeffreys medium, so that, at $q = 0$, the model reduces to an ordinary viscous fluid; the second parameter is the dynamic elasticity coefficient scaled with thermal energy.

For a magnetically hard particle, the energy U that enters Equation (4) reduces to the Zeeman interaction with uniform field $\vec{H}$:

$$U = -\left(\vec{e}\,\vec{H}\right) = -\mu H \left(\vec{e}\,\vec{h}\right), \quad \vec{h}^2 = 1. \tag{7}$$

In non-dimensional form, the reference magnitude of that energy is

$$\xi = \mu H/T.$$

It is easy to verify that the equilibrium solution of Equation (4) is given by an extended Boltzmann distribution

$$W_0(\vec{e}, \vec{Q}) \propto \exp\left[-\frac{1}{T}\left(U + \frac{Q^2}{2K} \right) \right]. \tag{8}$$

From Equation (8), it follows that the equilibrium state of the system in the absence of field $U = 0$ is isotropic. Note that, although a constant external field endows the system with uniaxial anisotropy, the phase variables $\vec{e}$ and $\vec{Q}$ remain statistically independent in that case as well.

In a dilute (the interparticle interaction is neglected) statistical ensemble of such particles, the magnetization and susceptibility are rendered by expressions

$$\vec{M} = n\mu \langle \vec{e} \rangle, \quad \chi_{ij} = \partial M_i / \partial H_j, \tag{9}$$

where n is the number concentration of particles, whereas angular brackets denote averaging with the distribution function from Equation (4).

3. Dynamic Susceptibility

The MRX relaxation function, i.e., the dependence $\Delta M(H, t)$, where ΔM is the projection of magnetization on the direction of the probing field $\vec{H}_1$, and H is the bias field strength (see Figure 1), could be derived by several ways. We chose the one where the relaxation function is obtained in terms of the dynamic magnetic susceptibility $\chi(\omega)$. The reason is two-fold. First, in the linear response approximation, the relation between the relaxation function and $\chi(\omega)$ has a simple form. Second, in Refs. [18,26], we have developed a workable technique to obtain the pertinent dynamic susceptibility for the case of zero external field. Using those results, one may skip a considerable part of lengthy calculations and just add only the extension allowing for the bias field. Besides that, in [18,26], while analyzing the results obtained from solving the kinetic equation (the analogue of Equation (4)) in an exact way, i.e., using a long multi-moment expansion, we have shown that, to get a plausibly accurate approximation, it suffices to truncate the infinite moment set to just the first two equations for the dynamic variables $\delta \vec{e} = \vec{e} - \langle \vec{e} \rangle_0$ and $\vec{P} = \left(\vec{Q} \times \vec{e} \right)$; here, angular brackets with index 0 denote averaging over the equilibrium distribution (8).

The aforementioned approximation is well known as the effective field model, and it has proven its usefulness for diverse problems of orientational kinetics of nanoparticles many times [12,27,28]. For the case of zero bias, the dynamic magnetic susceptibility in the effective field approximation is [28]:

$$\tilde{\chi}(\omega, \beta) = \frac{\chi(\omega, \beta)}{\chi_0} = \frac{1}{1 - i\omega\tau_D \left(1 + \dfrac{\frac{1}{2}\beta}{1 + \frac{1}{2}\beta/q - i\omega\tau_D} \right)}, \tag{10}$$

where $\chi_0 = n\mu^2 / 3T$ is the static magnetic susceptibility of an ensemble of noninteracting particles bearing magnetic moments μ.

As seen from Equation (10), under zero bias the dynamic susceptibility is isotropic, and the type of its frequency dependence is determined by the elasticity β and "maxwellity" q of the medium (corona). At high temperatures $\beta \ll 1$, the viscoelastic properties of the corona do not manifest themselves, and the susceptibility reduces to a plain Debye formula with the reference relaxation time τ_D. For a medium with high elasticity ($\beta \gg 1$), the dependence (10) has two maxima and may be with plausible accuracy presented as a superposition of two relaxation modes: slow (s) and fast (f) in the form

$$\tilde{\chi}(\omega, \beta) \simeq \frac{1 - 2/\beta}{1 - i\omega\tau_s} + \frac{2/\beta}{1 - i\omega\tau_f}, \quad \tau_s = \tau_D \left(1 + \frac{\frac{1}{2}\beta}{1 + \beta/2q} \right), \quad \tau_f = \frac{2\tau_D}{\beta}, \quad \beta \gg 1. \tag{11}$$

Equation (11) shows that, under enhancement of elasticity, the slow component grows, and its peak moves further to the low-frequency domain, which means gradual increase of its reference time. Concurrently, the contribution of the fast relaxation mode goes down monotonically with β.

In the presence of bias field $\vec{H}$ that makes the system uniaxially anisotropic, the susceptibility becomes a second-rank tensor that is diagonal in the coordinate frame whose Oz axis points along the bias field:

$$\tilde{\chi}_{ij}(\omega, \beta, \xi) = \tilde{\chi}_\perp \delta_{ij} + \left(\tilde{\chi}_\parallel - \tilde{\chi}_\perp \right) h_i h_j.$$

For the case of perturbation induced by a weak probing field $\vec{H}_1$ (in non-dimensional form, $\vec{\xi}_1$), the set of moment equations in the effective field approximation takes the form

$$\left(\frac{\partial}{\partial t} + \gamma_\alpha^{(e)}\right)\langle\delta\vec{e}_\alpha\rangle = \frac{1}{2}\vec{\xi}_1 f_\alpha + \frac{1}{2}\beta\langle\vec{P}\rangle, \tag{12}$$

$$\left[\frac{\partial}{\partial t} + \frac{1}{2}\beta(1+q^{-1}) + \gamma_\alpha^{(M)}\right]\langle\vec{P}\rangle = -\frac{1}{2}\vec{\xi}_1 f_\alpha + \gamma_\alpha^{(e)}\langle\delta\vec{e}_\alpha\rangle,$$

with the coefficients

$$\gamma_\perp^{(e)} = \gamma_\parallel^{(M)} = (1+c_2)/2(1-c_2), \quad \gamma_\parallel^{(e)} = (1-c_2)/2(c_2-c_1^2), \tag{13}$$

$$\gamma_\perp^{(M)} = (3-c_2)/2(1+c_2), \quad f_\perp = \frac{1}{2}(1+c_2), \quad f_\parallel = 1-c_2,$$

defined in terms of functions

$$c_k(\xi) = \int_{-1}^{1} x^k \exp(\xi x) dx \Big/ \int_{-1}^{1} \exp(\xi x) dx. \tag{14}$$

On solving Equations (12), one obtains the sought for dynamic susceptibility

$$\widetilde{\chi}_\alpha(\omega,\beta,\xi) = \frac{\chi_\alpha(\omega,\beta,\xi)}{\chi_0} = \frac{1}{1 - i\omega\dfrac{\tau_D}{\gamma_\alpha^{(e)}}\left(1 + \dfrac{\frac{1}{2}\beta}{\gamma_\alpha^{(M)} + \beta/2q - i\omega\tau_D}\right)}, \tag{15}$$

where index α assumes one of two possible values: $\perp$ or $\parallel$.

Presenting Function (15) as a sum of two relaxation modes—this time, for arbitrary values of parameters β and ξ and—one gets

$$\widetilde{\chi}_\alpha(\omega,\beta,\xi) = \frac{A_s^{(\alpha)}(\beta,\xi)}{1 - i\omega\tau_s^{(\alpha)}} + \frac{A_f^{(\alpha)}(\beta,\xi)}{1 - i\omega\tau_f^{(\alpha)}}, \quad \tau_{s,f} = \frac{\tau_D}{\lambda_{s,f}^{(\alpha)}}, \tag{16}$$

where the relaxation modes decrements $\lambda_{s,f}^{(\alpha)}$ are the roots of characteristic equation

$$\lambda^2 - \left[\Gamma_\alpha + \gamma_\alpha^{(e)}\right]\lambda + \gamma_\alpha^{(e)}\left(\Gamma_\alpha - \frac{1}{2}\beta\right) = 0, \tag{17}$$

with $\Gamma_\alpha = \gamma_\alpha^{(M)} + \frac{1}{2}\beta\left(1+q^{-1}\right)$. The explicit expressions for the mode amplitudes and decrements are

$$A_{s,f}^{(\alpha)}(\beta,\xi) = \frac{1}{2}\left[1 \pm \frac{\Gamma_\alpha - \gamma_\alpha^{(e)}}{\sqrt{\left(\Gamma_\alpha - \gamma_\alpha^{(e)}\right)^2 + 2\beta\gamma_\alpha^{(e)}}}\right], \quad \lambda_{s,f}^{(\alpha)} = \frac{1}{2}\left[\Gamma_\alpha + \gamma_\alpha^{(e)} \mp \sqrt{\left(\Gamma_\alpha - \gamma_\alpha^{(e)}\right)^2 + 2\beta\gamma_\alpha^{(e)}}\right]. \tag{18}$$

Let us consider the frequency dependence of the transverse ($\perp$) component of susceptibility (16). Under a weak constant field, it only slightly differs from the one rendered by Formula (10). However, in a medium with a pronounced elasticity, the fast mode might become dominating. Indeed, under a growing bias field, the decrement of the slow mode virtually does not change while $\lambda_f^{(\perp)}$ undergoes rapid increase, i.e., the relaxation time goes down. As a result, the spectrum as a whole shifts to the higher frequency range.

4. Magnetorelaxometry

4.1. Relaxation Functions

Consider an MRX experiment that is performed as follows. On the particle ensemble that dwells in equilibrium under a constant bias field $\vec{H}$, a weak perturbing field $\vec{H}_1$ is imposed transversely to $\vec{H}$. This induces the magnetization component $\Delta M = n\mu \langle \vec{\delta e}_\perp \rangle$ in the direction of $\vec{H}_1$. As soon as the magnetization attains its stationary value in the newly established equilibrium, the field $\vec{H}_1$ is turned off, and the process of relaxation of ΔM is recorded. This "transverse" configuration seems to be more preferable than measuring the longitudinal relaxation by varying $\vec{H}$ in small steps because the greater the relative magnitude of the relaxing perturbations (transverse vs. longitudinal), the stronger the bias.

According to the linear response theory (see [24], for example), the reaction to a stepwise turning off the field, is obtained from the dynamic susceptibility (15) by way of the integral relation

$$\widetilde{\chi}_\perp (\omega, \beta, \xi) = 1 + i\omega \int_0^\infty e^{i\omega t} R_\perp (t)\, dt, \tag{19}$$

where $R_\perp (t) = \langle \delta e_\perp \rangle / \langle \delta e_\perp \rangle_0$ is the relaxation function, i.e., the transverse dynamic magnetization normalized by its initial value.

Subjecting Formula (19) to inverse Fourier transformation, one arrives at the expression for relaxation function

$$R_\perp (t) = \frac{2}{\pi} \int_0^\infty \frac{\text{Im}\widetilde{\chi}_\perp (\omega, \beta, \xi)}{\omega} \cos \omega t\, d\omega. \tag{20}$$

As mentioned above, we take the stress relaxation time τ_M as the main characteristic of viscoelasticity of the macromolecular corona. If to describe the corona substance as a Jeffreys fluid, this time parameter is expressed in terms of the rheological coefficients as $\tau_M = \zeta_M / K = \eta_M / G$. Assuming that formation of the corona does not affect the viscosity η_N of the solvent, one finds that, for presenting the relaxation function $R_\perp (t)$, it is convenient to scale the time in units of $\tau_D = 3\eta_N V / T$. Moreover, the same quantity is appropriate as well for scaling the stress relaxation time τ_M. On that basis, we introduce the non-dimensional parameter

$$\kappa = \tau_M / \tau_D = 2\zeta_M T / \zeta_N K = 2q / \beta. \tag{21}$$

As seen, κ grows with the increase of "maxwellity" of the corona and/or of temperature and goes down with enhancement of the corona elasticity.

The plots illustrating the behavior of relaxation function $R_\perp (t)$ under variation of bias field H are presented in Figure 2. Their comparison reveals two main specific features. On the one hand, the more viscoelastic the corona, the slower the magnetic relaxation. On the other hand, the bias field, enhancing the effective restoring torque that acts on the particle, accelerates its orientational relaxation. Besides that, Figure 2 proves that magnetic relaxation, as soon as viscoelasticity (non-zero β) is introduced in the particle environment, is multi-mode, and formally the number of modes makes an infinite countable set. Meanwhile, in our model, this set is replaced just by two modes, slow and fast (see Equation (11)). However, as the detailed multi-mode analysis shows (would be presented elsewhere), this truncation does not affect the above-obtained essential qualitative conclusion: the considered relaxation process has two main reference time scales whose ranges differ by orders of magnitude. In logarithmic representation of Figure 2, the plot of $R_\perp (t)$ could be schematized as two quasi-straight lines—greater slope at short times and smaller slope at long times—connected by a crossover part. As seen from Figure 2—see curves *1* in all the panes—under zero bias, the crossover part is but weakly distinguishable. However, by using an appropriate bias field, the crossover part of $R_\perp (t)$ function could be enhanced and exposed to observation (see curves *4* in all the panes).

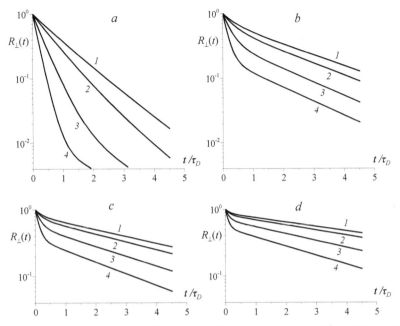

Figure 2. Relaxation functions for the "transverse" MRX; parameter $\kappa = 2 \times 10^3$ (**a**); 100 (**b**); 40 (**c**); 20 (**d**); the non-dimensional bias field is $\xi = 0.1$ (1), 2 (2), 5 (3), 10 (4).

4.2. Integral Relaxation Time

For studying of multi-mode processes, a useful and experimentally easily determined characteristic is the so-called integral relaxation time that is defined as the area under the $R(t)$ curve:

$$\tau_\alpha^{int} = \int_0^\infty R_\alpha(t)dt. \tag{22}$$

In the linear response approximation that we use here, the integral time expressed in terms of the dynamic susceptibility is

$$\tau_\alpha^{int} = \lim_{\omega \to 0}\left[\frac{\mathrm{Im}\tilde{\chi}_\alpha(\omega, \beta, \xi)}{\omega}\right]. \tag{23}$$

With Function (16), integral (23) is taken easily and yields

$$\tau_\alpha^{int} = \tau_D \frac{\Gamma_\alpha}{\gamma_\alpha^{(e)}(\Gamma_\alpha - \frac{1}{2}\beta)} = \tau_D \frac{\gamma_\alpha^{(M)} + \frac{1}{2}\beta(1 + q^{-1})}{\gamma_\alpha^{(e)}\left[\gamma_\alpha^{(M)} + \beta/2q\right]}. \tag{24}$$

This formula (with accuracy of the effective field approximation) in a qualitatively correct way describes the dependence of magnetization relaxation on all the relevant material parameters of the system. For example, for an isotropic system ($\xi = 0$) assuming that $\beta > 1$ and going back to dimensional time, one gets from Equation (23):

$$\tau^{int} = \tau_D \frac{1 + \frac{1}{2}\beta(1 + q^{-1})}{1 + \beta/2q} \simeq \tau_D \begin{cases} 1 + \frac{1}{2}\beta, & \beta \ll 2q, \\ 1 + q, & \beta \gg 2q. \end{cases} \tag{25}$$

As the second line of Equation (25) shows, the asymptotic value of the integral time is defined by the slow diffusion. However, to really attain this regime, one needs a system with very strong elasticity:

$\beta \gg 2q$. Even in dense polymer gels and "live polymers" where $q \sim 10^3 - 10^5$, one could hardly expect to match this condition. Under this limitation, one may always assume that the dependence of integral time (23) on the elasticity parameter is by and large linear.

In the presence of a bias field, full Expressions (13) for coefficients $\gamma_\alpha^{(e)}$ and $\gamma_\alpha^{(M)}$ should be used. Substitution of those in Equation (24) yields the formulas that enable one to estimate the effect of the bias field on the integral times of transverse and longitudinal relaxations of the dynamic magnetization:

$$\tau_\perp^{int} = 2\tau_D \frac{1-c_2}{1+c_2}\left(1 + \beta\frac{1+c_2}{3-c_2}\right), \quad \tau_\parallel^{int} = 2\tau_D \frac{c_2 - c_1^2}{1-c_2}\left(1 + \beta\frac{1-c_2}{1+c_2}\right), \quad \beta \ll 2q. \tag{26}$$

For a strong field ($\zeta \gg 1$), Equation (26) expands to simple asymptotic expressions

$$\tau_\perp^{int} = \tau_H\left[1 + \beta - \frac{2\beta}{\zeta}\right], \quad \tau_\parallel^{int} = \frac{1}{2}\tau_H\left(1 + \frac{1+\beta}{\zeta}\right), \quad \tau_H \equiv \frac{\zeta_N}{\mu H} = \frac{2\tau_D}{\zeta}. \tag{27}$$

Therefore, as it follows from Equation (27), under a strong bias, the initial stage of relaxation is mostly due to the forced field-induced rotation of the particles with reference time $\tau_\perp^{int} \sim \tau_H \sim 2\tau_D/\zeta$.

The overall behavior of the integral time is illustrated in Figure 3, where its dependence on the bias magnitude is given for different values of parameter $\kappa = \tau_M/\tau_D$. The values of integral time under zero bias obey very well Formula (25) for $\beta \ll q$: the integral time grows linearly with the elasticity parameter $\beta = K/T$. As already mentioned, this increase might halt only at $\beta \gg 2q$, where $\tau_\perp^{int} \simeq (q+1)\tau_D$, but that limit is practically unattainable. For the parameters used for plotting Figure 3, this upper limit would have ranged $\tau_\perp^{int}/\tau_D \simeq 100$.

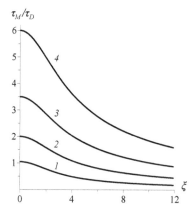

Figure 3. Dependence of the integral relaxation time on the strength of bias field for MRX in "transverse" configuration; the viscoelasticity parameter $\kappa = 2 \times 10^3$ (1), 100 (2), 40 (3), 20 (4); "maxwellity" of the corona is $q = 100$.

5. Conclusions

To summarise the results of the work, three essential points should be mentioned. First, the macromolecular corona covering the particles is considered as a viscoelastic entity. This approximation is qualitatively different from a customary approach where the effect of corona is reduced to just a change of the particle hydrodynamic diameter. Second, we study MRX that is performed in the presence of a bias field, so that the bias field strength is a controllable parameter of the experiment. Third, the main attention is focused on the MRX measured in the "transverse" geometry where the probing field is applied normally to the bias one. The advantage of that variant comes from a significant enhancement of the response in comparison with that obtained in "longitudinal" configuration.

Sensors **2018**, *18*, 1661

Therefore, the transverse configuration enables one to preserve the sensitivity of the method despite that in general the bias field diminishes the particle response to orientational perturbations. We are well aware that the considerations presented here are to a certain extent illustrative as they are obtained with a severely simplified kinetic description: the whole set of moment equations has been replaced by just a pair of them (effective field model). However, our goal is to show, in the first place, the qualitatively relevant features of the considered problem. The work on a complete solution of the kinetic equation is under way.

Author Contributions: V.V. and Y.R. have done all the work on the paper together: from conceiving the idea to writing of the final text.

Funding: The work was done under the auspices of RFBR grant 17-42-590504.

Conflicts of Interest: The authors declare no conflict of interest.

References

1. Sanson, C.; Diou, O.; Thévenot, J.; Ibarboure, E.; Soum, A.; Brûlet, A.; Miraux, S.; Thiaudière, E.; Tan, S.; Brisson, A.; et al. Doxorubicin loaded magnetic polymersomes: Theranostic nanocarriers for MR imaging and magneto-chemotherapy. *ACS Nano* **2011**, *5*, 1122–1140. [CrossRef] [PubMed]

2. Oliveira, H.; Pérez-Andrés, E.; Thevenot, J.; Sandre, O.; Berra, E.; Lecommandoux, S. Magnetic field triggered drug release from polymersomes for cancer therapeutics. *J. Control. Release* **2013**, *169*, 165–170. [CrossRef] [PubMed]

3. Li, Y.; Huang, G.; Zhang, X.; Li, B.; Chen, Y.; Lu, T.; Lu, T.J.; Xu, F. Magnetic hydrogels and their potential biomedical applications. *Adv. Funct. Mater.* **2013**, *23*, 660–672. [CrossRef]

4. Pankhurst, Q.A.; Thanh, N.K.T.; Jones, S.K.; Dobson, J. Progress in applications of magnetic nanoparticles in biomedicine. *J. Phys. D Appl. Phys.* **2009**, *42*, 224001. [CrossRef]

5. Dutz, S.; Hergt, R. Magnetic nanoparticle heating and heat transfer on a microscale: Basic principles, realities and physical limitations of hyperthermia for tumour therapy. *Int. J. Hyperth.* **2013**, *29*, 790–800. [CrossRef] [PubMed]

6. Sanchez, C.; El Hajj Diab, D.; Connord, V.; Clerc, P.; Meunier, E.; Pipy, B.; Payré, B.; Tan, R.P.; Gougeon, M.; Carrey, J.; et al. Targeting a G-protein-coupled receptor overexpressed in endocrine tumors by magnetic nanoparticles to induce cell death. *ACS Nano* **2014**, *8*, 1350–1363. [CrossRef] [PubMed]

7. Foroozandeh, P.; Aziz, A.A. Merging worlds of nanomaterials and biological environment: Factors governing protein corona formation on nanoparticles and its biological consequences. *Nanoscale Res. Lett.* **2015**, *10*, 221. [CrossRef] [PubMed]

8. Eberbeck, D.; Bergemann, C.; Wiekhorst, F.; Steinhoff, U.; Trahms, L. Quantification of specific bindings of biomolecules by magnetorelaxometry. *J. Nanobiotechnol.* **2008**, *6*. [CrossRef] [PubMed]

9. Wiekhorst, F.; Steinhoff, U.; Eberbeck, D.; Trahms, L. Magnetorelaxometry assisting biomedical applications of magnetic nanoparticles. *Pharm. Res.* **2012**, *29*, 1189–1202. [CrossRef] [PubMed]

10. Liebl, M.; Wiekhorst, F.; Eberbeck, D.; Radon, P.; Gutkelch, D.; Baumgartner, D.; Steinhoff, U.; Trahms, L. Magnetorelaxometry procedures for quantitative imaging and characterization of magnetic nanoparticles in biomedical applications. *Biomed. Eng.* **2015**, *60*, 427–443. [CrossRef] [PubMed]

11. Shliomis, M.I.; Stepanov, V.I. Theory of the dynamic susceptibility of magnetic fluids. *Adv. Chem. Phys.* **1994**, *78*, 1–30, ISBN 0-471-30312-1.

12. Raikher, Y.L.; Shliomis, M.I. Effective field method for orientational kinetics of magnetic fluids and liquid crystals. *Adv. Chem. Phys.* **1994**, *78*, 595–751, ISBN 0-471-30312-7.

13. Fortin, J.-P.; Wilhelm, C.; Servais, J.; Ménager, C.; Bacri, J.-C.; Gazeau, F. Size-sorted anionic iron oxide nanomagnets as colloidal mediators for magnetic hyperthermia. *J. Am. Chem. Soc.* **2007**, *129*, 2628–2635. [CrossRef] [PubMed]

14. Raikher, Y.L.; Stepanov, V.I. Physical aspects of magnetic hyperthermia: Low-frequency ac field absorption in a magnetic colloid. *J. Magn. Magn. Mater.* **2014**, *368*, 421–427. [CrossRef]

15. Oswald, P. *Rheophysics: The Deformation and Flow of Matter*; Cambridge University Press: Cambridge, UK, 2009; ISBN 978-0-521-88362-7.

16. Raikher, Y.L.; Rusakov, V.V. Brownian motion in a Jeffreys fluid. *J. Exp. Theor. Phys.* **2010**, *111*, 883–889. [CrossRef]

17. Raikher, Y.L.; Rusakov, V.V.; Perzynski, R. Brownian motion in a viscoelastic medium modelled by a Jeffreys fluid. *Soft Matter* **2013**, *9*, 10857–10865. [CrossRef]

18. Raikher, Y.L.; Rusakov, V.V.; Perzynski, R. Brownian motion in the fluids with complex rheology. *Math. Model. Nat. Phenom.* **2015**, *10*, 1–43. [CrossRef]

19. Ochab-Marcinek, A.; Wieczorek, S.A.; Ziębacz, N.; Hołyst, R. The effect of depletion layer on diffusion of nanoparticles in solutions of flexible and polydisperse polymers. *Soft Matter* **2012**, *8*, 11173–11179. [CrossRef]

20. Gomez-Solano, J.R.; Bechinger, C. Transient dynamics of a colloidal particle driven through a viscoelastic fluid. *New J. Phys.* **2015**, *17*, 103032. [CrossRef]

21. Pavlovsky, L.; Younger, J.G.; Solomon, M.J. In situ rheology of Staphylococcus epidermidis bacterial biofilms. *Soft Matter* **2013**, *9*, 122–132. [CrossRef] [PubMed]

22. Gardel, M.L.; Valentine, M.T.; Weitz, D.A. Microrheology. In *Microscale Diagnostic Techniques*; Breuer, K., Ed.; Springer: New York, NY, USA, 2005; ISBN 3-540-23099-8.

23. Happel, J.; Brenner, H. *Low Reynolds Number Hydrodynamics*; Prentice-Hall: Boston, MA, USA, 1965.

24. Coffey, W.T.; Kalmykov, Y.P. *The Langevin Equation*, 4rd ed.; World Scientific: Singapore, 2017; ISBN 978-981-3221-99-4.

25. Ilg, P.; Evangelopoulos, E.A.S. Magnetic susceptibility, nanorheology, and magnetoviscosity of magnetic nanoparticles in viscoelastic environments. *Phys. Rev. E* **2018**, *97*, 032610. [CrossRef] [PubMed]

26. Rusakov, V.V.; Raikher, Y.L. Magnetic response of a viscoelastic ferrodispertion: From a nearly Newtonian ferrofluid to a Jeffreys ferrogel. *J. Chem. Phys.* **2017**, *147*, 124903. [CrossRef] [PubMed]

27. Raikher, Y.L.; Stepanov, V.I. Nonlinear dynamic susceptibilities and field-induced birefringence in magnetic particle assemblies. *Adv. Chem. Phys.* **2004**, *129*, 419–588.

28. Raikher, Y.L.; Rusakov, V.V. Orientational Brownian motion in a viscoelastic medium. *Colloid J.* **2017**, *79*, 264–269. [CrossRef]

Article

Water-Based Suspensions of Iron Oxide Nanoparticles with Electrostatic or Steric Stabilization by Chitosan: Fabrication, Characterization and Biocompatibility

Galina V. Kurlyandskaya [1,2,*], Larisa S. Litvinova [3], Alexander P. Safronov [2,4],
Valeria V. Schupletsova [3], Irina S. Tyukova [2], Olga G. Khaziakhmatova [3], Galina B. Slepchenko [5],
Kristina A. Yurova [3], Elena G. Cherempey [5], Nikita A. Kulesh [2], Ricardo Andrade [6],
Igor V. Beketov [4] and Igor A. Khlusov [3,7]

[1] Departamento de Electricidad y Electrónica and BCMaterials, Universidad del País Vasco UPV-EHU, 48080 Bilbao, Spain
[2] Institute of Natural Sciences and Mathematics, Ural Federal University, Ekaterinburg 620002, Russia; safronov@iep.uran.ru (A.P.S.); tyukov.sergey@mail.ru (I.S.T.); kuleshnik@list.ru (N.A.K.)
[3] Laboratory of Immunology and Cell Biotechnology, I. Kant Baltic Federal University, Kaliningrad 23601, Russia; larisalitvinova@yandex.ru (L.S.L.); vshupletsova@mail.ru (V.V.S.); hazik36@mail.ru (O.G.K.); kristina_kofanova@mail.ru (K.A.Y.)
[4] Institute of Electrophysics, Ural Division RAS, Ekaterinburg 620016, Russia; igor.beketov@mail.ru
[5] Department of Physical and Analytical Chemistry, National Research Tomsk Polytechnic University, Tomsk 634050, Russia; slepchenkogb@mail.ru (G.B.S.); cherempey@mail.ru (E.G.C.)
[6] Advanced Research Facilities (SGIKER), Universidad del País Vasco UPV-EHU, 48080 Bilbao, Spain; ricardo.andrade@ehu.eus
[7] Department of Experimental Physics, National Research Tomsk Polytechnic University, Tomsk 634050, Russia; khlusov63@mail.ru
[*] Correspondence: galina@we.lc.ehu.es; Tel.: +34-9460-13237; Fax: +34-9460-13071

Received: 23 October 2017; Accepted: 10 November 2017; Published: 13 November 2017

Abstract: Present day biomedical applications, including magnetic biosensing, demand better understanding of the interactions between living systems and magnetic nanoparticles (MNPs). In this work spherical MNPs of maghemite were obtained by a highly productive laser target evaporation technique. XRD analysis confirmed the inverse spinel structure of the MNPs (space group Fd-3m). The ensemble obeyed a lognormal size distribution with the median value 26.8 nm and dispersion 0.362. Stabilized water-based suspensions were fabricated using electrostatic or steric stabilization by the natural polymer chitosan. The encapsulation of the MNPs by chitosan makes them resistant to the unfavorable factors for colloidal stability typically present in physiological conditions such as pH and high ionic force. Controlled amounts of suspensions were used for in vitro experiments with human blood mononuclear leukocytes (HBMLs) in order to study their morphofunctional response. For sake of comparison the results obtained in the present study were analyzed together with our previous results of the study of similar suspensions with human mesenchymal stem cells. Suspensions with and without chitosan enhanced the secretion of cytokines by a 24-h culture of HBMLs compared to a control without MNPs. At a dose of 2.3, the MTD of chitosan promotes the stimulating effect of MNPs on cells. In the dose range of MNPs 10–1000 MTD, chitosan "inhibits" cellular secretory activity compared to MNPs without chitosan. Both suspensions did not caused cell death by necrosis, hence, the secretion of cytokines is due to the enhancement of the functional activity of HBMLs. Increased accumulation of MNP with chitosan in the cell fraction at 100 MTD for 24 h exposure, may be due to fixation of chitosan on the outer membrane of HBMLs. The discussed results can be used for an addressed design of cell delivery/removal incorporating multiple activities because of cell capability to avoid phagocytosis by immune cells. They are also promising for the field of biosensor development for the detection of magnetic labels.

Keywords: magnetic biosensors; iron oxide magnetic nanoparticles; chitosan; ferrofluids; human blood mononuclear leukocytes; morphofunctional response

---

1. Introduction

Magnetic ferrofluids (FFs) have been attracting increasing interest for technological and biomedical applications in recent years. A FF is a stable suspension of colloidal magnetic nanoparticles (MNPs) dispersed in a carrier liquid [1].Biomedical applications demand magnetic MNPs in the form of water-based ferrofluids or ferrogels [2,3]. The list of proposed applications has extended rapidly, counting with uses such as drug delivery carriers, contrast media for magnetic resonance imaging, magnetic labels for magnetic biosensing, thermal ablation and hyperthermia, thermal activation for drug release and others [4–7]. Asystematic comparative analysis of the morphofunctional response of different kinds of living cells to the presence of MNPs is still lackingbut a better understanding of the interactions of the living systems with MNPs and search for synergetic combinations "cell type-MNPs" would be beneficial [8]. The progress in the fabrication of the biocomposites is especially important for the field of magnetic biosensing because further development of the compact analytical devices in which a magnetic transducer converts a magnetic field variation into a change of frequency, current, voltage requires thoroughly characterized biological samples with well-known amount and distribution of MNPs [9,10].

The necessary condition for the biomedical application of the suspension of MNPs is the maintenance of its colloidal stability in the neutral range of pH and high ionic strength, inherent to biological tissues. There are two main possibilities to obtain stable colloidal suspensions of deaggregated nanoparticles: electrostatic and steric stabilization [9]. An increase of the ionic strength of the biological medium makes the electrostatic stabilization of the suspensions less effective. In a number of recent studies the natural polymer chitosan has been discussed as an effective electrosteric stabilizer [11,12].

Chitosan is a polysaccharide produced by chemical modification of a natural polymer—chitin. The biocompatibility, biodegradability, immuno-stimulating features, high chemical reactivity of chitosan provides its extensive usage in a variety of biomedical and bioengineering applications [13]. Chitosan is a polyelectrolyte, whose typical molecular structure is shown in Figure 1.

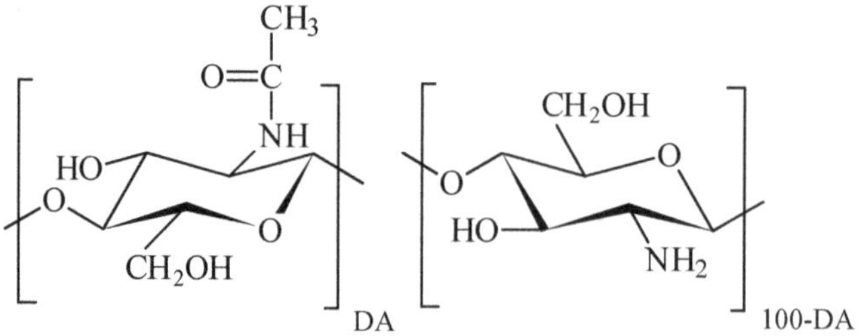

Figure 1. Chemical structure of chitosan monomer units: acetylated to the left, deacetylated to the right. DA is the degree of acetylation.

The macromolecular backbone is based on β-glucose units with hydroxyl residue in 2nd position substituted either by acetamide or by amine group. In the natural precursor of chitosan (chitin) all monomeric units contain acetamide moieties, which are partly or completely removed during chemical modification of chitin, which is known as deacetylation. The degree of acetylated units (DA, %) remained in the molecular structure is the characteristic feature of the chemical composition of chitosan, which governs its interaction with water and solutes. Amine groups in the deacetylated monomer units can be protonated and thus they can carry on positive electrical charge, which makes chitosan a polycation. The polycationic nature of chitosan favours its interaction with polyanionic biomacromolecules such as proteins and nucleic acids, which results in the formation of polyelecrolyte complexes. The formation and the stability of chitosan polyelectrolyte complexes depends on the DA, charge density, molecular weight, quality of a solvent, ionic strength, pH, and temperature [14]. The formation of chitosan—protein complexes takes place if both counterparts are ionized. Polyelectrolyte complexes with collagen, gelatin, and milk albumin are reported in literature [15].

Chitosan is widely tested as an efficient biocompatible stabilizer in the suspensions of nanoparticles of different chemical origin designed for biomedical applications [16]. The presence of both hydroxyl and amine residues in chitosan monomer unit enhance specific interaction of chitosan macromolecules with the surface of metal oxide MNPs and provide the stability of suspensions at low polymer concentration [17,18]. However, the majority of the reported data on the stabilization of the suspensions of MNPs by chitosan are related to acidic solutions in which chitosan is protonated and acts as a polycation. Meanwhile, the necessary condition for the biomedical application of the suspension of MNPs is the maintenance of its colloidal stability in the neutral range of pH and high ionic strength, inherent for biological tissues. In the neutral pH range as its monomer units are deprotonated, chitosan is not efficient as a stabilizer. In our recent works [19,20] we had introduced a route to overcome this limitation and provide the stabilization of iron oxide MNPs by chitosan in phosphate buffer saline (PBS, pH = 6.3), which is widely used in biomedical and bioengineering studies.

This approach is based on the well-known feature of colloidal suspensions that their stability depends not only on the composition but also on the prior history of their preparation. First, suspensions of iron oxide MNPs electrostatically stabilized by sodium citrate were prepared. Sodium citrate is a very good stabilizer for iron oxide MNPs in water due to the adsorption of citrate anions onto the MNPs' surface. It was shown earlier [21,22] that using citrate as an electrostatic stabilizer the suspension of individual iron oxide MNPs could be obtained. At the next step the electrostatically stabilized suspension was mixed with acidic chitosan solution. As chitosan was protonated and carried positive charges, it interacted with negatively charged MNPs electrostatically and adsorbed on their surface. Then PBS was added to deprotonate chitosan macromolecules and to make them collapse around MNPs. Thus the steric stabilization of MNPs resistant in high ionic strength was achieved. For the sake of comparison electro-statically stabilized water-based stable suspension of iron oxide nanoparticles from the same batch was also prepared using sodium citrate. We present our herein study on the fabrication, thorough characterization and biocompatibility of the suspensions of iron oxide MNPs electrostatically or sterically stabilized by chitosan tested for the case of human blood mononuclear leukocytes.

2. Experimental

2.1. Iron Oxide MNPs

Iron oxide MNPs were synthesized by laser target evaporation (LTE)—the method of high power physical dispersion of iron oxide in gas phase by laser irradiation. The details of the experimental procedure are given elsewhere [22,23]. LTE was performed using laboratory installation with Ytterbium (Yb) fiber laser with 1.07 μm wavelength. The laser operated in a pulsed regime with pulse frequency 4.85 KHz and pulse duration 60 μs. Average output power of irradiation was 212 W. The target pellet of 65 mm in diameter, 20 mm in height was pressed from commercial magnetite (Fe_3O_4) (Alfa Aesar, Ward Hill, MA, USA) powder (specific surface area 6.9 m^2/g). The target was evaporated in a mixture of N_2 and O_2 in the volume ratio 0.79:0.21.

The X-ray diffraction (XRD) studies were performed with a DISCOVER D8 diffractometer (Bruker, Billerica, MA, USA) using Cu-K$_\alpha$ radiation (λ = 1.5418 Å), a graphite monochromator and a scintillation detector. The MNPs were mounted on a zero background silicon wafer placed in a sample holder. A fixed divergence and antiscattering slit were used. The quantitative analysis was done using the TOPAS-3 software with Rietveld full-profile refinement [24]. Transmission electron microscopy (TEM) was performed to evaluate themorphology of the MNPs (JEM2100, JEOL, Tokyo, Japan). The specific surface area of MNPs (S_{sp}) was measured by the low-temperature adsorption of nitrogen (TriStar3000, Micromeritics, Norcross, GA, USA).

Magnetic measurements of the hysteresis loops (M(H)) and thermomagnetic curves (ZFC-FC, see [23] for details) were carried out with a MPMSXL-7 SQUID-magnetometer (Quantum Design, San Diego, CA, USA) in the ±70 Oe field range. Thermomagnetic zero field cooled/field cooled (ZFC-FC) curves were obtained for the field of 100 Oe (5–300 K temperature range) for air-dry MNPs and MNPs dried from ferrofluids. Ferrofluids and biological samples were dried prior to measurement and polymer container contributions were carefully subtracted.

2.2. Preparation of Ferrofluid and Suspension of MNPs Encapsulated by Chitosan

Before the preparation of the suspensions the iron oxide MNPs were dry-heat sterilized with a Binder FD53 device (Binder GmbH, Tuttlingen, Germany) at 180 °C for 1 h. Ferrofluid based on iron oxide MNPs was prepared in distilled water with the addition of an electrostatic stabilizer sodium citrate in 5 mM concentration. Suspension in an initial concentration 6% (by weight) was treated by ultrasound for 30 min using a CPX-750 processor (Cole-Parmer Instruments Corp., Vernon Hills, IL, USA) operated at 300 W power output. After the ultrasound treatment the suspension was centrifuged at 10,000 rpm for 5 min using Z383 centrifuge (Hermle, Hermle-labortechnik, Wehingen, Germany) equipped with a 218 rotor. Such a prepared suspension (electrostatically stabilized water-based ferrofluid) was designated FF1.

Encapsulation of MNPs by chitosan was performed using commercial chitosan SK-1000 (AlexinChem, Tula, Russia) with MW = 4.4×10^5 and DA = 30%. Chitosan was dissolved in 0.2M HCl to prepare 2% stock solution; the solution was kept 48 h at 25 °C for equilibration. Then chitosan stock solution was mixed with ferrofluid based on MNPs in 1:3 volume ratio and vigorously stirred. Then the pH of suspension was adjusted by the dropwise addition of PBS solution (in 1:2 volume ratios to suspension) under vigorous stirring. Such a prepared suspension (at a time electrostatically and sterically stabilized water-based ferrofluid) was designated the name FF2.

Hydrodynamic diameters of the MNPs and their aggregates were measured by dynamic light scattering (Zeta Plus particle size analyzer, Brookhaven Brookhaven Instruments Corp., Holtville, NY, USA). The electrokinetic zeta-potentials of the ferrofluids were measured by electrophoretic light scattering with the same instrument.

2.3. Cell Culture

The study of real biological samples (cells, unicelular organisms or tissues) has certain testing limitations when physical methods are employed. These limitations are caused by the wide variety of complex processes that exist in the living systems and the wide variety of morphologies of each particular sample. To make biophysical research withMNPs more effective, one must select model systems, with well known or most stable/typical properties. One of such model systems is human blood mononuclear leukocytes (HBMLs). HBMLs of health volunteer (Permit No. 4 from 23.10.2013 of Local Ethics Committee of Innovation park of Immanuel Kant Baltic Federal University, Russian Federation) were collected by venous blood gradient ($\varrho^* = 1.077$) Ficoll-Paque Premium (Sigma-Aldrich, St. Louis, MO, USA) centrifugation at 1500 rpm for 10 min. The HBMLs were twice washed by phosphate-buffered saline (pH 7.2) and resuspended in complete culture medium consisting of 90% RPMI-1640 (Sigma-Aldrich, St. Louis, MO, USA), 10% inactivated (for 30 min at 56 °C) fetal bovine serum (Sigma), and 0.3 mg/mLL-glutamine (Sigma). Viability was 90–95% for living cells unstained by 0.4% trypan blue.

The HBMLs suspension was added into 24-well plates (Orange Scientific, Braine-l'Alleud, Belgium) with a final concentration of 2×10^6 viable cells per 1 mL of nutrient medium containing

RPMI-1640 (Sigma-Aldrich, St. Louis, MO, USA), 10% inactivated (for 30 min at 56 °C) fetal bovine serum (Sigma-Aldrich, St. Louis, MO, USA), 50 mg/L gentamicin (Invitrogen, Glasgow, UK) and freshly added L-glutamine sterile solution in a final concentration of 0.3 mg/mL (Sigma-Aldrich, St. Louis, MO, USA). Iron oxide maximum tolerated dose (MTD) was calculated on the basis of iron ions: one MTD of iron in water was equal to 0.3 mg/L. Then FF1 or FF2 suspensions of magnetic nanoparticles were immediately added to cell culture in final concentrations of 2.3, 10, 100 or 1000 MTDs. Both suspensions were pipetted fivetimes. The control cell culture (0 MTD) had nether MNPs nor chitosan solution. The cell cultures were incubated for 24 h in atmosphere of 95% air and 5% CO_2 at 37 °C. Preliminary study showed no sign of chitosan toxic effect on HBMLs culture. After incubation, the cell suspension was centrifuged at 1300 rpm for 10 min. The obtained cell pellet was used to measure cell death, membrane antigen presentation, transmission electron microscopy (TEM), and energy dispersive X-ray fluorescence analysis. The cell culture supernatants were employed to measure cytokine concentrations.

2.4. Measurement of Cell Death

Calculations of cell concentration and viability were conducted with a Countess TM Automated Cell Counter (Invitrogen, Carlsbad, CA, USA) using 0.4% trypan blue solution (Invitrogen, Carlsbad, CA, USA). To this end, the cells to be tested were transferred in a volume of 12.5 μL to an immunological plate, and 12.5 μL of stain was added. Viable cells were unstained by 0.4% trypan blue.

2.5. Cytokine Profile in the Cell Culture

Cytokines are small peptide information molecules. Cytokines have a molecular mass not exceeding 30 kD. Their main producers are lymphocytes. Cytokines regulate intercellular and intersystem interactions, determine cell survival, stimulation or suppression of their growth, differentiation, functional activity and apoptosis, and also ensure the compatibility of the action of the immune, endocrine and nervous systems under normal conditions and in response to pathological effects. Cytokines are active in very low concentrations. Their biological effect on cells is realized through interaction with a specific receptor located on the cell cytoplasmic membrane. The formation and secretion of cytokines occurs briefly and it is strictly regulated. In the present work, to measure the spontaneous secretion of interleukins (IL-2, IL-4, IL-6, IL-8, and IL-10),granulocyte-macrophage colony-stimulating factor (GM-CSF), interferon-gamma (IFN_γ), and tumor necrosis factor alpha (TNF_α) in the supernatants (intracelular fluids), a flow cytometry (FC) was performed. The FC procedure was conducted according to the instructions of the manufacturer of the cytokine assay system (Bio-Plex Pro Human Cytokine 8-Plex Panel, Bio-Rad, Hercules, CA, USA) using an automated processing system (Bio-Plex Protein Assay System, Bio-Rad, USA). The concentration of each cytokine was expressed in pg/mL.

2.6. Cellular Immunophenotype Detection

The cellular antigen profile was analyzed using a method based on the interaction between specific monoclonal antibodies (mAbs; see below) and clustering determinants on the cell surface according to the manufacturer's instructions. After culturing, the cells were washed with phosphate-buffered saline (pH 7.2), and a single-cell suspension in a volume of 10 μL was mixed 1:1 with a single standard mAb against CD45, CD3, CD4, CD8, CD25 (Abcam, Cambridge, UK), CD28, CD45RO, CD45RA, CD71, or CD95 (e-Bioscience, San Diego, CA, USA). The mAbs were labeled with fluorescein isothiocyanate (FITC), allophycocyanin (APC), phycoerythrin (PE), or peridinin chlorophyll protein (PerCP). After 30 min of incubation with the labeled mAb, the cells were assayed using a MACS Quant flow cytometer (Miltenyi Biotec, Teterow, Germany). Measurements of the orange/green/red fluorescence parameters were performed at the gate of the analyzed cells, and the numbers of cells presenting the studied antigenic determinants were calculated. The cytometric results were examined using KALUZA Analysis Software (Beckman Coulter, Indianapolis, IN, USA).

2.7. Electrochemical Testing

Concentrations of iron oxide MNPs in biological liquids were confirmed by stripping voltammetry (SV) of iron ions [25].Calculated maximum tolerated doses in medium were compared to those estimated by SVas described in [12].

2.8. Transmission Electron Microscopy of Cells

The protocol to visualize cells under TEM was follows. Cell culture was washed in PBS at 37 °C and pre-fixed in 10 mL of 0.5% glutaraldehyde in Sörenson buffer at room temperature for 15 min. After pre-fixation, cells were decanted into a 15 mLFalcon tube and centrifuged at 1500 rpm for 10 min. As the next step the supernatant was removed and freshly made 2% glutaraldehyde in Sörenson buffer added to the pellet. After 2 h fixation the pellet was spun at 2000 g for further compaction. Samples were fixed in 1% Osmium Tetroxide in Sörenson buffer, dehydrated, embedded in Epon Polarbed resin and cut as 70 nm ultrathin sections for TEM studies (EM208S, Philips, Moosseedorf, Germany).

To compare iron oxide MNPs distribution in different cells the primary culture of post-natal adipose-derived multipotent mesenchymal stromal cells (AMMSCs) was prepared from human fate tissue after processing of lipoaspirates (Permission No. 4 from 23.10.2013 of Local Ethics Committee of Innovation park of Immanuel Kant Baltic Federal University). AMMSCs culturing and TEM was prepared as described in our previous work [12].

2.9. Statistical Analysis

The results were analyzed using STATISTICA software for Windows 10.0. The following distribution parameters were calculated: the median (Me), the 25% quartile (Q1) and the 75% quartile (Q3). The Mann-Whitney U-test (P_U) was performed, and differences were considered significant at $p < 0.05$. The relationship between the studied parameters was established via regression analyses. The coefficients (r) were kept at a significance level greater than 95%.

2.10. Energy Dispersive X-ray Fluorescence Analysis of Cells

Total reflection X-ray fluorescent spectrometry (TXRF) [26] is a relatively new method of elemental analysis that can be applied to samples in the form of thin layer including dried drops of homogenized suspensions of fine particles or a thin layer of whole cells. Although the TXRF method without a pressure controlled chamber proved to be limited for low Z elements quantification (such as carbon, nitrogen, and oxygen) [27,28], it can be successfully applied for the determination of elements with higher-energy characteristic lines. All TXRF measurements were carried out by a Nanohunter spectrometer (Rigaku, Tokyo, Japan). For every experiment the same parameters were used: exposure time of 500 s, angle of 0.05°, X-ray tube withCu anode as a primary beam source. Samples were dried at 50 °C in a Rigaku Ultra dry chamber at normal pressure. For all calculations of iron concentration, we neglected the matrix effects assuming thin film sample geometry.

3. Results and Discussion

Figure 2a shows a typical TEM microphotograph of iron oxide MNPs synthesized by the LTE technique. The MNPs were non-agglomerated and their shape was close to spherical. Only few of the particles appeared to be hexagonal or having hexagonal corners. Weighted particle size distribution (PSD) (Figure 2a inset) was lognormal, with a median value of 26.8 nm and dispersion 0.362. The specific surface area of MNPs was 78 m^2/g. The surface average diameter of MNPs, calculated from this value using the equation $d_s = 6/(\varrho \times S_{sp})$ (ϱ = 4.6 g/cm^3 being iron oxide density) was 16.7 nm. It was in good agreement with the value d_s = 15.9 nm, obtained using PSD with aforementioned parameters. XRD plot of iron oxide MNPs is given in Figure 2b.

The crystalline structure of MNPs corresponded to the inverse spinel lattice with a space group Fd3m. The lattice period (a) was found a = 0.8358 nm, which was larger than that for iron oxide (γ-Fe$_2$O$_3$, a = 0.8346 nm) but lower than that for magnetite (Fe$_3$O$_4$, a = 0.8396 nm) [29]. Based on the dependence between the lattice period of the spinel cell and the effective state of oxidation of Fe the composition of MNPs was defined. It contained 76% of γ-Fe$_2$O$_3$ and 24% of Fe$_3$O$_4$.

Average hydrodynamic diameters of aggregates in suspensions were monitored by the dynamic light scattering (Brookhaven Zeta Plus). The concentration of ferrofluid after centrifuging was 5.0% of MNPs by weight for FF1 suspension. The average hydrodynamic diameter of MNPs in suspension was 56 nm.

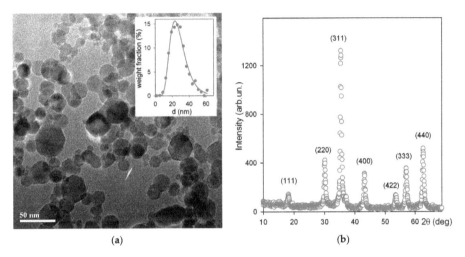

Figure 2. TEM image of iron oxide MNPs (JEOL JEM2100 operating at 200 kV). Inset: particle size distribution (number averaged) obtained by the graphical analysis of TEM images (2160 particles) (**a**). XRD plot for iron oxide MNPs (Bruker D8 DISCOVER) Cu-K$_\alpha$ radiation (λ = 1.5418 Å), a graphite monochromator and a scintillation detector (**b**).

Magnetic measurements confirmed that MNPs were close to a superparamagnetic state with low coercivity (H$_c$) at 20 °C (Figure 3a). Following H$_c$ values were obtained for air-dried MNPs—30 Oe, FF1 MNPs—5 Oe and FF2 MNPs—5 Oe. It is noteworthy that the saturation was not totally reached at a maximum available field of 70 kOe. The saturation magnetization (M$_s$) value of about 70 emu/g for MNPs is in good agreement with about 20 nm sized MNPs of maghemite [22,30]. The coercivity decrease down to about 5 Oe and Ms decrease for FF1 and FF2 MNPs were consistent with the separation process during the suspension fabrication and the fact of the reduction of the mass of magnetic material per unit mass due to sodium citrate and chitosan incorporation. Although analysis of the transition between SPM behavior and blocked state was not the subject of the presents study in order to demonstrate very roughly the features of magnetic interactions we show the results of ZFC-FC measurements for air-dry LTE MNPs and MNPs obtained by drying the FF1 ferrofluid. For used concentrations MNPs are supposed to be weakly interacting but the ensemble can contain coarse particles which contribute much to coercivity and complex shape of ZFC–FC thermomagnetic curves (Figure 3d). Even so these curves additionally verify the MNP superparamagnetic behavior in the ensemble under study. Thus, the blocking temperature of about 150 K typical for superparamagnetics can be seen. It is interesting that the parameters of ZFC–FC thermomagnetic curves for MNPs produced from dried aqueous suspension FF1, are very close to those for air—dry MNPs. The latter is explained by a small original size of the MNPs, i.e., their separation during the preparation of magnetic suspensions has no considerable effect on the size distribution parameters.

The FF2 suspensions of iron oxide MNPs for biocompatibility studies were prepared through several steps schematically shown in Figure 4a. First, a water suspension of individual MNPs electrostatically stabilized by adsorbed surface layer of citrate anions was prepared. Its zeta-potential was –48 mV, which provided efficient mutual repellence of negatively charged MNPs and prevented their aggregation. The next step was the treatment of MNPs with the acidic solution of chitosan, which resulted in the adsorption of chitosan polycations on the negatively charged surface of MNPs. The adsorption of chitosan was not, however, purely electrostatic, because it resulted in the inversion of the surface charge of the MNPs. Their zeta-potential after the adsorption step was +52 mV, which was very close to the zeta-potential of the chitosan solution (+50 mV). It means that the total positive charge of adsorbed chitosan exceeds the negative charge located at the surface of MNPs. Most likely the bonding of chitosan macromolecules to the surface stems from the molecular interactions due to amine and hydroxyl groups in monomer units (see Figure 1). The final step of preparation was the adjustment of pH to neutral level by PBS, which caused the de-ionization of chitosan and the contraction of its coils around the MNPs. The details and the physicochemical background of such an encapsulation of iron oxide MNPs by chitosan was given in our previous studies [19,20].

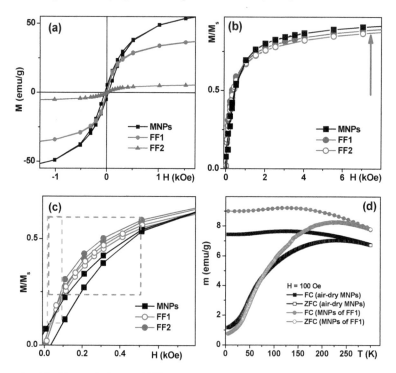

Figure 3. (**a**)—Hysteresis loops of LTE MNPs, FF-1, FF-2 suspensions at room temperature. Recalculated M/M_s hysteresis loops for M_s—being a saturation magnetization values corresponding to each particular case (**b**,**c**), red arrow shows the direction of the increase of the magnetization; (**c**)—dashed lines show two important intervals for possible applications: light green—magnetic biosensing interval; green—cell use as native microcapsules for targeted delivery in a supplement of nanomedicine, theranostics, cell technologies, and regenerative medicine; (**d**)—ZFC-FC curves for air dry LTE MNPs and MNPs of FF1.

Figure 4b shows PSD (obtained by DLS) of species in the suspension of iron oxide MNPs ecapsulated by chitosan in comparison with PSD in the precursor suspension of MNPs stabilized by citrate and with PSD in the solution of chitosan. One can see, that in the case of PSD of the suspension FF1 the median value of the diameter was about 50 nm. It is higher than the weight average value of the diameter determined in the graphical analysis of TEM images (about 30 nm). The difference, however, is not large and may likely be attributed to the thickness of solvating shells and the thickness of the double electric layer at the surface of the particles, which provide the stability of their suspension. The median diameter of PSD for the solution of chitosan and for the suspension of the encapsulated MNPs is much larger. In both cases it is around 300 nm. The obtained value for the chitosan solution is much higher than that anticipated based on the molecular parameters of chitosan.

The characteristic dimension of the polymeric chain in the solution is the mean square end-to-end distance for the macromolecular coil, which can be estimated using the following equation [31]:

$$\left\langle R^2 \right\rangle = NA^2 \tag{1}$$

where N is the number of statistical segments in the macromolecule and A is the length of the segment.

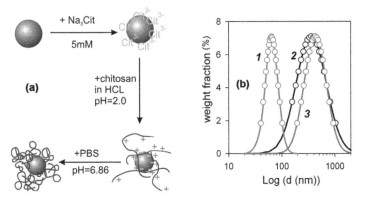

Figure 4. (a)—Scheme of encapsulation of iron oxide MNPs by chitosan. (b)—PSD (obtained by DLS) in suspensions at different steps of the encapsulation: 1—suspension FF1 of MNPs in 5 mM citrate—solution of chitosan taken for encapsulation—suspension FF2 of encapsulated MNPs.

The length of the segment for chitosan varies in different reports [32], instead the value 10 nm might be taken as a fair estimation. It corresponds to ca. 20 monomer units in the segment. The average total number of monomer units in chitosan used in the study can easily be evaluated as a ratio of the molecular weight of chitosan to the formula molecular weight of the unit (148). Thisgives ca. 3000 units in chitosan and, hence, ca. 150 segments in the macromolecule. The calculation of end-to-end distance based on these values gives ca. 150 nm, which is two times lower than the median diameter of PSD for chitosan solution. It means that the solution of chitosan does not contain individual macromolecules but their aggregates. This result is consistent with published reports on the association and aggregation in chitosan solution [33,34]. According to them association is inevitable and occurs even in filtered dilute solutions being kept for some period of time in isothermal conditions. We had obtained the mean value of the diameter of 80 nm for the acidic solutions of chitosan with 0.002 g/L concentration filtered through Wattman 0.1 μm filter. Meanwhile, the associates were re-established after the three-day period of storage of the solution at 25 °C. It is obvious from Figure 4 that the diameter of encapsulated MNPs is the same as the diameter of chitosan aggregates. It means that the encapsulation of individual MNPs by chitosan can possibly be achieved only in very dilute solution; but in the solutions of finite concentration the process of interaction among chitosan and MNPs rather ends up in the incorporation of MNPs into the associates (aggregates) of chitosan macromolecules.

Figure 5a shows the influence of pH on the zeta-potential and on the mean hydrodynamic diameter of aggregates in the suspension of the encapsulated MNPs. Zeta-potential is almost constant in the pH range up to pH = 4 and gradually diminishes with the further increase of pH. It reaches values below +20 mV in the physiologically relevant range of pH = 6.5–7.5. The decrease of zeta-potential happens due to the deprotonation of amine groups in the chitosan macromolecule according to the reaction:

$$Chit\text{-}NH_3^+Cl^- + OH^- => Chit\text{-}NH_2 + H_2O + Cl^-$$

As the positive electric charge of chitosan macromolecules decreases, the electrostatic factor of stability of the suspension diminishes. In principle, it is unfavorable to the stability, as it is known from the theoretical consideration in terms of the Derjaguin-Landau-Verwey-Overbeek (DLVO) approach that the zeta-potential should maintain the value above 30 mV irrespective of the sign [35] to provide the stability of the suspension against the aggregation due to the attractive Van-der-Waals forces between MNPs. Meanwhile, the suspension of encapsulated MNPs retains its stability, and the mean hydrodynamic diameter of chitosan associates does not increase substantially. This effect is totally due to the steric repulsion among chitosan aggregates with embedded iron oxide MNPs. This repulsion stems from the osmotic forces, which appear if the polymeric shells of aggregates overlap. In this case the local concentration of segments increases in the overlapping region and it causes the incoming osmotic flux of water, which prevents interpenetration of aggregates and keeps them apart.

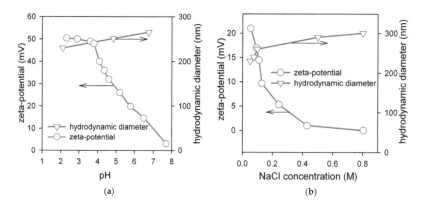

(a)   (b)

Figure 5. (**a**)—pH dependence of the zeta-potential and of the mean hydrodynamic diameter of aggregates in the suspension FF2 of the encapsulated iron oxide MNPs. (**b**)—Dependence of the zeta-potential and of the mean hydrodynamic diameter of aggregates in the FF2 suspension of the encapsulated MNPs at pH = 6.5.

Figure 5b shows the influence of salt (NaCl) concentration on the zeta-potential and on the mean hydrodynamic diameter of aggregates in the suspension of the encapsulated MNPs at pH = 6.5. The zeta-potential of the suspension decreases substantially if salt is added due to the contraction of the double electric layer on the surface of chitosan aggregates. At NaCl concentration 0.4 M zeta-potential vanishes to zero. It means that the electrostatic factor of the stability of the suspension is effectively eliminated by the ionic force of the salt. Meanwhile, the level of aggregation indicated by mean hydrodynamic diameter of chitosan associates retains almost the same. Certainly, it is the result of the efficient steric stabilization of the encapsulated MNPs in FF2 suspension.

In general, the dependences given in Figure 5 show that the encapsulation of LTE iron oxide MNPs by chitosan makes them resistant to unfavorable factors for the colloidal stability such as pH and high ionic force, which are typically present under physiological conditions. It makes the suspensions of MNPs encapsulated by chitosan promising candidates for the biocompatibility testing. Therefore, as the next step biological experiments were provided for comparative analysis of the biocompatibility of the FF1 and FF2 suspensions for the case of human blood mononuclear leukocytes.

According to the experimental data, intact HBMLs express a wide range of membrane markers in a 24-h in vitro culture. 99% of the cells express CD45CD3 T-lymphocyte antigens, predominantly (55%) of CD45RA$^+$ naive (antigen non-activated) CD4$^+$ T-helper/inducers (67%) [36,37]. Membrane activation and co-stimulatory molecules CD25, CD28, CD71 and CD95 are present for CD45CD3$^+$ cells in 9%; 79%; 2.5% and 15% of cases, respectively. CD45RO isoform of the transmembrane antigen expressed in vitro on activated T-lymphocytes and/or T-memory cells was detected on 35% for CD45CD3$^+$ cells [37,38] (Table 1).

Table 1. Molecular determinants of HBMLs after 24 h of in vitro culture, Me (Q1–Q3).

% CD45$^+$ Cells, Bearing Membrane Markers, $n = 4$ *								
CD3	CD4	CD8	CD25	CD95	CD28	CD71	RA	RO
99 (98–99)	67 (65–68)	21 (19–23)	9 (9–10)	15 (14–17)	79 (78–79)	2.5 (2–4)	55 (54–55)	35 (34–36)

* Note: n is the number of blood samples tested.

24-h culture of the viable HBMLs (median of living cells was 93.5%, Table 2) spontaneously secreted into the intercellular fluid the spectrum of immunomodulating cytokines and chemokines with pro-inflammatory (IL-2, IL-8, TNFα, GM-CSF, IFNγ) and anti-inflammatory activity (IL-4, IL-10) [39] capable of modulating the in vitrosurvival, proliferation, differentiation and maturation of HBMLs through autocrine/paracrine signaling pathways. It should be emphasized specially that both tested suspensions (without and with chitozan) did not cause cell death in 24-h culture in vitro (Table 2), which implies a non-toxic but irritating effect of massive doses of MNPs on the secretory function of HBMLs.

In the 2.3–1000 maximum tolerated dose (MTD) range, FF1 suspensions (without chitosan) produced a dose-dependent statistically significant increase (2–47 times with respect to the baseline) of the secretion of all cytokines studied (Table 3). Significantly, the concentration of IL-10 in the intercellular fluid increased in an exponential dependence on the dose of MNPs without chitosan (FF1 suspension) (Figure 5).

Table 2. Percentage of viable HBMLs (according to ISO 10993-5 test with 0.4% trypan blue) after 24-h co-cultivation with iron oxide MNPs from FF1 or FF2, Me (Q1–Q3).

Dose of MNPs in MTD	Amount of Viable Cells (%), $n = 4$ *	
	For FF1	For FF2
0	93.5 (88.5–95.5)	
2.3	95 (94.5–96)	95 (93–96)
10	94.5 (93.5–96)	93.5 (89–95)
100	95 (95–96)	94 (92.5–96.5)
1000	94.5 (93–97.5)	93.5 (91.5–96)

* Note: n is the number of samples tested in each group. Maximum tolerated dose (MTD) was 0.3 mg/L iron ions in water solution.

We observed the most indicative difference in dynamics of changes in suspensions of MNPs with and without chitosan in the range of 10–1000 MTD for the case of IL-10. With chitosan, a decrease in cytokine secretion in this range of low doses very pronounced. At the same time, the relative increase in the secretory activity of HBMLs in the range of 100–1000 MTDs was slowed down with the dependence on the plateau of the classical S-shaped dose-response curve (Figure 5 and Table 3).

Dispersing LTE iron oxide MNPs in a chitosan matrix changed the type of the dose-response curve for the secretion of biological molecules by HBMLs culture. Figure 5 shows the summary of these results: curve 1 (corresponding to the FF1 suspension without chitosan) was well fitted by exponent $y = 2.13e^{1.06x}$, $R^2 = 0.96$; curve 2 (corresponding to the FF2 suspension with chitosan) was well fitted by 3-degree polynomial dependence with a high degree of approximation: $y = 6.5x^3 + 51.9x^2 + 124.1x - 73$, $R^2 = 0.99$ and curve 3 showing correlation between calculated dependence of iron concentration in MTDs and in parts per million (ppm) was well fitted by exponent $y = 0.0119e^{1.9785x}$, $R^2 = 0.99$, wherein, the value of 2.3 MTDs of MNPs with chitosan statistically increased the concentrations of all tested cytokines, compared to both levels—the baseline level and the corresponding dose of MNPs without chitosan (Table 3).

Table 3. The concentration of cytokines in supernatants of HBMLs after 24 h of co-cultivation with nanoparticles of iron oxides at various doses, Me (Q1–Q3).

MTD	The Concentration of Cytokines, pg/mL, $n = 3$							
	IL-2	IL-4	IL-6	IL-8	IL-10	GM-CSF	IFNγ	TNFa
0	4.23 (3.87–4.51)	1.26 (0.73–1.98)	49.0 (43.5–54.3)	2345 (2189–2416)	6.27 (5.99–8.55)	10.61 (10.23–13.40)	66.2 (59.1–68.9)	48.3 (44.3–56.4)
2.3 FF1	**6.50** (6.23–6.56)	1.82 (1.70–1.86)	**56.3** (55.7–57.0)	**3560** (3507–3651)	**14.28** (13.2–15.32)	**17.83** (16.22–18.21)	**97.6** (95.4–100.2)	**70.0** (69.3–70.2)
FF2	**9.20 *** (9.03–10.11)	**2.65 *** (2.57–2.76)	**58.0 *** (57.7–58.5)	**4598 *** (4570–4610)	**17.56 *** (16–09–1821)	**25.6 *** (24.2–26.0)	**137.7 *** (135.2–138.2)	**165.2 *** (152.3–167.4)
10 FF1	**10.04** (9.8–11.2)	**2.92** (2.82–3.01)	**60.6** (60.1–61.0)	**4494** (4489–4500)	**52.74** (48.12–56.33)	**26.88** (25.3–26.9)	**146** (132.8–147.5)	**216.4** (210.2–219.0)
FF2	**6.19 *** (6.09–7.11)	**1.85 *** (1.8–1.95)	**55.6 *** (55.0–56.1)	**2905 *** (2800–3100)	**10.4 *** (10.18–11.1)	**25.9 *** (25.53–26.33)	**110.1 *** (108.2–112.6)	**229.3 *** (220.3–230.3)
100 FF1	**12.8** (12.08–13.01)	**3.42** (3.32–3.56)	**63.1** (62.5–63.4)	**4778** (4770–4790)	**267.6** (259.3–280.9)	**30.12** (29.11–31.24)	**197.7** (189.0–2017)	**278.3** (262.0–280.7)
FF2	**3.8 *** (3.63–4.01)	**1.35 *** (1.29–1.4)	**50.7 *** (50.0–51.0)	**2900 *** (2895–3000)	**5.96 *** (4.91–6.11)	**16.92 *** (16.8–17.22)	**79.5 *** (78.1–81.1)	**78.3 *** (76.2–81.8)
1000 FF1	**14.23** (13.12–14.98)	**3.82** (3.75–4.01)	**62.6** (62.1–63.0)	**4998** (4980–5000)	**295.5** (285.3–302.3)	**31.76** (30.12–32.12)	**200.9** (198.9–202.2)	**318.2** (315.2–323.4)
FF2	**10.11 *** (9.4–11.01)	**2.98 *** (2.81–3.01)	**60.6 *** (60.0–60.9)	**5280 *** (5257–5310)	**60.8 *** (58.2–61.1)	**39.79 *** (37.8–40.22)	**161.6 *** (159.8–162.1)	**239.2 *** (232.3–246.5)

Note: n is the number of samples tested in each group. Almost all values at doses 2.3–1000 MTD (bold font) are above the corresponding control samples (0 MTD); (*)—values for FF2 suspension are higher or lower than corresponding value for FF1 suspension without chitosan; $p < 0.05$ according to the Mann-Whitney test.

On the contrary, in the range of 10 to 100 MTDs of MNPs with chitosan leveled the stimulating effect of MNPs on the secretory activity of HBMLs. At 100 MTDs, a decrease in the concentrations of IL-2, IL-4, IL-6, IL-10 cytokines reached the baseline level. At 1000 MTDs, the failure of secretion caused by MNPs plus chitosan was replaced by an overshoot (Table 3), which in only two cases (IL-8, GM-CSF) significantly exceeded the corresponding values reached with the suspension FF1 without chitosan. The obtained data made it possible to formulate a working hypothesis:

(1) on the uneven distribution of different doses of MNPs in the chitosan matrix;
(2) the predominantly retarding effect of chitosan on the irritating effect of high (10–1000 MTDs) doses of magnetic maghemite nanoparticles.

At that, 100 MTDs turned out to be a critical (reference) point for research. To test the hypothesis, additional experiments with cell culture (TEM of biological samples, energy dispersive X-ray fluorescence analysis) were carried out. Calculated MTDs of MNPs in the medium corresponded to those estimated by stripping voltammetry. Thus, the electrochemical analysis of supernatants providing for the complete dissolution of MNPs during sample preparation showed the result of three measurements of the concentration of iron 33 ± 11 mg/kg at 100 MTDs and 2.9 ± 0.9 mg/kg at 10 MTDs. It is to be recalled that the maximum tolerated dose (MTD) is 0.3 mg/L iron ions in water solution.

Energy distribution X-ray fluorescence analysis of cells and supernatants showed (Table 4) the uneven distribution of MNPs in the cell fraction and intercellular fluids (supernatants) in low-speed centrifugation of HBMLs. According to Table 4, the measured iron concentration in the suspension of MNPs without chitosan at a maximum of 1000 MTDs was significantly lower than the calculated one presented in Figure 6. In contrast, the measured concentration of MNPs with chitosan at 100 MTDs (Table 4), primarily in the cell fraction, was significantly (3–10 times) higher than the calculated (Figure 6 and Table 4). On one hand, we should make a special remark related to the low solubility of iron oxide nanoparticles in the time being tested (24 h) [40] and therefore the existing ambiguity in determining the MTDs in the case of MNPs.

Table 4. The distribution of MNPs in the HBMLs cellular sediment and intercellular fluid (supernatants) according to energy dispersive X-ray fluorescence analysis, Me (Q1–Q3).

MTDs (ppm)	Cellular Sediments, n = 2–3		Supernatants, n = 3–4	
	For FF1	For FF2	For FF1	For FF2
0 (control)	0.6 (0.4–1.9); n = 3		1.45 (1.15–1.75); n = 4	
2.3 (0.69)	0.6–1.3	0.6–1.36	1.8 (1.2–2.6)	1.0 (0.15–1.2)
10 (3.0)	1.3	2.4–32	2.2 (1.2–3.0)	1.0 (0.9–5.3)
100 (30.0)	15.5–114.9	200–431	17.65 (15.95–19.10)	92.0 (45.7–97.0)
1000 (300.0)	32.8–38.8	307–408	43.25 (35.45–51.35)	5.7 (4.55–13.9)

Note: n is the number of samples tested in each group.

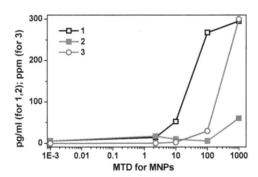

Figure 6. An example of modulating effect of suspensions on cytokine (here, IL-10), concentrations are given for 24-h HBMLs culture. Curve 1 corresponds to the FF1 suspension without chitosan, curve 2 corresponds to the FF2 suspension with chitosan and curve 3 shows correlation between calculated dependence of iron concentration in MTDs in ppm.

Strictly speaking the maximum tolerated dose definition is given for the ionic form of iron: 0.3 mg/L of Fe ions in water. Nanoparticles and, in a broader sense, colloidal particles, strictly speaking, cannot be considered a soluble form. The particles are a separate insoluble phase with an interface with the aqueous medium. Although some iron solubility on the boundary of the particle is possible, but most likely it does not exceed value around one MTD. The observed high error in the iron concentration definition by TXRF analysis (30–300 ppm) may be a consequence of quite non-uniform iron distribution in the biosample when non-dissolved composites were measured. With the low solubility of iron oxide nanoparticles in the time being tested, probably an instrumental approach to determination of high doses of MNPs similar to soluble forms of iron is not entirely applicable. On the other hand, chitosan seems to have contributed to the accumulation of MNPs in the cell fraction at high iron concentrations after centrifugation as compared with supernatants (Table 4). Therefore, TEM studies of HBMLs fractions incorporated into resin were studied in order to determine the possible localization of nanoparticles with respect to cells at 100 MTDs grows conditions.

Analysis of TEM images showed rather dispersed intracellular distribution of individual MNPs in HBMLs cellswithout the formation of conglomerates in both tested cases (with and without chitosan). Figure 7 demonstrates some representative examples, including general view of control culture grown without suspension of MNPs but in the same conditions. Due to the visually small penetration of MNPs into the HBMLs, we failed to make a comparative quantitative assessment of their intracellular concentrations. In addition, in the case of MNPs without chitosan (FF1), nanoparticle agglomerates were not detected in the intercellular environment (supernatants), that is individual MNPs were dispersed outside the cells, and only very few on the cytoplasmic membrane and inside the cells. At the same time, MNPs with chitosan (FF2) outside the cells were concentrated in the form of agglomerates in the chitosan matrix fixed as separate "flakes" to the outside of the cytoplasmic membrane of HBMLs. Apparently, chitosan prevented the free penetration of MNPs into HBMLs. On the other hand, it caused uneven distribution of MNPs, which contributed to an excess of the calculated dose of 100 MTDs when non-dissolved composites were measured.

Recently we have performedin vitro experiments with human adipose-derived mesenchymal stem cells (AMMSCs) underthe same conditions: stable colloidal suspensions using electrostatic or steric (by chitosan) stabilization of iron oxide MNPs obtained by LTE for 100 MTD MNPs concentration and after 24 h contact with suspension [12]. Intracellular MNPs inclusions were clearly observed in mainly inside organelles of AMMSCs contacted with 100 MTD of FF1. MNPs aggregates were noted inside endosomes, contoured the hydrolytic vesicles/secretory granules and outer membrane of mitochondria and only single inclusions were freely situated in cellular cytoplasm. Mitochondria looked like myelin-like bodies with hyperelectron-dense inclusions between membranous layers. In the case of FF2 suspension hyperelectron-dense aggregated MNPs They seemed to enter into the cells to a lesser extent in comparison with FF-1 MNPs: very few inclusions contoured only the hydrolytic vesicles/secretory granules. For sake of comparison Figure 8 shows AMMSCs images for the same conditions as those used for the Figure 7 cases.

Figure 8f shows very interesting case—MNPs of FF2 suspension as the aggregates in chitozan matrix. Very roughly one can estimate the average size of the aggregate to be the order of 50 to 200 nm and containing approximately 20 to 100 nanoparticles. Aggregates have large variety of shapes but they do not tend to be very spherical but rather characterized by the presence of various chains of different length. Interestingly, the average size of the aggregates observed by TEM was close to the calculated of end-to-end distance for the segment for chitosan (150 nm), which is two times lower than the median diameter of PSD for chitosan solution. In the case of the MNPs aggregates inside the AMMSCs, the average size of the aggregate concept is not applicable as the size of the MNPs aggregates predetermined by the size of the cell organelles. In both cases (HBMLs and AMMSCs) the size of the main organelles like mitochondria was much larger comparing with the size of typical MNPs aggregates for FF2.

It is important to mention that T-lymphocytes (CD45+CD3+ cells) account more than for 95% of all mononuclear cells (Table 1). The difference between T-lymphocytes and the AMMSCs is that T-lymphocytes are not phagocytic cells. Previously we observed that AMMSCs can accumulate tested MNPs in endosomes, secretory granules and mitochondria (Figure 8). As a consequence, our suppositions based on the obtained data were: (1) the existence of uneven distribution of MNPs in chitosan due to agglomeration of the latter with cell membranes, (2) the existence of the predominantly inhibitory effect of chitosan for the irritating effect of high (10–1000) MTDs of magnetic maghemite nanoparticles on cytokine secretion. Perhaps the observed phenomenon is caused by a decrease in the penetration of MNPs into cells and enveloping cells with chitosan flakes, which can lead to a decrease in the free secretory cell surface or the fixation of cytokines in chitosan matrix. "Enveloping" effect is an interesting phenomenon, because usually the main attention is attracted to the phenomenon of the irritating effect of the nanoparticles. In practical terms, this may find application in the future when developing biotechnological methods restriction of complications of cellular allergic reactions similar to hypersensitivity of delayed type, for example, with bronchial asthma (as a spray) or dermatitis. The question of why this effect manifests itself, mainly at 100 MTDs, remains open and extremely interesting for interdisciplinary research. It is possible that maghemite MNPs passively does not enter the cells due to special combination of the factors between others the zeta potential and magnetic interactions may play important roles.

As we have confirmed the possibility of either massive incorporation of LTE MNPs into AMMSCs (FF1) or their adhesion toward the cell membrane (HBMLs, AMMSCs for FF2) one can discuss the possibility of their magnetic filed assisted application. To make the comparison easier let us re-calculate M/M_s hysteresis loops for Ms—being a saturation magnetization for each particular case (Figure 3). Figure 3a shows that the saturation magnetization of air-dry MNPs is the highest one and M_s for FF2 MNPs is the lowest as one could expect taking into account the fabrication process and inevitable separation step resulted in the removal of large MNPs. At the same time in low field interval below 1 kOe, which is most reasonable for applications, one can observe that FF1 and FF2 MNPs reach the higher moment comparing with air-dry in the same low field. This is clear advantage for such application as magnetic biosensing (magnetic fields below 100 Oe [12]). The second interval for possible applications (100–500 Oe) is connected to idea of the cell use as native microcapsules for targeted delivery in a supplement of nanomedicine and theranostics, cell technologies, and regenerative medicine [41–43].

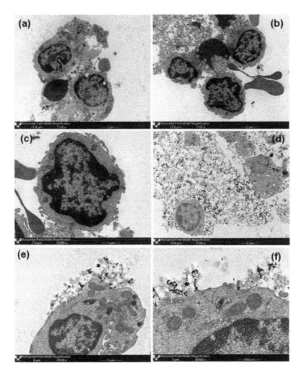

Figure 7. TEM of HBMLs mononuclear blood leukocytes. (**a**)—control culture grown without MNPs. HBMLs mononuclear blood leukocytes after 24-h contact with MNPs of suspension of iron oxide MNPs without chitosan (FF1) at a dose of 100 MTDs: (**b**)—magnification ×7100, (**c**)—magnification ×14,000. HBMLs after 24-h contact with MNPs of suspension of iron oxide MNPs with chitosan (FF2) at a dose of 100 MTDs, magnifications: (**d**)—×7100, (**e**)—×18000, (**f**)—×36000.

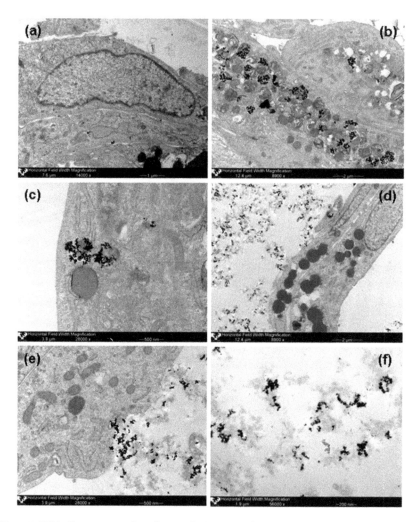

Figure 8. TEM of human mesenchymal stem cells. (**a**)—control AMMSCs culture grown without MNPs. AMMSCs cells after 24-h contact with MNPs of suspension of iron oxide MNPs without chitosan (FF1) at a dose of 100 MTDs: (**b**)—magnification ×8900, (**c**)—magnification ×28,000. AMMSCs cells after 24-h contact with iron oxide MNPs of suspension with chitosan (FF2) at a dose of 100 MTDs: (**d**)—magnification ×8900, (**e**)—magnification ×28,000. Aggregates of LTE MNPs encapsulated by chitozan (**f**)—magnification ×56,000.

Recently we have shown that the presence of LTE MNPs changes the physical properties of ferrogels synthesized by radical polymerization of acrylamide and their biocompatibility [43]. We found that the gradual increase of MNPs concentration in the gel network resulted in the significant increase of the negative value of electrical potential and adhesion index for both the human dermal fibroblasts and the human peripheral blood leucocytes. Ferrimagnetic MNPs affect hemopoietic and stromal cells and promote cellular adhesion and formation of cell-to-cell contacts along the magnetic field lines. Thisimplies cell use as native microcapsules for targeted delivery in a supplement of nanomedicine and theranostics, cell technologies, and regenerative medicine. From viewpoint of biomedical applications, the inclusion of small amount of LTE MNPs into the polymer network seems to be very positive step which significantly enhances the mechanical and electrical properties of ferrogels, and improves biocompatibility of these systems. Although, this point was not yet investigated, one can expect certain influence of the applied magnetic field on the adhesion process. Here we can propose to plan the experiments with LTE MNPs loaded cells (AMMSCs cells after 24-contact with MNPs of suspension of iron oxide MNPs without chitosan (FF1) or of HBMLs mononuclear blood leukocytes. after 24-contact with LTE MNPs of suspension of iron oxide MNPs with chitosan (FF2) at a dose of 100 MTDs).

Even more sophisticated applications can be thought. The majority of the present day in vitro studies are focused on the understanding of one particular culture. In the case of HBMLs and MMSC cells studied in similar conditions we found clear difference in MNPs internacionalization process for FF1 case. This means that LTE MNPs will affect specifically on HBMLs if both cells are exposed together, the situation which seems to be realistic from practical point of view.

One of therapeutical problems of MNPs applications is rapid recognition of MNPs by the immune system [44]. LTE MNPs encapsulation by MMSC cells can be a good solution for certain stage of the treatment especially taking into account the possibility to manipulate living cells anchored with MNPs by a magnetic field. As for therapeutically purposes like hyperthermia or thermal ablation the excess of MNPs is always necessary there is a request for new techniques of removal of the excess of MNPs after the therapy. One can propose low invasive cathetering of ferrogel pads with adhered cell culture toward the region of the MNPs excess (which may be concentrated by the external magnetic field application) after the pad cells loading with MNPs pad can be removed providing removal of the MNPs excess. Of course, these hypothetic scenarios require special long series of investigation steps but the general directions seems to be promising.

4. Conclusions and Outlook

Magnetic iron oxide nanoparticles were obtained by a highly productive laser target evaporation technique. XRD analysis confirmed their inverse spinel structure (space group Fd-3m). According to TEM, the shape of MNPs was close to spherical. The analysis of TEM images revealed that the ensemble obeyed lognormal size distribution with amedian value of 26.8 nm and 0.362 dispersion. Stabilized water-based suspensions with MNPs were fabricated using electrostatical or sterical stabilization by chitosan. It was shown that the encapsulation of iron oxide MNPs by chitosan makes them resistant to unfavorable factors for the colloidal stability such as pH and high ionic force, which are typically present under physiological conditions. This makes the suspensions of MNPs encapsulated by chitosan promising candidates for biocompatibility testing.

Controlled amounts of suspensions were used for in vitro experiments with human blood mononuclear leukocytes in order to study their morphofunctional response in the wide Fe concentration range. Special efforts were made in order to provide complete physical and physical-chemical characterization of obtained MNPs and suspensions. For sake of comparison the results obtained in the present study were compared with our previous results of the study of similar suspensions with human mesenchymal stem cells.

Both types of suspensions (with and without chitosan) enhance the secretion of cytokines by a 24-h culture of HBMLs (predominantly T-lymphocytes) compared to a control without nanoparticles. At a dose of 2.3, the MTD of chitosan promotes the stimulating effect on cells; in the dose range of MNPs 10–1000 MTDs, chitosan "inhibits" cellular secretory activity compared to "pure" MNPs. Both suspensions did not cause cell death by necrosis, hence, the secretion of cytokines is due to the enhancement of the functional activity of HBMLs. Increased accumulation of MNPs with chitosan in the cell fraction at 100 MTDs for 24 h exposure, according to energy dispersive X-ray fluorescence analysis, may be due to fixation of chitosan on the outer membrane of HBMLs. Within the cells, the distribution pattern of MNPs is similar.

The discussed results can be used for an addressed design of cell delivery and removal systems incorporating multiple activities and functions because of cell capability to avoid phagocytosis by immune cells as opposed to individual MNPs in the blood streem. They are also promising for the field of magnetic biosensor development. The possibility to incorporate living cells anchored with MNPs manipulation step using external magnetic field is also promising.

Acknowledgments: This work was supported in part within framework of a program for increasing competitiveness and with financial support from the Program for Organization of Research (20.4986.2017/6.7) of Immanuel Kant Baltic Federal University; in the framework of the state task of the Ministry of Education and Science of Russia 3.6121.2017/8.9; Russian Federation state task project 0389-2014-0002, RFBR grant 16-08-00609 and by the ELKARTEK grant KK-2016/00030 of the Basque Country Government. Selected studies were made at SGIKER Common Services of UPV-EHU and URFU Common Services. We thank Iulia P. Novoselova, Anatoly I. Medvedev, Aidar M. Murzakaev and Aitor Larrañaga for special support.

Author Contributions: G.V.K., L.S.L., A.P.S. and I.A.K. conceived and designed the experiments; A.P.S., V.V.S., I.S.T., O.G.K., G.B.S., K.A.Y., E.G.C., N.A.K., R.A., I.V.B., G.V.K. performed the experiments; G.V.K., L.S.L., A.P.S., V.V.S., N.A.K. and I.A.K. analyzed the data; A.P.S., I.A.K., G.V.K. and L.S.L. wrote the manuscript. All authors discussed the results and implications, and commented on the manuscript at all stages. All authors read and approved the final manuscript.

Conflicts of Interest: The authors declare no conflict of interest.

References

1. Tartaj, P.; Del Puerto Morales, M.; Veintemillas-Verdaguer, S.; González-Carreño, T.; Serna, C.J. The preparation of magnetic nanoparticles for applications in biomedicine. *J. Phys. D Appl. Phys.* **2003**, *36*, R182–R197. [CrossRef]

2. Roca, A.G.; Costo, R.; Rebolledo, A.F.; Veintemillas-Verdaguer, S.; Tartaj, P.; González-Carreño, T.; Morales, M.P.; Serna, C.J. Progress in the preparation of magnetic nanoparticles for applications in biomedicine. *J. Phys. D Appl. Phys.* **2009**, *42*, 224002. [CrossRef]

3. Kurlyandskaya, G.V.; Fernandez, E.; Safronov, A.P.; Svalov, A.V.; Beketov, I.V.; Burgoa Beitia, A.; Garcıa-Arribas, A.; Blyakhman, F.A. Giant magnetoimpedance biosensor for ferrogel detection: Model system to evaluate properties of natural tissue. *Appl. Phys. Lett.* **2015**, *106*, 193702. [CrossRef]

4. Bucak, S.; Yavuzturk, B.; Sezer, A.D. Magnetic nanoparticles: Synthesis, surface modification and application. In *Recent Advances in Novel Drug Carrier Systems*; Sezer, A.D., Ed.; InTech: Rijeka, Croatia, 2012; pp. 165–200.

5. Coisson, M.; Barrera, G.; Celegato, F.; Martino, L.; Vinai, F.; Martino, P.; Ferraro, G.; Tiberto, P. Specific absorption rate determination of magnetic nanoparticles through hyperthermia measurements in non-adiabatic conditions. *J. Magn. Magn. Mater.* **2016**, *415*, 2–7. [CrossRef]

6. Kurlyandskaya, G.V.; Fernandez, E.; Safronov, A.P.; Blyakhman, F.A.; Svalov, A.V.; Burgoa Beitia, A.; Beketov, I.V. Magnetoimpedance biosensor prototype for ferrogel detection. *J. Magn. Magn. Mater.* **2017**, *441*, 650–655. [CrossRef]

7. Moroz, P.; Jones, S.K.; Gray, B.N. Status of hyperthermia in the treatment of advanced liver cancer. *J. Surg. Oncol.* **2001**, *77*, 259–269. [CrossRef] [PubMed]

8. Morrison, I.D.; Ross, S. *Colloidal Dispersions: Suspensions, Emulsions and Foams*; Wiley: New York, NY, USA, 2002.

9. Baselt, D.R.; Lee, G.U.; Natesan, M.; Metzger, S.W.; Sheehan, P.E.; Colton, R.J. A biosensor based on magnetoresistance technology. *Biosens. Bioelectron.* **1998**, *13*, 731–739. [CrossRef]

10. Kurlyandskaya, G.V.; Levit, V.I. Advanced materials for drug delivery and biosensorsbased on magnetic label detection. *Mater. Sci. Eng. C* **2007**, *27*, 495–503. [CrossRef]

11. Kuo, C.H.; Liu, Y.C.; Chang, C.M.J.; Chen, J.H.; Chang, C.; Shieh, C.J. Optimum conditions for lipase immobilization on chitosan-coated Fe_3O_4 nanoparticles. *Carbohydr. Polym.* **2012**, *87*, 2538–2545. [CrossRef]

12. Kurlyandskaya, G.V.; Novoselova, I.P.; Schupletsova, V.V.; Andrade, R.; Dunec, N.A.; Litvinova, L.S.; Safronov, A.P.; Yurova, K.A.; Kulesh, N.A.; Dzyuman, A.N.; et al. Nanoparticles for magnetic biosensing systems. *J. Magn. Magn. Mater.* **2017**, *431*, 249–254. [CrossRef]

13. Muzzarelli, R.A.A.; Muzzarelli, C. Chitin and chitosan hydrogels. In *Handbook of Hydrocolloids*; Phillips, G.O., Williams, P.A., Eds.; CRC Press: Boca Raton, FL, USA, 2009; pp. 849–888.

14. Il'ina, A.V.; Varlamov, V.P. Chitosan-dased polyelectrolyte complexes: A review. *Appl. Biochem. Microbiol.* **2005**, *41*, 5–11. [CrossRef]

15. Krayuhina, M.A.; Samoylova, N.A.; Yamskov, I.A. Chitosan polyelectrolyte complexes: Formations, properties and applications. *Uspekhi Khimii* **2008**, *77*, 854–869.

16. Spizzo, F.; Sgarbossa, P.; Sieni, E.; Semenzato, A.; Dughiero, F.; Forzan, M.; Bertani, R.; Del Bianco, L. Synthesis of ferrofluids made of iron oxide nanoflowers: Interplay between carrier fluid and magnetic properties. *Nanomaterials* **2017**, *7*, 373. [CrossRef] [PubMed]

17. Qu, J.; Liu, G.; Wang, Y.; Hong, R. Preparation of Fe_3O_4—Chitosan nanoparticles used for hypothermia. *Adv. Powder Technol.* **2010**, *21*, 421–427. [CrossRef]

18. Ravi Kumar, M.N.V.; Muzzarelli, R.A.A.; Muzzarelli, C.; Sashiwa, H.; Domb, A.J. Chitosan chemistry and pharmaceutical perspectives. *Chem. Rev.* **2004**, *104*, 6017–6084. [CrossRef] [PubMed]

19. Tyukova, I.S.; Safronov, A.P.; Kotel'nikova, A.P.; Agalakova, D.Y. Electrostatic and steric mechanisms of iron oxide nanoparticle sol stabilization by chitosan. *Polym. Sci. Ser. A* **2014**, *56*, 498–504. [CrossRef]

20. Safronov, A.P.; Beketov, I.V.; Tyukova, I.S.; Medvedev, A.I.; Samatov, O.M.; Murzakaev, A.M. Magnetic nanoparticles for biophysical applications synthesized by high-power physical dispersion. *J. Magn. Magn. Mater.* **2015**, *383*, 281–287. [CrossRef]

21. Beketov, I.V.; Safronov, A.P.; Medvedev, A.I.; Alonso, J.; Kurlyandskaya, G.V.; Bhagat, S.M. Iron oxide nanoparticles fabricated by electric explosion of wire: Focus on magnetic nanofluids. *AIP Adv.* **2012**, *2*, 022154. [CrossRef]

22. Safronov, A.P.; Beketov, I.V.; Komogortsev, S.V.; Kurlyandskaya, G.V.; Medvedev, A.I.; Leiman, D.V.; Larrañaga, A.; Bhagat, S.M. Spherical magnetic nanoparticles fabricated by laser target evaporation. *AIP Adv.* **2013**, *3*, 52135. [CrossRef]

23. Novoselova, I.P.; Safronov, A.P.; Samatov, O.M.; Beketov, I.V.; Medvedev, A.I.; Kurlyandskaya, G.V. Water based suspensions of iron oxide obtained by laser target evaporation for biomedical applications. *J. Magn. Magn. Mater.* **2016**, *415*, 35–38. [CrossRef]

24. Rietveld, H.M. A profile refinement method for nuclear and magnetic structures. *J. Appl. Crystallogr.* **1969**, *2*, 65–71. [CrossRef]

25. Dubova, N.M.; Slepchenko, G.B.; Khlusov, I.A.; Kho, S.L. The study of iron-based nanoparticles stability in biological fluids by stripping voltammetry. *Procedia Chem.* **2015**, *15*, 360–364. [CrossRef]

26. Szoboszlai, N.; Polgári, Z.; Mihucz, V.G.; Záray, G. Recent trends in total reflection X-ray fluorescence spectrometry for biological applications. *Anal. Chim. Acta* **2009**, *633*, 1–18. [CrossRef] [PubMed]

27. Polgári, Z.; Szoboszlai, N.; Mihucz, V.G.; Záray, G. Possibilities and limitations of the total reflection X-ray fluorescence spectrometry for the determination of low Z elements in biological samples. *Microchem. J.* **2011**, *99*, 339–343. [CrossRef]

28. Kulesh, N.A.; Novoselova, I.P.; Safronov, A.P.; Beketov, I.V.; Samatov, O.M.; Kurlyandskaya, G.V.; Morozova, M.V.; Denisova, T.P. Total reflection X-ray fluorescence spectroscopy as a tool for evaluation of iron concentration in ferrofluids and yeast samples. *J. Magn. Magn. Mater.* **2016**, *415*, 39–44. [CrossRef]

29. Pearson, W.B. *Handbook of Lattice Spacing Structures of Metals and Alloys*; Pergamon Press: London, UK, 1958; p. 1044.

30. O'Handley, R.C. *Modern Magnetic Materials*; John Wiley & Sons: New York, NY, USA, 1972; p. 740.

31. Rubinstein, M.; Colby, R.H. *Polymer Physics*; Oxford University Press: New York, NY, USA, 2003; p. 59.

32. Rinaudo, M. Chitin and Chitosan: Properties and Applications. *Prog. Polym. Sci.* **2006**, *31*, 603–632. [CrossRef]

33. Philippova, O.E.; Korchagina, E.V. Chitosan and its hydrophobic derivatives: Preparation and aggregation in dilute aqueous solution. *Polym. Sci. Ser. A* **2012**, *54*, 552–572. [CrossRef]

34. Popa-Nita, S.; Alcouffe, P.; Rochas, C.; Laurent, D.; Domard, A. Continuum of structural organization from chitosan solutions to derived physical forms. *Biomacromolecules* **2010**, *11*, 6–12. [CrossRef] [PubMed]

35. Kosmulski, M. *Chemical Properties of Material Surfaces*; CRC Press: Boca Raton, FL, USA, 2001; p. 248.

36. Gallagher, P.F.; Fazekas de St. Groth, B.; Miller, J.F. CD4 and CD8 Molecules can Physically Associate with the same T-cell Receptor. *PNAS* **1989**, *86*, 10044–10048. [CrossRef] [PubMed]

37. Law, H.K.W.; Tu, W.; Liu, E.; Lau, Y.L. Insulin-like growth factor I promotes cord blood T cell maturation through monocytes and inhibits their apoptosis in part through interleukin-6. *BMC Immunol.* **2008**, *9*, 74–86. [CrossRef] [PubMed]

38. Akbar, A.N.; Terry, L.; Timms, A.; Beverley, P.C.; Janossy, G. Loss of CD45R and gain of UCHL1 reactivity is a feature of primed T cells. *J. Immunol.* **1988**, *140*, 2171–2178. [PubMed]

39. Khalaf, H.; Jass, J.; Olsson, P.E. Differential cytokine regulation by NF-kappaB and AP-1 in Jurkat T-cells. *BMC Immunol.* **2010**, *11*, 26. [CrossRef] [PubMed]

40. Khlusov, I.A.; Zagrebin, L.V.; Shestov, S.S.; Itin, V.I.; Sedoi, V.S.; Feduschak, T.A.; Terekhova, O.G.; Magaeva, A.A.; Naiden, E.P.; Antipov, S.A.; et al. Colony-forming activity of unipotent hemopoietic precursors under the effect of nanosized ferrites in a constant magnetic field in vitro. *Bull. Exp. Biol. Med.* **2008**, *145*, 151–157. [CrossRef] [PubMed]

41. Pavlov, A.M.; De Geest, B.G.; Louage, B.; Lybaert, L.; De Koker, S.; Koudelka, Z.; Sapelkin, A.; Sukhorukov, G.B. Magnetically engineered microcapsules as intracellular anchors for remote control over cellular mobility. *Adv. Mater.* **2013**, *25*, 6945–6950. [CrossRef] [PubMed]

42. Pavlov, A.M.; Gabriel, S.A.; Sukhorukov, G.B.; Gould, D.J. Improved and targeted delivery of bioactive molecules to cells with magnetic layer-by-layer assembled microcapsules. *Nanoscale* **2015**, *7*, 9686–9693. [CrossRef] [PubMed]

43. Blyakhman, F.A.; Safronov, A.P.; Zubarev, A.Y.; Shklyar, T.F.; Makeyev, O.G.; Makarova, E.B.; Melekhin, V.V.; Larrañaga, A.; Kurlyandskaya, G.V. Polyacrylamide ferrogels with embedded maghemite nanoparticles for biomedical engineering. *Results Phys.* **2017**. [CrossRef]

44. Grossman, J.H.; McNeil, S.E. Nanotechnology in cancer medicine. *Phys. Today* **2012**, *65*, 38–42. [CrossRef]

Review

Aptamer-Modified Magnetic Beads in Biosensing

Harshvardhan Modh, Thomas Scheper and Johanna-Gabriela Walter *

Institute of Technical Chemistry, Leibniz University of Hannover, Hannover 30167, Germany;
modh@iftc.uni-hannover.de (H.M.); scheper@iftc.uni-hannover.de (T.S.)
* Correspondence: walter@iftc.uni-hannover.de; Tel.: +49-511-762-2955

Received: 24 February 2018; Accepted: 26 March 2018; Published: 30 March 2018

Abstract: Magnetic beads (MBs) are versatile tools for the purification, detection, and quantitative analysis of analytes from complex matrices. The superparamagnetic property of magnetic beads qualifies them for various analytical applications. To provide specificity, MBs can be decorated with ligands like aptamers, antibodies and peptides. In this context, aptamers are emerging as particular promising ligands due to a number of advantages. Most importantly, the chemical synthesis of aptamers enables straightforward and controlled chemical modification with linker molecules and dyes. Moreover, aptamers facilitate novel sensing strategies based on their oligonucleotide nature that cannot be realized with conventional peptide-based ligands. Due to these benefits, the combination of aptamers and MBs was already used in various analytical applications which are summarized in this article.

Keywords: aptamer; magnetic beads; analytical applications; electrochemical assays; optical assays; point-of-care-testing

1. Introduction

Aptamers are synthetic single-stranded (ss) DNA (deoxyribonucleic acid) or RNA (ribonucleic acid) molecules, which specifically bind to their target molecules with high affinity. They are selected by an iterative in vitro process termed systematic evolution of ligands by exponential enrichment (SELEX). In recent years, aptamers have been developed against a broad range of target molecules, including metal ions, small molecules, peptides, proteins, and even complex targets such as whole cells. Due to various advantages of aptamers, such as their animal free and cost effective production, high temperature stability, chemical stability, target versatility, and high affinity and selectivity for their targets, aptamers are appealing alternatives to antibodies (AB) for use in analytical applications [1].

The specific advantages offered by aptamers include the easy modification with functional groups resulting in the possibility to control the orientation of the aptamer after immobilization. This controlled orientation facilitates high activity of immobilized aptamers, which is beneficial in their analytical applications [2]. In addition, aptamers can undergo considerable structural changes while interacting with target molecules. These changes have been extensively studied and exploited for the development of novel assays including target-induced structural switching (TISS) and target-induced dissociation (TID) of complementary oligonucleotides (Figure 1). These possibilities are specific for aptamers and allow for the design of sensing strategies even in cases, where conventional strategies, such as sandwich assays are not applicable. Moreover, aptamers can be regenerated and aptamer-modified sensors can be reused [3,4].

Recently, evolving analytical techniques and improved use of established methods have begun to incorporate micro- or nano-sized magnetic beads (MBs) [5]. The specific properties of MBs, such as colloidal stability of magnetic nanoparticles, homogenous size distribution, high and uniform magnetite content, a fast response to applied magnetic field, and presence of surface functional groups are essential for their analytical applications. The superparamagnetic properties of MBs permit the easy isolation of analytes from complex matrices, such as biological and environmental samples, by attaching specific ligands on the surface of MBs and separating the MBs-bound analytes with the aid of an external magnetic field. In order to provide different specificities, MBs can be functionalized with various reactive groups, such as amines, carboxyls, epoxyls, and tosyls, which can be used to immobilize high affinity ligands such as aptamers, proteins, antibodies, etc. according to required applications [6]. Suspended modified MBs with immobilized ligands can be highly recommended for the detection of analytes in complex matrices and the isolation of targets from larger volumes [7]. In some applications, the magnetic properties of the MBs, such as magnetic-relaxation switch, have been used for signal generation [8,9]. Lately, magnetic separation processes have been introduced in biotechnology for the purification of proteins [10–16], protein digestion [17–19], separation of cells [20,21], and analytical applications [18,19,22–24].

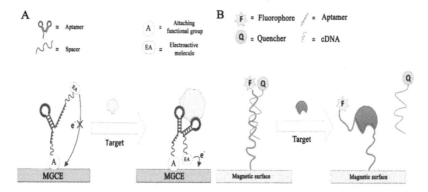

Figure 1. (**A**) Target-induced structure switching (TISS) type of assay. Here, the interaction between an aptamer and a target molecule leads to change in the conformation of the aptamer. The conformational changes can be exploited for signal generation, e.g., by using an electroactive molecule (EA) fused to the aptamer. In the figure, MGCE is a magnetic glass carbon electrode; (**B**) Target-induced dissociation (TID) type of assay. Here, the aptamer is hybridized with a complementary oligonucleotide (cDNA). The interaction between a target molecule and an aptamer leads to release of the cDNA sequence from the aptamer. The release of the cDNA can provide different types of signals in different assay formats, in the given example FRET-is used for signal generation. Adapted from [3] with permission. Copyright 2014, De Gruyter.

The combination of MBs with aptamers opens up new possibilities in a number of applications, such as sample preparation, wastewater treatment, water purification, disease therapy, disease diagnosis including magnetic resonance imaging, cell labelling and imaging, and biosensors. Within this review article, first a brief introduction of magnetic beads and their applications will be given in Section 2. In Section 3, the use of aptamer-modified MBs in biosensing will be described and recent examples will be highlighted.

2. Magnetic Beads and Their Biological Applications

Within this section magnetic beads will first be briefly introduced together with potential strategies to modify their surfaces. Consequently, the use of magnetic beads in some of the important biological applications such as separating biomolecules from complex matrices will be described.

2.1. Magnetic Beads and Their Modification

In 1792, William Fullarton described the separation of iron material with a magnet in a patent, which paved the way for applications of magnetic fields in separation techniques [6]. Initially, magnetic properties of sediments were used for separation. In 1852, a company from New York separated magnetite from apatite and later on, magnetite was separated from iron from brass fillings, turnings, metallic iron from furnace products and plain gauge etc. Gradually, magnetic separation evolved into complex and diverse commercial applications. In 1950, introduction of high gradient magnetic separation (HGMS) systems allowed faster and broad magnetic separation applications [25]. Towler et al. [26] reported the use of micron sized magnetite particles with adsorbent, manganese dioxide, on the surface to recover radium, lead and polonium from seawater samples. Remarkably, Safarikova et al. introduced silanized magnetite particles (blue magnetite [27]) in magnetic solid-phase extraction (MSPE) for the first time to preconcentrate organic compounds [28] prior to analysis. Nowadays, magnetic separations have been widely used for protein purification [10–16], separation and purification of cells [20,21], and analytical applications [18,19,22–24].

In recent applications, MBs are largely composed of a magnetic core, a surface coating, and specific binding ligands at the surface (Figure 2). Generally, magnetic cores can be composed from various materials, which exhibit magnetic properties. Largely, they consist of either pure metals (e.g., Co, Fe, and Ni) or their oxides. In addition, transition-metal-doped oxides and metal alloys, including $CoPt_3$, FeCo, and FePt, are also good candidates. Among these magnetic materials, particularly iron oxides such as magnetite (Fe_3O_4) and maghemite (γ-Fe_2O_3) are considered to be the most attractive candidates for biological applications, owing to their strong magnetic property and biocompatibility [6]. U.S. Food and Drug Administration (FDA) and the European Medicines Agency (EMA) have approved the use of iron oxide MBs as magnetic resonance imaging (MRI) contrast agents [29]. Several approaches are available to synthesize the iron oxides. One widely used approach is co-precipitation from iron (III) chloride and iron (II) chloride solutions in the presence of aqueous ammonia solution [20,30,31]. Other methods are available including the extraction of bacterial magnetic particles (BMPs) from the flagella of magnetic bacteria from marine sediments [32,33].

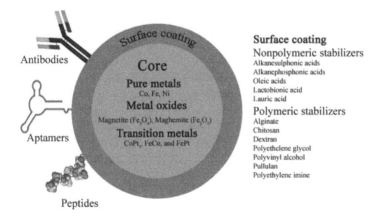

Figure 2. Composition of magnetic beads (MBs) used in analytical applications.

Finely divided iron is highly reactive toward oxidizing agents in the presence of water or humid air. Thus, surface coating of magnetic particles is required to obtain physically and chemically stable systems. Such stabilization can be achieved by surface coating of the magnetic particles in numerous ways (Figure 2). Surface coating can be performed by using stabilising surface coating material, encapsulation into polymeric shells and into lipososmes [34].

Surface coating is primarily necessary to stabilize the newly formed surface of the particles and to prevent aggregation of the particles. Surface stabilization is generally achieved using nonpolymeric stabilizers based on organic monomers such as alkanesulphonic and alkanephosphonic acids, or phosphonates; oleic acid, lactobionic acid, lauric acid, or polymeric stabilizers, i.e., alginate, chitosan, dextran, polyethylene glycol, polyvinyl alcohol, pullulan, or polyethylene imine (Figure 2) [6]. Encapsulation into polymeric shells improves the water dispersibility and chemical and physical stability of the magnetic particles. In addition, the polymeric shell can provide a basis for conjugation of the magnetic particles to the targeting ligands by providing functional groups such as amine or carboxyl groups on the surface. Several ligands are available for the analytical purpose; the most popular among them are antibodies (ABs), aptamers, and peptides. The commonly used approach to attach ABs to MBs is the coupling of amine groups of the AB to carboxylated MBs via ethyl (dimethylaminopropyl) carbodiimide/N-hydroxysuccinimide (EDC/NHS) chemistry, but this can lead to random orientation of AB on the MBs. Different strategies have been attempted to avoid random orientation. For example, protein A can be used in order to immobilize the AB on the MB in a controlled orientation [35].

Currently, aptamers are also becoming popular for analytical applications. Several strategies are used for immobilising aptamers on the MBs. One of the popular strategies is to introduce amine groups on one terminus of the aptamer to allow immobilization to carboxylated MBs via EDC/NHS coupling. In contrast to AB immobilization, this results in highly oriented immobilization of aptamers, since only the terminal amine group can participate in coupling procedure. It is also possible to use streptavidin coated MBs and biotin-labelled aptamer for the attachment. Most important advantage here is the orientation of the aptamer, which can be easily controlled, as aptamers are chemically synthesized. During chemical synthesis, various modifications can be attached at defined positions within the aptamer sequence.

2.2. Separation of Biomolecules Using Modified MBs

The separation of biomolecules such as whole cells, proteins and peptides, and mRNA from complex matrices is very challenging. In this context, MBs modified with specific ligands allow to isolate the target molecule using strong magnets. This MB-based sample pre-treatment also allows to increase the concentration of target molecules in case of low concentration of the target molecules. In addition, time-consuming sample pre-treatment procedures like centrifugation, filtration and solid-phase extraction can be avoided [10]. In the following paragraphs, some exemplary applications of MBs to separate the biomolecules will be briefly discussed.

2.2.1. Cell Separations

MB-based separation of cells from complex mixtures has become a popular tool and a valuable alternative to fluorescence activated cell sorting. In this process, specific aptamers or antibodies are anchored on the surface of magnetic particles. As example, Herr et al. immobilised an aptamer against leukemia cells on MBs. In this work, it was possible to specifically recognize the cells from complex mixtures including whole blood samples [36]. In a similar work by Zamay et al., MBs modified with aptamer against lung adenocarcinoma cells were used to separate circulating tumor cells (CTC) from human blood [37]. Interestingly, the applications of aptamer-modified MBs have been also integrated with microfluidic device to isolate cancer cell subpopulations [38]. Magnetic cell sorting (MACS) is used with antibodies since a long time [39]. The US FDA has also approved the Cellsearch® system as in vitro diagnostic system for the detection of CTC in the clinic [40].

2.2.2. mRNA Isolation

The isolation of mRNA using MBs is based on A-T pairing. Short sequence of dT (normally $dT_{(25)}$) can be covalently attached to MBs, which will hybridise to dA-tail of mRNA and the isolation of mRNA can be possible within 15 min [41]. This technique eliminates the cumbersome steps used in traditional methods of mRNA isolation such as use of centrifugation and membrane-based spin columns. Importantly, MB-based isolation of mRNA fulfils the demand of automated systems and microfluidic devices, thereby promoting fast processing and high sample throughput [42]. The possibility of high throughput analysis has facilitated the identification of genetic aberrations in cancer cells [43], understanding of biochemical pathways [44] and phylogenetic studies [45].

2.2.3. Protein and Peptide Enrichment

Traditionally, proteins are purified using expensive liquid chromatography systems, centrifuges, filters and other equipment. In addition, this purification process requires several steps and there is significant loss of protein/peptide at each step of purification. Purification using MBs reduces the amount of handling step and all the step can be done in a single test tube, which results in higher efficiency and reduced risk of contamination [46]. In some cases where intracellular proteins are targeted, it is even possible to combine the disruption of cells and separation of the protein form the complex mixture and thus shorten the total purification time. For example, Ni^{2+} or Co^{2+} coated MBs can be used to easily purify His-tagged proteins from cells [6]. Alternatively, aptamer-modified MBs can be used for the isolation of His-tagged proteins [47] or other proteins [48].

These methods can be used to isolate and enrich proteins prior to analysis. For example, immunomagnetic assays rely on MBs modified with antibodies (AB). Enzyme-linked immunosorbent assay (ELISA) is a gold-standard method for the detection of the protein in complex mixture. In immunomagnetic assays, the capturing antibody is immobilized on the MBs, which can reduce the incubation time and efficiency of the assay, as MBs remain suspended throughout the procedure [49]. Morozov et al. have developed an immunomagnetic assay for the detection of streptavidin in a microfluidic device. This has been highly successful in reducing the assay time (three minutes) and better sensitivity (2×10^{-17} M) of the assay [50]. While Section 2 provided a brief overview on the general biological applications of magnetic beads, the next section will focus on the analytical applications of aptamer-modified magnetic beads.

3. Aptamer-Modified Magnetic Beads in Analytical Applications

In aptamer-modified MB-based assays, aptamers are used as binding ligands and MBs are mostly used for the separation of the analyte from complex matrices. MBs have also been coupled with antibodies, but aptamer-based assays offer prominent advantages such as high stability, broad dynamic range, prolonged shelf life, and low cross reactivity [51]. Moreover, aptamers are synthesized chemically, which facilitates straightforward and highly controlled modification with functional groups and different labels [52]. Moreover, the use of aptamers facilitates new assay designs that cannot be realized by using Abs. Aptamers can undergo significant conformational change upon binding to the target molecule. This can be exploited in target-induced structure switching (TISS)-based assays. Another assay format specific for aptamers is based on target-induced dissociation (TID) of complementary oligonucleotides. In case of TID, aptamers ability to hybridize with complementary sequences are used. Here, the helix structure formed between aptamer and its complementary sequence can be easily dissociated by competitive binding of the aptamer with its target (Figure 1). TISS as well as TID can be used for signal generation in various sensing strategies and are especially useful in cases were conventional formats such as sandwich assays are not applicable [3,53,54].

As shown in Table 1, a growing number of research groups are already using aptamer-modified MBs in analytical applications using various sensor designs. In the following sections, the applications have been broadly divided in the electrochemical, optical, piezoelectric and PCR-based assays.

Table 1. Examples of coupling magnetic beads in aptamer-based analytical applications.

Method	Analytes	Detection Limit	Reference
Electrochemical			
Voltammetric			
Differential pulse voltammetry (DPV)	Human activated protein C	2.35 µg mL^{-1}	[55]
DPV	Thrombin	5.5 fM	[56]
DPV	Thrombin	5 nM	[57]
DPV	Human liver hepatocellular carcinoma cells (HepG2)	15 cells mL^{-1}	[58]
DPV	Platelet derived growth factor BB (PDGF BB)	0.22 fM	[59]
DPV	Adenosine	0.05 nM	[60]
DPV	Hg^{2+}	0.33 nM	[61]
Squarewave voltammetry (SWV)	Tumor necrosis factor-alpha (TNF-α)	10 pg mL^{-1}	[62]
SWV	Ochratoxin A	0.07 pg mL^{-1}	[63]
Potentiometric			
Potentiometric carbon-nanotube aptasensor	Variable surface glycoprotein from African Trypanosomes	10 pM	[64]
Direct Potential Measurement	*Listeria monocytogenes*	10 cfu mL^{-1}	[65]
Chronopotentiometry	*Vibrio alginolyticus*	10 cfu mL^{-1}	[66]
Impedimetric			
Electrochemical impedance spectroscopy	*Salmonella*	25 cfu mL^{-1}	[67]
Impedimetric microfluidic analysis	Protein Cry1Ab	0.015 nM	[68]
Microfluidic impedance device	Thrombin	0.01 nM	[69]
Electrogenerated Chemiluminescence			
Electrochemiluminescence resonance energy transfer system	β-amyloid	4.2 × 10^{-6} ng mL^{-1}	[70]
Ratiometric electrochemiluminescence	Cancer cells	150 cells mL^{-1}	[71]
Optical			
Fluorescence			
Signal-on fluorescent aptasensor	Ochratoxin A	20 pg mL^{-1}	[72]
Aptamer-conjugated upconversion nanoprobes assisted by magnetic separation	Circulating tumour cells	20 cells mL^{-1}	[73]
Enzyme-linked aptamer assay	Oxytetracycline	0.88 ng mL^{-1}	[74]
Colorimetric			
Colorimetric assay (Methylene Blue-based)	Hg(II)	0.7 nM	[75]
Chemiluminescence			
Chemiluminescent	Hepatitis B Virus	0.1 ng mL^{-1}	[76]
Chemiluminescence (integrated microfluidic system)	Glycated haemoglobin	0.65 g dL^{-1} for HbA1c and 8.8 g dL^{-1} for Hb	[77]
Surface enhanced Raman scattering			
Molecular embedded SERS aptasensor	Aflatoxin B1	0.0036 ng mL^{-1}	[78]
Universal SERS aptasensor	Aflatoxin B1	0.54 pg mL^{-1}	[79]
Induced Target-Bridged Strategy	platelet derived growth factor BB	3.2 pg mL^{-1}	[80]
Piezoelectric			
Quartz crystal microbalance sensor	*Salmonella enterica*	100 cfu mL^{-1}	[81]
Magnet-quartz crystal microbalance system	Acute leukemia cells	8 × 10^3 cells mL^{-1}	[82]
PCR-based assays			
Apta-qPCR	ATP	17 nM	[54]
Apta-qPCR	Ochratoxin A	0.009 ng mL^{-1}	[53]
Rolling circle amplification	Cocaine	0.48 nM	[83]
Micromagnetic aptamer PCR	PDGF-BB	62 fM	[84]
Real-time PCR	*Escherichia coli*	100 cfu mL^{-1}	[85]
Magnetic relaxation			
Magnetic nanosensors	CCRF-CEM cell	40 cells mL^{-1}	[8]
Magnetic relaxation switch	*Pseudomonas aeruginosa*	50 cfu mL^{-1}	[9]

3.1. Electrochemical

Since the first use of electrochemical (EC) assay for the detection of glucose by Clark and Lyons in 1963 [86], the applications have evolved in different kind of assays including ABs and aptamer-based assays. EC assays are generally rapid, highly sensitive, cost-effective and easy to miniaturize, which is highly attractive for the development of modern bioassays. In aptamer-based electrochemical assays, the change in electrochemical signals (current, voltage, and impedance) due to interaction of analytes and aptamers is measured [87].

EC assays can be broadly classified as amperometric, potentiometric, voltammetric, impedimetric, and electrogenerated chemiluminescence (ECL) assays, according to their working principles. In Table 1, recent work on aptamer-modified MBs in different electrochemical assays is summarized.

3.1.1. Voltammetric Assays

In voltammetric assays, a specific potential is applied to a working electrode in comparison to a reference electrode. The electrochemical reduction or oxidation at the surface of the working electrode results in the generation of a current. Here, amperometric assays are included as a subclass of voltammetric assays as, in amperometric assays also, a constant potential is applied on the working and reference electrode, and the change in current is measured over a period of time. In case of voltammetric assays, a potential range is applied and the changes in both, current and potential are observed. In both assays, the observed change in current is proportional to the concentration of the analyte. Different voltammetric modes are available such as cyclic voltammetry (CV), differential pulse voltammetry (DPV), squarewave voltammetry (SWV), and alternating current voltammetry (ACV) [88]. CV is majorly used to evaluate the electrode surface including purity, stability, and reproducibility of the electrode and to examine the aptamer immobilization on the electrode surface since it allows to check the redox behaviour over a wide potential range [58,88,89].

3.1.2. Differential Pulse Voltammetry (DPV)

Due to its sensitivity and high selectivity, DPV is preferred in analytical applications. In DPV, different electroactive labels are used to generate the signal including small molecules and enzymes, such as horseradish peroxidase (HRP), alkaline phosphatise (AP), and glucose oxidase (GOD), etc. Among these enzyme labels, HRP is one of the most commonly used. For example, Zhao et al. [56] developed a highly sensitive and selective assay for the detection of thrombin (Figure 3A). In this assay, the capture probe was prepared to capture thrombin from the sample solution by immobilizing a thrombin aptamer-1 on MB-AuNP. The detection probe was prepared from another thrombin aptamer-2, horseradish peroxidase (HRP), thiolated chitosan (CS) nanoparticle and gold nanoparticle (CS-AuNP-HRP-Apt2). Presence of thrombin resulted in formation of the sandwich structure of MB-AuNP-Apt1/thrombin/Apt2-HRP-AuNP-CS. The sandwich structures were captured on the surface of a screen printed carbon electrode (SPCE) by a magnet located at the edge of SPCE. Due to the presence of HRP within the sandwich structure, the oxidation of hydroquinone (HQ) with H_2O_2 was dramatically accelerated. The observed electrochemical signal was proportional to the concentration of thrombin in the samples. A similar assay was also developed by Sun et al. [58] for the detection of human liver hepatocellular carcinoma cells (HepG2).

Centi et al. [57] have developed an electrochemical sandwich assay coupled to magnetic beads for the detection of thrombin in plasma. In this work, a microfluidic device was developed for the detection. The electrodes were screen-printed on a magnetic bar. Two different aptamers against thrombin were used where the first aptamer sequence was immobilised on MBs, to capture thrombin from the samples, using streptavidin-biotin interactions. The second aptamer was linked to alkaline phosphatase. The presence of thrombin in the sample resulted in current generation at the electrode, as more product was formed in the presence of alkaline phosphatase.

Zheng et al. [59] have developed an assay based on aptamer-conjugated to methylene blue for the detection of human platelet-derived growth factor BB (PDGF-BB). In this work, boronic acid modified SiO$_2$@Fe$_3$O$_4$@PDA@AuNP composite was used to capture PDGF-BB from the sample. An aptamer against PDGF was linked to SiO$_2$-methylene blue sphere, where methylene blue was used for its high electron-transfer efficiency. Due to the magnetic property of the electrode, the SiO$_2$@Fe$_3$O$_4$@PDA@AuNP/PDGF/aptamer-SiO$_2$-methylene blue could be easily enriched on the electrode surface. Later on, the electrochemical responses could be detected to quantify PDGF.

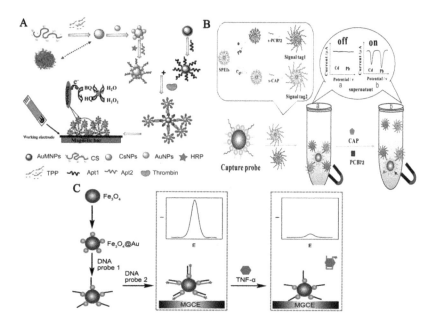

Figure 3. Utilization of MBs in aptamer-based electrochemical assays. (**A**) Using an electric signal mediator. Here, the electroactive molecules (HRP) were brought close to the electrode using aptamer-modified MBs. Reproduced with permission from [56]. Copyright 2012, Royal Society of Chemistry; (**B**) Signal-on type of electrochemical assay. The interactions between aptamers and the target molecules (Chloramphenicol and PCB 72) lead to generation of electrochemical signal. Reproduced with permission [90]. Copyright 2015, Elsevier; (**C**) Signal off type of electrochemical assay. In this type of assay, the interaction between the aptamer and the target molecule leads to reduction in electric signal. Reproduced with permission from [62]. Copyright 2017, Springer.

While in the previous examples, proteins were detected with aptamer-modified MBs, also small molecules are suitable targets, which can be detected e.g., by using the aptamer-specific TISS and TID-based assays. In this context, a TID-based label-free assay was developed by Yang et al. [60] for the detection of adenosine using thionine (Th) for generation of the electrochemical signal. In this work, adenosine aptamer was immobilised on MBs and an oligonucleotide complementary to the target-binding site of the aptamer (abbreviated as cDNA) was hybridized to the immobilized aptamer. Addition of adenosine resulted in the formation of aptamer-adenosine complex and the release of cDNA from the aptamer due to TID. The released cDNA was captured on the sensing electrode through DNA hybridization. As the cDNA is modified with thiol groups at the 5′ termini, AuNP can attach to the cDNA via the formation of S-Au bonding. Subsequently, the electroactive molecules, thionine, are adsorbed on the surfaces of the AuNPs and result in signal generation.

A TISS-based assay in which the structural changes of the aptamer were exploited for signal generation was developed by Wu et al. [61] for the detection of Hg^{2+} using streptavidin modified magnetic beads (Fe_3O_4-SA) and thionine, as electron mediator. In this work, Streptavidin-modified MBs (MB-SA) were immobilized onto the glassy carbon electrode (GCE) and provided magnetic character to the electrode. Then biotin-labelled aptamer against Hg^{2+} was immobilized to the electrode via SA-biotin interaction. Addition of Hg^{2+} resulted in a stable folded structure of thymine (T)-Hg^{2+}-T where Th can easily intercalate. The detection of Hg^{2+} was achieved by recording the DPV signal of Th.

3.1.3. Squarewave Voltammetry (SWV)

SWV is the popular pulse technique and widely considered for automatous and kinetic studies complementary to cyclic voltammetry. Here, TID-based assays are very popular and the interaction of the aptamer with the target molecule can result in electrical signal gain (signal-on) or signal suppression (signal-off).

Yan et al. [90] have developed an assay for the simultaneous detection of two different molecules, chloramphenicol and polychlorinated biphenyls-72 using a signal-on mechanism (Figure 3B). In this work, aptamers were immobilised on MBs and hybridised with a cDNA sequences attached to CdS or PbS QDs as electrochemical signal tracers. Binding of chloramphenicol caused the release of CdS QDs and binding of polychlorinated biphenyls-72 caused the release of PbS QDs. The released CdS and PbS QDs were simultaneously detected through the square wave voltammetry (SWV), which can switch the signals of the biosensor to "on" state. Hao et al. [63] have also developed an assay based on signal-on approach for the detection of ochratoxin A (OTA). In this work, addition of OTA caused the release of CdTe QDs modified with cDNA, from the MB-modified aptamers. After magnetic separation of aptamer-modified MBs, CdTe QDs remaining in the supernatant were dissolved by HNO_3 and the concentration of Cd ions, which was directly proportional to the concentration of OTA, was detected by SWV.

Miao et al. [62] have developed a signal-off assay for the detection of TNF-α (Figure 3C). In this simple assay, methylene blue-tagged aptamer was immobilised on magnetic glassy carbon electrode (MGCE) using a cDNA. Addition of the target molecule resulted in release of methylene blue-tagged aptamer, resulting in decrease in electrochemical signal by SWV.

3.1.4. Potentiometric Assays

In potentiometric sensors, the change in electric potential between two electrodes is detected by a field-effect transistor (FET) [91]. Here, the indicator electrode reports change in electric potential according to analyte concentration and the reference electrode provides constant electric potential.

Recently, Zhao et al. [66] have developed a potentiometric sensor for the detection of *Vibrio alginolyticus*, which is an opportunistic marine pathogen and can cause otitis, wound infection, and chronic diarrhoea in mammals. In this work, a cDNA sequences were immobilized on the surface of the magnetic beads using streptavidin-biotin interaction. The aptamer and H1/H2 (two different oligonucleotides) hybridize successively with the cDNA to form the DNA structure-modified magnetic beads. The resulted DNA structure can interact with protamine (polycation) due to electrostatic interactions, which can be detected by the polycation-sensitive electrode. When a sample containing *Vibrio alginolyticus* was added, the aptamers interact with the target due to its high affinity with the target. Consequently, the DNA structure disassembled and a reduction in potential was observed which was proportional the target concentration. A similar assay was also used for the detection of small molecules such as bisphenol A [92].

3.1.5. Impedimetric Assays (EIS)

In impedimetric assays, electrochemical impedance spectroscopy (EIS) is popular due to its high sensitivity. EIS involves the analysis of the resistive and capacitive properties, which are based on the perturbation of a system at equilibrium by a small amplitude of excitation signal. EIS allows rapid and accurate detection of the small changes along the electrode by a transducer. The signal is enhanced by an amplifier, which makes EIS appealing in analytical applications. In addition, EIS is a simple technique and the detection in EIS does not require attachment of capture molecules to the electrodes and thus allows label-free detection.

Wang et al. [69] have developed a microfluidic analysis system assay for the detection of thrombin using aptamer-modified magnetic separation. In this work, thrombin aptamer-modified MBs were used to capture and separate the target protein from serum. Later on, the bound complex was injected into the microfluidic flow cell for impedance measurement. Similar analysis system was also developed by Jin et al. [68] for the detection of Cry1Ab protein to detect genetically modified crops. Another interesting impedimetric assay based on TID mechanism was developed by Lee et al. [93] for the detection of prostate-specific antigen (PSA). Combining PSA aptamer-modified magnetic nanoparticles with rolling circle amplification (RCA) has provided a better sensitivity of 0.74 pg mL^{-1} PSA in human serum.

3.1.6. Electrogenerated Chemiluminescence

Electrogenerated chemiluminescence (ECL), also called electrochemiluminescence, refers to the emission of light via electron transfer reactions from electrochemically generated reagents. ECL combines the sensitivity and wide dynamic range from chemiluminescence (CL) with the advantages offered by electrochemical methods, such as simplicity, stability, and facility to be miniaturized [94,95]. Lately, ECL has been widely accepted in different analytical applications including fundamental studies to detecting trace amount of target molecules. Among different luminophores, use of luminol, quantum dots (QDs) and ruthenium(II) complexes, have been widely employed in ECL assays [96].

Ke et al. [70] have developed an assay for the detection of β-amyloid (Aβ) using a sandwich-type ECL sensing platform. In this work, Ru(bpy)$_3$$^{2+}$ was used as ECL donor and gold nanorods (GNRs) were used as ECL acceptor. Here, resonance energy transfer (RET) donor nanohybrids were prepared with mesoporous carbon nanospheres (MCNs)@nafion/Ru(bpy)$_3$$^{2+}$/Aβ antibody. After incubation with target Aβ protein and GNRs-attached aptamer, prominent decrease in ECL signal was observed due to the quenching effect between Ru(bpy)$_3$$^{2+}$ and GNRs. This innovative approach performed well with sensitivity of 4.2 fg mL^{-1} in real Alzheimer's patient cerebrospinal fluid samples.

In a TID-based approach, Wang et al. [71] have developed a novel ECL sensing system for the detection of HL-60 cancer cells. Here, Ag-polyamidoamine (PAMAM) was prepared and functionalized with cDNA and bio-bar-code DNA (bbcDNA). The prepared composite was hybridized with the aptamer-modified MBs. Addition of HL-60 cancer cells resulted in the release of cDNA-Ag-PAMAM composite in the supernatant. For the detection, an oligonucleotide complementary to cDNA was immobilised on the electrode surface resulting in hybridization of released cDNA-Ag-PAMAM.

3.2. Optical

Optical assays have been widely used due to their specific advantages such as high sensitivity, quick response, high signal-to-noise ratio, reduced cost of manufacture, and relatively simple operation. Aptamers are preferred ligands in optical assays due to flexibility in modification with various fluorophores and other labels. As already described for electrochemical assays, application of aptamer-modified MBs can be highly advantageous due to easy separation of the target molecule from the complex matrices, which offers high signal-to-noise ratio during measurement. The assays can be classified based on the detection principle including fluorescence, colorimetry, chemiluminescence, surface plasmon resonance (SPR), and Raman scattering.

3.2.1. Fluorescence Based Assays

Aptamer-based fluorescence assays can be mainly divided into labelled and label-free assays. Easy modification of aptamers with fluorophores and quenchers during chemical synthesis facilitates the design of various assays.

Luo et al. [97] developed a nicking enzyme assisted signal amplification (NEASA)-based assay relying on TID mechanism (Figure 4A). In their work, ampicillin aptamer was immobilised on MBs and attached to a complementary sequence (cDNA) through Watson-Crick base pairing. Addition of ampicillin resulted in release of cDNA which can bind to Taqman probe having fluorescent and quencher probes at opposite end. The nicking enzyme cleaved the Taqman probe only when it was bound to cDNA. The decrease in fluorescence signal was proportional to the concentration of ampicillin.

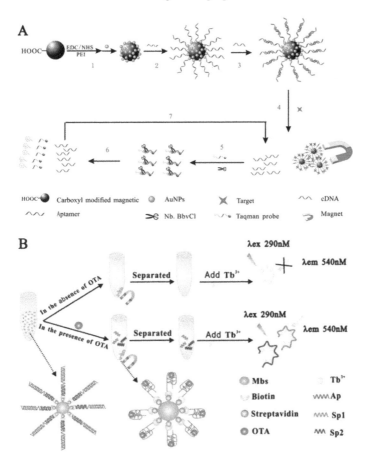

Figure 4. Fluorescence-based assays. (**A**) Combining fluorophores and quencher molecules. Here the interaction between the aptamer and the target molecule leads to the release of quencher molecule and the increase of fluorescence signal. Reproduced with permission from [97]. Copyright 2017, Elsevier; (**B**) Label-free assay. Being oligonucleotides, aptamers can specifically interact with dyes specific for ssDNA or dsDNA. In this example, Tb^{3+} was used which interacts specifically with ssDNA (cDNA), which was released due to TID from ochratoxin A (OTA) aptamer. Reproduced with permission from [72]. Copyright 2013, Elsevier.

Upconversion nanoparticles (UCNPs), nanocrystals containing lanthanide ions, emerged as an important fluorophor, as they lack autofluorescence and their use result in high signal-to-noise ratio. In addition, the optical properties of UCNP can be tuned with different lanthanide dopants such as Er^{3+}, Tm^{3+}, and Ho^{3+} [98,99]. Fang et al. [73] developed an assay for the detection of circulating tumor cells (CTC) using UCNPs. In their work, UCNPs were modified with the aptamer and biotinylated-PEG. Aptamer was used to recognize CTC and biotinylated-PEG was used to attach UCNP to the MBs. Here, whole blood samples were mixed with the modified UCNPs and it was possible to detect as low as 10 cells into 0.5 mL of whole blood samples. Efforts have been also made to further increase the sensitivity of fluorescence-based assays. In this context, Wang et al. [100] used RuBpy-doped silica nanoparticles (RSiNPs), which are highly photostable and provide significant enhancement in fluorescent signal when compared with single RuBpy dye molecules. In this work, aptamer-modified MBs were used to separate thrombin from human serum.

Aptamer-modified MBs have been also used in enzyme-linked aptamer sandwich assays. John Bruno et al. [101] used aptamer-modified MBs to capture *Campylobacter jejuni* from the samples. Later on, a second aptamer modified with QDs was introduced. The bound complex was brought on the photo detector using external magnet. The whole detection procedure could be finished in 15 min and resulted in high sensitivity. Similar detection principle was also used by Hao et al. for the detection of thrombin [102].

To reduce the labelling cost and reducing the effect of labelling on aptamer conformation, label free assays have attracted big attention. Being DNA sequences, aptamers can also bind to DNA binding chemicals, such as crystal violet [103], SYBR Green I (SGI) [104,105], 4′,6-diamidino-2-phenylindol (DAPI) [106], malachite green [107], OliGreen [108] and terbium (III) (Tb^{3+}). Zhang et al. [72] developed a label-free fluorescent aptasensor based on the Tb^{3+}, structure-switching of anti-OTA aptamer and MBs for the detection of ochratoxin A in wheat (Figure 4B).

3.2.2. Colorimetric Assay

Colorimetric assays are widely used in analytical application due to simplicity of the measurement. Among aptamer-based colorimetric assays, AuNP-based assays are widely used. MBs-based separation in these assays provide better signal-to-noise ratio and sensitivity.

For example, Liang et al. [109] reported an aptamer-protein interaction-induced aggregation assay for the detection of thrombin. In their work, two different aptamers against human α-thrombin were immobilised on nanoroses (MB-AuNP core-shell structure in a flowerlike shape). Addition of human α-thrombin in solution resulted in aggregation of nanoroses and thus in a characteristic change in UV-Vis absorption spectra of the colloid. An interesting assay was also developed by Wang et al. [75] for the detection of Hg(II) based on hybridization chain reaction (HCR). In this work, aptamer against Hg(II) were immobilized on the MB-AuNP. HCR process is inhibited in the presence of Hg(II) enabling less methylene blue to intercalate into the dsDNA structure.

The use of peroxidase-like activity of magnetic nanoparticles is exploited in different colorimetric assays. Kim et al. [110] reported an assay for the detection of metal ions. Here, the aptamers were adsorbed on the positively charged surface of MBs, which reduces the catalytic activity of MBs. Addition of the target molecules released the adsorbed aptamers from MBs surface and MBs recover the peroxidase-like activity. Similar assays were also developed for the detection of ochratoxin A in cereal samples [111] and thrombin in blood plasma [112].

3.2.3. Chemiluminescence Assays

In chemiluminescence (CL) assays, the light emitted during a chemical reaction is detected. The unique characteristics of CL include high sensitivity, wide dynamic ranges and simple instrumentation. In case of CL, an excitation light-source is not required which is highly cost effective.

A label-free CL detection of adenosine in human serum was realized by Yan et al. [113]. In their work, 3,4,5-trimethoxyl-phenylglyoxal (TMPG) was used as the signalling molecule for CL, as TMPG can intercalate with guanine (G) nucleobases. Firstly, the cDNA was immobilized on the surface of MBs and hybridized with a G-rich adenosine aptamer. Addition of adenosine containing sample caused the release of the aptamer from MBs modified with cDNA and a decrease in CL signal was observed, which was proportional to adenosine concentration.

In contrast, HRP-based catalysis can be widely used for the generation of CL signal. As an example for CL assays using HRP, Li et al. [114] immobilized aptamers directed against cocaine on the surface of AuNP-functionalized MBs (MB-AuNP). Therefore, aptamers were hybridized with the cDNA immobilized on the double-functional AuNP modified with HRP (HRP-AuNP). When cocaine was introduced, a dissociation of the aptamer was achieved due to binding of the aptamer to cocaine. Consequently, HRP-AuNP were eluted from the MB-AuNP due to target-induced dissociation (TID). The recorded CL signals were proportional to the concentration of cocaine (Figure 5).

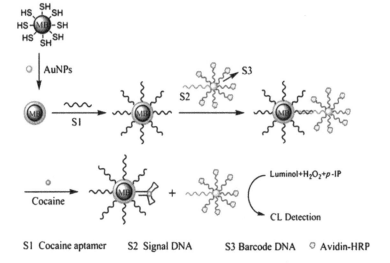

S1 Cocaine aptamer   S2 Signal DNA   S3 Barcode DNA   Avidin-HRP

Figure 5. Chemiluminescence assay. This assay is based on TID. Here, the interaction between the aptamer and the target molecule (cocaine) caused the release of cDNA attached to HRP-modified AuNPs. Released HRP generated chemiluminescence signal which was proportional to cocaine concentration. Reproduced with permission from [114]. Copyright 2011, Springer.

3.2.4. Surface-Enhanced Raman Scattering-Based Assays

Surface-enhanced Raman scattering (SERS) relies on the principle that the Raman scattering intensity of molecules will be greatly improved after their adsorbtion onto the metal surface.

Quansheng et al. [78] developed a SERS assay for ultrasensitive detection of aflatoxin B1 (AFB1) detection in peanut oil (Figure 6). In this study, AFB1 aptamers were immobilised on the MBs and cDNA was immobilised on gold nanorods (cDNA-AuNRs). Presence of AFB1 resulted in the release of cDNA-AuNRs and decrease in SERS signal was observed. In an interesting work, aptamer-conjugated magnetic beads were used for the separation of circulating tumor cells from whole blood samples and the tumor cells were detected using surface-enhanced Raman scattering imaging [115].

Yoon et al. [116] developed SERS-based magnetic aptasensors for the detection of thrombin in serum samples. In this work, two different aptamers against thrombin were used for the detection. One aptamer was immobilised on MBs and second aptamer was immobilised on AuNP-coated with Raman reporter molecules, X-rhodamine-5-(and -6)-isothiocyanate (XRITC). Addition of thrombin resulted in the formation of sandwich aptamer complexes and an increase of SERS signal according to thrombin concentration in the sample was observed.

3.3. Piezoelectric Assays

Since the discovery of piezoelectric effect in 1880 by the Curie brothers, the piezoelectric effect has been very popular in analytical applications. Lately, the progress made in the fields of microelectronics and microfluidics further promotes the development of label-free piezoelectric assays [117]. Particularly, quartz-crystal microbalance (QCM)-based assays have become popular in analytical applications of aptamers. Utilizing MBs in these assays is highly advantageous due to their inherent piezoelectric properties, and potential to concentrate the analyte molecules at the QCM surface [118].

Ozalp et al. [81] developed a QCM biosensor for the detection of *Salmonella* cells in food samples. Here, aptamer-modified MBs were, firstly, used to capture the target pathogens from the food samples. The magnetically separated pathogens were detected by QCM sensor and 100 cfu mL^{-1} *Salmonella* cells could be detected in milk samples. In a similar assay, Pan et al. [82] detected leukemia cells in complex matrices. In this work, aptamer-modified MBs were used to capture leukemia cells from the biological sample and to approximate them to the QCM sensor using an external magnet.

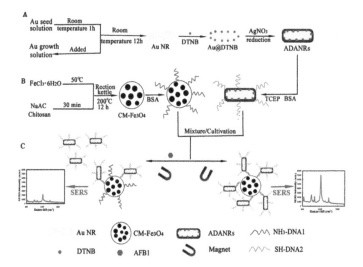

Figure 6. Surface-enhanced Raman scattering-based assays. (**A**) Immobilization of aflatoxin B1 (AFB1) aptamer on gold nanorods (AuNRs). (**B**) Immobilization of cDNA on chitosan-modified MBs. (**C**) Schematic representation of AFB1 measurement. Here, the binding of AFB1 induced the release of cDNA and, in turn, AuNRs from the MBs and a decrease in SERS signal was observed. Reproduced with permission from [78]. Copyright 2018, Elsevier.

An interesting approach was used by Song et al. [119] for the detection of ATP via DNAzyme-activated and aptamer-based target-triggered circular amplification. In this work, AuNPs were used for mass amplification and captured on the modified gold electrode. The amplification scheme involved circular nucleic acid strand-displacement polymerization, aptamer binding strategy and DNAzyme signal amplification. Presence of ATP resulted in a two-cycle amplification process, triggered by the aptamer recognition of a target molecule.

3.4. PCR-Based Assays

Being oligonucleotide sequences, aptamers can be easily amplified and quantified using qPCR with high sensitivity and reproducibility. Recently, different assays based on this property of aptamers have been developed including Apta-qPCR, micromagnetic aptamer PCR (MAP), and assays involving

PCR-based amplification strategies as loop-mediated isothermal amplification (LAMP), rolling circle amplification, and isothermal signal amplification, proximity ligation assays, and nuclease protection assays. In these assays, MBs provides opportunity to separate target-bound and unbound aptamers, which is very important to get minimum background and high signal-to-noise ratios and in turn high sensitivity and reproducibility.

Our group [53] developed an Apta-qPCR assay for the detection of ochratoxin A in beer samples (Figure 7). In this work, ochratoxin A aptamer was hybridized to a corresponding cDNA, which was immobilized on MBs. Addition of the target molecules caused TID of aptamer from the MBs and the released aptamers were quantified using qPCR. This assay was able to detect 0.009 ng mL^{-1} OTA in beer samples. Similar assay was also used for the quantification of ATP present in HeLa cell lysate with the sensitivity of 17 nM ATP [54].

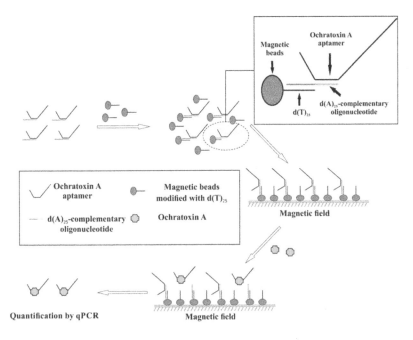

Figure 7. Apta-qPCR. This assay is based on TID, where the interaction of the target molecules (ochratoxin A) caused the release of aptamer from the cDNA-modified MBs. The released aptamers were quantified using qPCR. Reproduced with permission from [53]. Copyright 2017, Wiley.

Csordas et al. [84] developed an interesting concept for the detection of PDGF-BB using combination of antibody and aptamer in MAP. In this work, high-gradient magnetic field sample preparation was integrated within a microfluidic device with aptamer-based real-time PCR readout. Antibody-modified MBs were used for capturing PDGF-BB and an aptamer against PDGF-BB, which was modified with flanking PCR primer sequences, was added after washing. The bound aptamers were quantified using qPCR.

Ozalp et al. [85] developed a qPCR-based assay where aptamer-modified MBs were used to preconcentrate the *Escherichia coli* or *Salmonella typhimurium*. Later on, bacterial genomes were extracted which was quantified using qPCR. A similar assay was also developed by Feng et al. [120] for the detection of *Listeria monocytogenes* using a LAMP assay.

4. Conclusions and Future Trends

Aptamers are attractive bioreceptors in analytical applications due to their small size, animal free- and cost effective production, high stability (especially DNA aptamers), target versatility, high binding affinity and selectivity for their target molecules. In addition, several properties of aptamers including ease of chemical modification, measurable structural changes induced upon interaction of the aptamer with the target molecule, and the potential to amplify aptamers via PCR are advantageous in comparison to other binding ligands such as ABs. Due to these significant advantages of aptamers, they have been widely used for the detection of different analytes ranging from metal ions, small molecules, proteins to whole cells in diverse assay formats. In this review article, the focus was put on the magnetic bead-based analytical applications of aptamers. Utilization of MBs in aptamer-based applications allows to rapidly detect the analyte in the complex matrices with high signal-to-noise ratio. Recent developments in the synthesis of MBs resulted in MBs with better homogenous size distribution, high and uniform magnetite content, and a fast response to applied magnetic field, as well as high colloidal stability of magnetic nanoparticles.

In this review, different assay formats have been discussed where MBs were coupled with aptamers for the analytical applications. In many of the applications, MBs help to separate the target molecule from complex matrices. In some applications the aptamer-modified beads are also enriched directly on the sensor surface, thereby representing a surface modification used to immobilize the target. In few applications, magnetic properties of the MBs, such as magnetic-relaxation switch, have been used for signal generation.

Application of aptamer in analytical techniques is still in development phase, as many commercial applications use ABs. Slowly but steadily, aptamers are developed against a range of molecules. For some of them the development of other ligands, such as antibodies, is not easy, e.g., due to low immunogenicity or high toxicity. Moreover, aptamers seem to be especially advantageous for the detection of small molecules. In this context, TISS and TID mechanism provide the possibility to design assays that can detect small molecules, while other strategies like sandwich assays are not suitable for detection of small molecules. These advantages of aptamers and their combination with those of MBs, can further boost the development of new analytical procedures. The use of MBs in these assays can result in rapid detection of the target molecules even within complex matrices with no need for time-consuming sample pre-treatment procedures. Taking together the strengths of aptamers and MBs can therefore be especially advantageous in the development of POCT, where complex samples have to be analyzed within minutes.

Acknowledgments: German Academic Exchange Service (DAAD) is acknowledged for the financial support to Harshvardhan Modh. The publication of this article was funded by the Open Access fund of Leibniz Universität Hannover.

Author Contributions: Harshvardhan Modh and Johanna-Gabriela Walter wrote the manuscript. Thomas Scheper supervised the preparation of the manuscript.

Conflicts of Interest: The authors declare no conflict of interest.

References

1.   Sun, H.; Zu, Y. A highlight of recent advances in aptamer technology and its application. *Molecules* **2015**, *20*, 11959–11980. [CrossRef] [PubMed]
2.   Witt, M.; Walter, J.-G.; Stahl, F. Aptamer microarrays—Current status and future prospects. *Microarrays* **2015**, *4*, 115–132. [CrossRef] [PubMed]
3.   Walter, J.-G.; Heilkenbrinker, A.; Austerjost, J.; Timur, S.; Stahl, F.; Scheper, T. Aptasensors for small molecule detection. *Z. Naturforsch. B* **2012**, *67*, 976–986. [CrossRef]
4.   Urmann, K.; Walter, J.-G.; Scheper, T.; Segal, E. Label-free optical biosensors based on aptamer-functionalized porous silicon scaffolds. *Anal. Chem.* **2015**, *87*, 1999–2006. [CrossRef] [PubMed]
5.   Rocha-Santos, T.A.P. Sensors and biosensors based on magnetic nanoparticles. *TRAC Trend Anal. Chem.* **2014**, *62*, 28–36. [CrossRef]

6. Aguilar-Arteaga, K.; Rodriguez, J.A.; Barrado, E. Magnetic solids in analytical chemistry: A review. *Anal. Chim. Acta* **2010**, *674*, 157–165. [CrossRef] [PubMed]

7. Kudr, J.; Klejdus, B.; Adam, V.; Zitka, O. Magnetic solids in electrochemical analysis. *TRAC Trend Anal. Chem.* **2017**, *98*, 104–113. [CrossRef]

8. Bamrungsap, S.; Chen, T.; Shukoor, M.I.; Chen, Z.; Sefah, K.; Chen, Y.; Tan, W. Pattern recognition of cancer cells using aptamer-conjugated magnetic nanoparticles. *ACS Nano* **2012**, *6*, 3974–3981. [CrossRef] [PubMed]

9. Jia, F.; Xu, L.; Yan, W.; Wu, W.; Yu, Q.; Tian, X.; Dai, R.; Li, X. A magnetic relaxation switch aptasensor for the rapid detection of Pseudomonas aeruginosa using superparamagnetic nanoparticles. *Microchim. Acta* **2017**, *184*, 1539–1545. [CrossRef]

10. Khng, H.P.; Cunliffe, D.; Davies, S.; Turner, N.A.; Vulfson, E.N. The synthesis of sub-micron magnetic particles and their use for preparative purification of proteins. *Biotechnol. Bioeng.* **1998**, *60*, 419–424. [CrossRef]

11. Liao, M.-H.; Chen, D.-H. Fast and efficient adsorption/desorption of protein by a novel magnetic nano-adsorbent. *Biotechnol. Lett.* **2002**, *24*, 1913–1917. [CrossRef]

12. Bucak, S.; Jones, D.A.; Laibinis, P.E.; Hatton, T.A. Protein separations using colloidal magnetic nanoparticles. *Biotechnol. Prog.* **2003**, *19*, 477–484. [CrossRef] [PubMed]

13. Shao, D.; Xu, K.; Song, X.; Hu, J.; Yang, W.; Wang, C. Effective adsorption and separation of lysozyme with PAA-modified Fe_3O_4@silica core/shell microspheres. *J. Colloid Interface Sci.* **2009**, *336*, 526–532. [CrossRef] [PubMed]

14. Oktem, H.A.; Bayramoglu, G.; Ozalp, V.C.; Arica, M.Y. Single-step purification of recombinant thermus aquaticus DNA polymerase using DNA-aptamer immobilized novel affinity magnetic beads. *Biotechnol. Prog.* **2007**, *23*, 146–154. [CrossRef] [PubMed]

15. Shukoor, M.I.; Natalio, F.; Tahir, M.N.; Ksenofontov, V.; Therese, H.A.; Theato, P.; Schröder, H.C.; Müller, W.E.; Tremel, W. Superparamagnetic γ-Fe_2O_3 nanoparticles with tailored functionality for protein separation. *Chem. Commun.* **2007**, 4677–4679. [CrossRef] [PubMed]

16. Sun, Y.; Ding, X.; Zheng, Z.; Cheng, X.; Hu, X.; Peng, Y. A novel approach to magnetic nanoadsorbents with high binding capacity for bovine serum albumin. *Macromol. Rapid Commun.* **2007**, *28*, 346–351. [CrossRef]

17. Jeng, J.; Lin, M.F.; Cheng, F.Y.; Yeh, C.S.; Shiea, J. Using high-concentration trypsin-immobilized magnetic nanoparticles for rapid in situ protein digestion at elevated temperature. *Rapid Commun. Mass Spectrom.* **2007**, *21*, 3060–3068. [CrossRef] [PubMed]

18. Li, Y.; Xu, X.; Deng, C.; Yang, P.; Zhang, X. Immobilization of trypsin on superparamagnetic nanoparticles for rapid and effective proteolysis. *J. Proteome Res.* **2007**, *6*, 3849–3855. [CrossRef] [PubMed]

19. Lin, S.; Yao, G.; Qi, D.; Li, Y.; Deng, C.; Yang, P.; Zhang, X. Fast and efficient proteolysis by microwave-assisted protein digestion using trypsin-immobilized magnetic silica microspheres. *Anal. Chem.* **2008**, *80*, 3655–3665. [CrossRef]

20. Chen, W.; Shen, H.; Li, X.; Jia, N.; Xu, J. Synthesis of immunomagnetic nanoparticles and their application in the separation and purification of $CD34^+$ hematopoietic stem cells. *Appl. Surf. Sci.* **2006**, *253*, 1762–1769. [CrossRef]

21. Antoine, J.-C.; Rodrigot, M.; Avrameas, S. Lymphoid cell fractionation on magnetic polyacrylamide-agarose beads. *Immunochemistry* **1978**, *15*, 443–452. [CrossRef]

22. Krogh, T.N.; Berg, T.; Højrup, P. Protein analysis using enzymes immobilized to paramagnetic beads. *Anal. Biochem.* **1999**, *274*, 153–162. [CrossRef]

23. Gatto-Menking, D.L.; Yu, H.; Bruno, J.G.; Goode, M.T.; Miller, M.; Zulich, A.W. Sensitive detection of biotoxoids and bacterial spores using an immunomagnetic electrocheminescence sensor. *Biosens. Bioelectron.* **1995**, *10*, 501–507. [CrossRef]

24. Guesdon, J.-L.; Avrameas, S. Magnetic solid phase enzyme-immunoassay. *Immunochemistry* **1977**, *14*, 443–447. [CrossRef]

25. Yavuz, C.T.; Prakash, A.; Mayo, J.; Colvin, V.L. Magnetic separations: From steel plants to biotechnology. *Chem. Eng. Sci.* **2009**, *64*, 2510–2521. [CrossRef]

26. Towler, P.H.; Smith, J.D.; Dixon, D.R. Magnetic recovery of radium, lead and polonium from seawater samples after preconcentration on a magnetic adsorbent of manganese dioxide coated magnetite. *Anal. Chim. Acta* **1996**, *328*, 53–59. [CrossRef]

27. Šafařík, I.; Šafaříková, M.; Vrchotová, N. Study of sorption of triphenylmethane dyes on a magnetic carrier bearing an immobilized copper phthalocyanine dye. *Collect. Czech. Chem. Commun.* **1995**, *60*, 34–42. [CrossRef]

28. Šafaříková, M.; Šafařík, I. Magnetic solid-phase extraction. *J. Magn. Magn. Mater.* **1999**, *194*, 108–112. [CrossRef]

29. Hsing, I.; Xu, Y.; Zhao, W. Micro-and nano-magnetic particles for applications in biosensing. *Electroanalysis* **2007**, *19*, 755–768. [CrossRef]

30. Quy, D.V.; Hieu, N.M.; Tra, P.T.; Nam, N.H.; Hai, N.H.; Thai Son, N.; Nghia, P.T.; Anh, N.T.V.; Hong, T.T.; Luong, N.H. Synthesis of silica-coated magnetic nanoparticles and application in the detection of pathogenic viruses. *J. Nanomater.* **2013**, *2013*. [CrossRef]

31. Chen, C.-T.; Chen, Y.-C. Fe$_3$O$_4$/TiO$_2$ core/shell nanoparticles as affinity probes for the analysis of phosphopeptides using TiO$_2$ surface-assisted laser desorption/ionization mass spectrometry. *Anal. Chem.* **2005**, *77*, 5912–5919. [CrossRef] [PubMed]

32. Blakemore, R. Magnetotactic bacteria. *Science* **1975**, *190*, 377–379. [CrossRef] [PubMed]

33. Matsunaga, T.; Maeda, Y.; Yoshino, T.; Takeyama, H.; Takahashi, M.; Ginya, H.; Aasahina, J.; Tajima, H. Fully automated immunoassay for detection of prostate-specific antigen using nano-magnetic beads and micro-polystyrene bead composites, 'Beads on Beads'. *Anal. Chim. Acta* **2007**, *597*, 331–339. [CrossRef] [PubMed]

34. Canfarotta, F.; Piletsky, S.A. Engineered magnetic nanoparticles for biomedical applications. *Adv. Healthc. Mater.* **2014**, *3*, 160–175. [CrossRef] [PubMed]

35. Paleček, E.; Fojta, M. Magnetic beads as versatile tools for electrochemical DNA and protein biosensing. *Talanta* **2007**, *74*, 276–290. [CrossRef] [PubMed]

36. Herr, J.K.; Smith, J.E.; Medley, C.D.; Shangguan, D.; Tan, W. Aptamer-conjugated nanoparticles for selective collection and detection of cancer cells. *Anal. Chem.* **2006**, *78*, 2918–2924. [CrossRef] [PubMed]

37. Zamay, G.S.; Kolovskaya, O.S.; Zamay, T.N.; Glazyrin, Y.E.; Krat, A.V.; Zubkova, O.; Spivak, E.; Wehbe, M.; Gargaun, A.; Muharemagic, D. Aptamers selected to postoperative lung adenocarcinoma detect circulating tumor cells in human blood. *Mol. Ther.* **2015**, *23*, 1486–1496. [CrossRef] [PubMed]

38. Labib, M.; Green, B.; Mohamadi, R.M.; Mepham, A.; Ahmed, S.U.; Mahmoudian, L.; Chang, I.-H.; Sargent, E.H.; Kelley, S.O. Aptamer and antisense-mediated two-dimensional isolation of specific cancer cell subpopulations. *J. Am. Chem. Soc.* **2016**, *138*, 2476–2479. [CrossRef] [PubMed]

39. Miltenyi, S.; Müller, W.; Weichel, W.; Radbruch, A. High gradient magnetic cell separation with MACS. *Cytom. Part A* **1990**, *11*, 231–238. [CrossRef] [PubMed]

40. Hassan, E.M.; Willmore, W.G.; DeRosa, M.C. Aptamers: Promising tools for the detection of circulating tumor cells. *Nucl. Acid Ther.* **2016**, *26*, 335–347. [CrossRef] [PubMed]

41. Karrer, E.E.; Lincoln, J.E.; Hogenhout, S.; Bennett, A.B.; Bostock, R.M.; Martineau, B.; Lucas, W.J.; Gilchrist, D.G.; Alexander, D. In situ isolation of mRNA from individual plant cells: Creation of cell-specific cDNA libraries. *Proc. Natl. Acad. Sci. USA* **1995**, *92*, 3814–3818. [CrossRef] [PubMed]

42. Rodriguez, I.R.; Chader, G.J. A novel method for the isolation of tissue-specific genes. *Nucl. Acids Res.* **1992**, *20*, 3528. [CrossRef] [PubMed]

43. Maher, C.A.; Kumar-Sinha, C.; Cao, X.; Kalyana-Sundaram, S.; Han, B.; Jing, X.; Sam, L.; Barrette, T.; Palanisamy, N.; Chinnaiyan, A.M. Transcriptome sequencing to detect gene fusions in cancer. *Nature* **2009**, *458*, 97. [CrossRef] [PubMed]

44. Rogers, S.; Macheda, M.L.; Docherty, S.E.; Carty, M.D.; Henderson, M.A.; Soeller, W.C.; Gibbs, E.M.; James, D.E.; Best, J.D. Identification of a novel glucose transporter-like protein—GLUT-12. *Am. J. Physiol. Endocrinol. Metab.* **2002**, *282*, E733–E738. [CrossRef] [PubMed]

45. Helmkampf, M.; Bruchhaus, I.; Hausdorf, B. Phylogenomic analyses of lophophorates (brachiopods, phoronids and bryozoans) confirm the Lophotrochozoa concept. *Proc. R. Soc. Lond. B Biol. Sci.* **2008**, *275*, 1927–1933. [CrossRef] [PubMed]

46. Franzreb, M.; Siemann-Herzberg, M.; Hobley, T.J.; Thomas, O.R. Protein purification using magnetic adsorbent particles. *Appl. Microbiol. Biotechnol.* **2006**, *70*, 505–516. [CrossRef] [PubMed]

47. Kökpinar, Ö.; Walter, J.G.; Shoham, Y.; Stahl, F.; Scheper, T. Aptamer-based downstream processing of his-tagged proteins utilizing magnetic beads. *Biotechnol. Bioeng.* **2011**, *108*, 2371–2379. [CrossRef] [PubMed]

48. Lönne, M.; Bolten, S.; Lavrentieva, A.; Stahl, F.; Scheper, T.; Walter, J.-G. Development of an aptamer-based affinity purification method for vascular endothelial growth factor. *Biotechnol. Rep.* **2015**, *8*, 16–23. [CrossRef] [PubMed]

49. Song, F.; Zhou, Y.; Li, Y.; Meng, X.; Meng, X.; Liu, J.; Lu, S.; Ren, H.; Hu, P.; Liu, Z. A rapid immunomagnetic beads-based immunoassay for the detection of β-casein in bovine milk. *Food Chem.* **2014**, *158*, 445–448. [CrossRef] [PubMed]

50. Morozov, V.N.; Groves, S.; Turell, M.J.; Bailey, C. Three minutes-long electrophoretically assisted zeptomolar microfluidic immunoassay with magnetic-beads detection. *J. Am. Chem. Soc.* **2007**, *129*, 12628–12629. [CrossRef] [PubMed]

51. Modh, H.B.; Bhadra, A.K.; Patel, K.A.; Chaudhary, R.K.; Jain, N.K.; Roy, I. Specific detection of tetanus toxoid using an aptamer-based matrix. *J. Biotechnol.* **2016**, *238*, 15–21. [CrossRef] [PubMed]

52. Ilgu, M.; Nilsen-Hamilton, M. Aptamers in analytics. *Analyst* **2016**, *141*, 1551–1568. [CrossRef] [PubMed]

53. Modh, H.; Scheper, T.; Walter, J.G. Detection of ochratoxin A by aptamer-assisted real-time PCR-based assay (Apta-qPCR). *Eng. Life Sci.* **2017**, *17*, 923–930. [CrossRef]

54. Modh, H.; Witt, M.; Urmann, K.; Lavrentieva, A.; Segal, E.; Scheper, T.; Walter, J.G. Aptamer-based detection of adenosine triphosphate via qPCR. *Talanta* **2017**, *172*, 199–205. [CrossRef] [PubMed]

55. Erdem, A.; Congur, G. Voltammetric aptasensor combined with magnetic beads assay developed for detection of human activated protein C. *Talanta* **2014**, *128*, 428–433. [CrossRef] [PubMed]

56. Zhao, J.; Lin, F.; Yi, Y.; Huang, Y.; Li, H.; Zhang, Y.; Yao, S. Dual amplification strategy of highly sensitive thrombin amperometric aptasensor based on chitosan-Au nanocomposites. *Analyst* **2012**, *137*, 3488–3495. [CrossRef] [PubMed]

57. Centi, S.; Tombelli, S.; Minunni, M.; Mascini, M. Aptamer-based detection of plasma proteins by an electrochemical assay coupled to magnetic beads. *Anal. Chem.* **2007**, *79*, 1466–1473. [CrossRef] [PubMed]

58. Sun, D.; Lu, J.; Zhong, Y.; Yu, Y.; Wang, Y.; Zhang, B.; Chen, Z. Sensitive electrochemical aptamer cytosensor for highly specific detection of cancer cells based on the hybrid nanoelectrocatalysts and enzyme for signal amplification. *Biosens. Bioelectron.* **2016**, *75*, 301–307. [CrossRef] [PubMed]

59. Zheng, J.; Zhang, M.; Guo, X.; Wang, J.; Xu, J. Boronic acid functionalized magnetic composites with sandwich-like nanostructures as a novel matrix for PDGF detection. *Sens. Actuators B Chem.* **2017**, *250*, 8–16. [CrossRef]

60. Yang, C.; Wang, Q.; Xiang, Y.; Yuan, R.; Chai, Y. Target-induced strand release and thionine-decorated gold nanoparticle amplification labels for sensitive electrochemical aptamer-based sensing of small molecules. *Sens. Actuators B Chem.* **2014**, *197*, 149–154. [CrossRef]

61. Wu, D.; Wang, Y.; Zhang, Y.; Ma, H.; Pang, X.; Hu, L.; Du, B.; Wei, Q. Facile fabrication of an electrochemical aptasensor based on magnetic electrode by using streptavidin modified magnetic beads for sensitive and specific detection of Hg^{2+}. *Biosens. Bioelectron.* **2016**, *82*, 9–13. [CrossRef] [PubMed]

62. Miao, P.; Yang, D.; Chen, X.; Guo, Z.; Tang, Y. Voltammetric determination of tumor necrosis factor-α based on the use of an aptamer and magnetic nanoparticles loaded with gold nanoparticles. *Microchim. Acta* **2017**, *184*, 3901–3907. [CrossRef]

63. Hao, N.; Jiang, L.; Qian, J.; Wang, K. Ultrasensitive electrochemical Ochratoxin A aptasensor based on CdTe quantum dots functionalized graphene/Au nanocomposites and magnetic separation. *J. Electroanal. Chem.* **2016**, *781*, 332–338. [CrossRef]

64. Zelada-Guillén, G.A.; Tweed-Kent, A.; Niemann, M.; Göringer, H.U.; Riu, J.; Rius, F.X. Ultrasensitive and real-time detection of proteins in blood using a potentiometric carbon-nanotube aptasensor. *Biosens. Bioelectron.* **2013**, *41*, 366–371. [CrossRef] [PubMed]

65. Ding, J.; Lei, J.; Ma, X.; Gong, J.; Qin, W. Potentiometric aptasensing of Listeria monocytogenes using protamine as an indicator. *Anal. Chem.* **2014**, *86*, 9412–9416. [CrossRef] [PubMed]

66. Zhao, G.; Ding, J.; Yu, H.; Yin, T.; Qin, W. Potentiometric aptasensing of *Vibrio alginolyticus* Based on DNA nanostructure—Modified magnetic beads. *Sensors* **2016**, *16*, 2052. [CrossRef] [PubMed]

67. Jia, F.; Duan, N.; Wu, S.; Dai, R.; Wang, Z.; Li, X. Impedimetric salmonella aptasensor using a glassy carbon electrode modified with an electrodeposited composite consisting of reduced graphene oxide and carbon nanotubes. *Microchim. Acta* **2016**, *183*, 337–344. [CrossRef]

68. Jin, S.; Ye, Z.; Wang, Y.; Ying, Y. A novel impedimetric microfluidic analysis system for transgenic protein Cry1Ab detection. *Sci. Rep.* **2017**, *7*, 43175. [CrossRef] [PubMed]

69. Wang, Y.; Ye, Z.; Ping, J.; Jing, S.; Ying, Y. Development of an aptamer-based impedimetric bioassay using microfluidic system and magnetic separation for protein detection. *Biosens. Bioelectron.* **2014**, *59*, 106–111. [CrossRef] [PubMed]

70. Ke, H.; Sha, H.; Wang, Y.; Guo, W.; Zhang, X.; Wang, Z.; Huang, C.; Jia, N. Electrochemiluminescence resonance energy transfer system between GNRs and Ru(bpy)$_3^{2+}$: Application in magnetic aptasensor for β-amyloid. *Biosens. Bioelectron.* **2018**, *100*, 266–273. [CrossRef] [PubMed]

71. Wang, Y.-Z.; Hao, N.; Feng, Q.-M.; Shi, H.-W.; Xu, J.-J.; Chen, H.-Y. A ratiometric electrochemiluminescence detection for cancer cells using g-C$_3$N$_4$ nanosheets and Ag-PAMAM-luminol nanocomposites. *Biosens. Bioelectron.* **2016**, *77*, 76–82. [CrossRef] [PubMed]

72. Zhang, J.; Zhang, X.; Yang, G.; Chen, J.; Wang, S. A signal-on fluorescent aptasensor based on Tb^{3+} and structure-switching aptamer for label-free detection of ochratoxin A in wheat. *Biosens. Bioelectron.* **2013**, *41*, 704–709. [CrossRef] [PubMed]

73. Fang, S.; Wang, C.; Xiang, J.; Cheng, L.; Song, X.; Xu, L.; Peng, R.; Liu, Z. Aptamer-conjugated upconversion nanoprobes assisted by magnetic separation for effective isolation and sensitive detection of circulating tumor cells. *Nano Res.* **2014**, *7*, 1327–1336. [CrossRef]

74. Lu, C.; Tang, Z.; Liu, C.; Kang, L.; Sun, F. Magnetic-nanobead-based competitive enzyme-linked aptamer assay for the analysis of oxytetracycline in food. *Anal. Bioanal. Chem.* **2015**, *407*, 4155–4163. [CrossRef]

75. Wang, L.; Liu, F.; Sui, N.; Liu, M.; William, W.Y. A colorimetric assay for Hg(II) based on the use of a magnetic aptamer and a hybridization chain reaction. *Microchim. Acta* **2016**, *183*, 2855–2860. [CrossRef]

76. Xi, Z.; Huang, R.; Li, Z.; He, N.; Wang, T.; Su, E.; Deng, Y. Selection of HBsAg-specific DNA aptamers based on carboxylated magnetic nanoparticles and their application in the rapid and simple detection of hepatitis B virus infection. *ACS Appl. Mater. Interfaces* **2015**, *7*, 11215–11223. [CrossRef] [PubMed]

77. Chang, K.-W.; Li, J.; Yang, C.-H.; Shiesh, S.-C.; Lee, G.-B. An integrated microfluidic system for measurement of glycated hemoglobin Levels by using an aptamer-antibody assay on magnetic beads. *Biosens. Bioelectron.* **2015**, *68*, 397–403. [CrossRef] [PubMed]

78. Chen, Q.; Yang, M.; Yang, X.; Li, H.; Guo, Z.; Rahma, M. A large Raman scattering cross-section molecular embedded SERS aptasensor for ultrasensitive Aflatoxin B1 detection using CS-Fe$_3$O$_4$ for signal enrichment. *Spectrochim. Acta Part A Mol. Biomol. Spectrosc.* **2018**, *189*, 147–153. [CrossRef] [PubMed]

79. Yang, M.; Liu, G.; Mehedi, H.M.; Ouyang, Q.; Chen, Q. A universal sers aptasensor based on DTNB labeled GNTs/Ag core-shell nanotriangle and CS-Fe$_3$O$_4$ magnetic-bead trace detection of Aflatoxin B1. *Anal. Chim. Acta* **2017**, *986*, 122–130. [CrossRef] [PubMed]

80. He, J.; Li, G.; Hu, Y. Aptamer recognition induced target-bridged strategy for proteins detection based on magnetic chitosan and silver/chitosan nanoparticles using surface-enhanced Raman spectroscopy. *Anal. Chem.* **2015**, *87*, 11039–11047. [CrossRef] [PubMed]

81. Ozalp, V.C.; Bayramoglu, G.; Erdem, Z.; Arica, M.Y. Pathogen detection in complex samples by quartz crystal microbalance sensor coupled to aptamer functionalized core—Shell type magnetic separation. *Anal. Chim. Acta* **2015**, *853*, 533–540. [CrossRef] [PubMed]

82. Pan, Y.; Guo, M.; Nie, Z.; Huang, Y.; Pan, C.; Zeng, K.; Zhang, Y.; Yao, S. Selective collection and detection of leukemia cells on a magnet-quartz crystal microbalance system using aptamer-conjugated magnetic beads. *Biosens. Bioelectron.* **2010**, *25*, 1609–1614. [CrossRef] [PubMed]

83. Ma, C.; Wang, W.; Yang, Q.; Shi, C.; Cao, L. Cocaine detection via rolling circle amplification of short DNA strand separated by magnetic beads. *Biosens. Bioelectron.* **2011**, *26*, 3309–3312. [CrossRef] [PubMed]

84. Csordas, A.; Gerdon, A.E.; Adams, J.D.; Qian, J.; Oh, S.S.; Xiao, Y.; Soh, H.T. Detection of proteins in serum by micromagnetic aptamer PCR (MAP) technology. *Angew. Chem. Int. Ed.* **2010**, *49*, 355–358. [CrossRef] [PubMed]

85. Ozalp, V.C.; Bayramoglu, G.; Kavruk, M.; Keskin, B.B.; Oktem, H.A.; Arica, M.Y. Pathogen detection by core—Shell type aptamer-magnetic preconcentration coupled to real-time PCR. *Anal. Biochem.* **2014**, *447*, 119–125. [CrossRef] [PubMed]

86. Wang, J. Electrochemical glucose biosensors. *Chem. Rev.* **2008**, *108*, 814–825. [CrossRef] [PubMed]

87. Han, K.; Liu, T.; Wang, Y.; Miao, P. Electrochemical aptasensors for detection of small molecules, macromolecules, and cells. *Rev. Anal. Chem.* **2016**, *35*, 201–211. [CrossRef]

88. Meirinho, S.G.; Dias, L.G.; Peres, A.M.; Rodrigues, L.R. Voltammetric aptasensors for protein disease biomarkers detection: A review. *Biotechnol. Adv.* **2016**, *34*, 941–953. [CrossRef] [PubMed]

89. Feng, L.; Zhang, Z.; Ren, J.; Qu, X. Functionalized graphene as sensitive electrochemical label in target-dependent linkage of split aptasensor for dual detection. *Biosens. Bioelectron.* **2014**, *62*, 52–58. [CrossRef] [PubMed]

90. Yan, Z.; Gan, N.; Wang, D.; Cao, Y.; Chen, M.; Li, T.; Chen, Y. A "signal-on"aptasensor for simultaneous detection of chloramphenicol and polychlorinated biphenyls using multi-metal ions encoded nanospherical brushes as tracers. *Biosens. Bioelectron.* **2015**, *74*, 718–724. [CrossRef] [PubMed]

91. Bakker, E.; Pretsch, E. Nanoscale potentiometry. *TRAC Trends Anal. Chem.* **2008**, *27*, 612–618. [CrossRef] [PubMed]

92. Ding, J.; Gu, Y.; Li, F.; Zhang, H.; Qin, W. DNA nanostructure-based magnetic beads for potentiometric aptasensing. *Anal. Chem.* **2015**, *87*, 6465–6469. [CrossRef] [PubMed]

93. Lee, C.-Y.; Fan, H.-T.; Hsieh, Y.-Z. Disposable aptasensor combining functional magnetic nanoparticles with rolling circle amplification for the detection of prostate-specific antigen. *Sens. Actuators B Chem.* **2018**, *255*, 341–347. [CrossRef]

94. Palchetti, I.; Mascini, M. Electrochemical nanomaterial-based nucleic acid aptasensors. *Anal. Bioanal. Chem.* **2012**, *402*, 3103–3114. [CrossRef] [PubMed]

95. Zhou, Y.; Yan, D.; Wei, M. A 2D quantum dot-based electrochemiluminescence film sensor towards reversible temperature-sensitive response and nitrite detection. *J. Mater. Chem. C* **2015**, *3*, 10099–10106. [CrossRef]

96. Liu, Z.; Qi, W.; Xu, G. Recent advances in electrochemiluminescence. *Chem. Soc. Rev.* **2015**, *44*, 3117–3142. [CrossRef] [PubMed]

97. Luo, Z.; Wang, Y.; Lu, X.; Chen, J.; Wei, F.; Huang, Z.; Zhou, C.; Duan, Y. Fluorescent aptasensor for antibiotic detection using magnetic bead composites coated with gold nanoparticles and a nicking enzyme. *Anal. Chim. Acta* **2017**, *984*, 177–184. [CrossRef] [PubMed]

98. Haase, M.; Schäfer, H. Upconverting nanoparticles. *Angew. Chem. Int. Ed.* **2011**, *50*, 5808–5829. [CrossRef] [PubMed]

99. Liu, Y.; Tu, D.; Zhu, H.; Chen, X. Lanthanide-doped luminescent nanoprobes: Controlled synthesis, optical spectroscopy, and bioapplications. *Chem. Soc. Rev.* **2013**, *42*, 6924–6958. [CrossRef] [PubMed]

100. Wang, W.; Xu, D.-D.; Pang, D.-W.; Tang, H.-W. Fluorescent sensing of thrombin using a magnetic nano-platform with aptamer-target-aptamer sandwich and fluorescent silica nanoprobe. *J. Lumin.* **2017**, *187*, 9–13. [CrossRef]

101. Bruno, J.G.; Phillips, T.; Carrillo, M.P.; Crowell, R. Plastic-adherent DNA aptamer-magnetic bead and quantum dot sandwich assay for Campylobacter detection. *J. Fluoresc.* **2009**, *19*, 427. [CrossRef] [PubMed]

102. Hao, L.; Zhao, Q. Using fluoro modified RNA aptamers as affinity ligands on magnetic beads for sensitive thrombin detection through affinity capture and thrombin catalysis. *Anal. Methods* **2016**, *8*, 510–516. [CrossRef]

103. Jin, Y.; Bai, J.; Li, H. Label-free protein recognition using aptamer-based fluorescence assay. *Analyst* **2010**, *135*, 1731–1735. [CrossRef] [PubMed]

104. Tan, Y.; Zhang, X.; Xie, Y.; Zhao, R.; Tan, C.; Jiang, Y. Label-free fluorescent assays based on aptamer—Target recognition. *Analyst* **2012**, *137*, 2309–2312. [CrossRef] [PubMed]

105. McKeague, M.; Velu, R.; Hill, K.; Bardóczy, V.; Mészáros, T.; DeRosa, M.C. Selection and characterization of a novel DNA aptamer for label-free fluorescence biosensing of ochratoxin A. *Toxins* **2014**, *6*, 2435–2452. [CrossRef] [PubMed]

106. Zhu, Z.; Yang, C.; Zhou, X.; Qin, J. Label-free aptamer-based sensors for L-argininamide by using nucleic acid minor groove binding dyes. *Chem. Commun.* **2011**, *47*, 3192–3194. [CrossRef] [PubMed]

107. Babendure, J.R.; Adams, S.R.; Tsien, R.Y. Aptamers switch on fluorescence of triphenylmethane dyes. *J. Am. Chem. Soc.* **2003**, *125*, 14716–14717. [CrossRef] [PubMed]

108. Huang, C.-C.; Chang, H.-T. Aptamer-based fluorescence sensor for rapid detection of potassium ions in urine. *Chem. Commun.* **2008**, *12*, 1461–1463. [CrossRef] [PubMed]

109. Liang, G.; Cai, S.; Zhang, P.; Peng, Y.; Chen, H.; Zhang, S.; Kong, J. Magnetic relaxation switch and colorimetric detection of thrombin using aptamer-functionalized gold-coated iron oxide nanoparticles. *Anal. Chim. Acta* **2011**, *689*, 243–249. [CrossRef] [PubMed]

110. Kim, Y.S.; Jurng, J. A simple colorimetric assay for the detection of metal ions based on the peroxidase-like activity of magnetic nanoparticles. *Sens. Actuators B Chem.* **2013**, *176*, 253–257. [CrossRef]

111. Wang, C.; Qian, J.; Wang, K.; Yang, X.; Liu, Q.; Hao, N.; Wang, C.; Dong, X.; Huang, X. Colorimetric aptasensing of ochratoxin A using Au@Fe$_3$O$_4$ nanoparticles as signal indicator and magnetic separator. *Biosens. Bioelectron.* **2016**, *77*, 1183–1191. [CrossRef] [PubMed]

112. Zhang, Z.; Wang, Z.; Wang, X.; Yang, X. Magnetic nanoparticle-linked colorimetric aptasensor for the detection of thrombin. *Sens. Actuators B Chem.* **2010**, *147*, 428–433. [CrossRef]

113. Yan, X.; Cao, Z.; Kai, M.; Lu, J. Label-free aptamer-based chemiluminescence detection of adenosine. *Talanta* **2009**, *79*, 383–387. [CrossRef] [PubMed]

114. Li, Y.; Ji, X.; Liu, B. Chemiluminescence aptasensor for cocaine based on double-functionalized gold nanoprobes and functionalized magnetic microbeads. *Anal. Bioanal. Chem.* **2011**, *401*, 213–219. [CrossRef]

115. Sun, C.; Zhang, R.; Gao, M.; Zhang, X. A rapid and simple method for efficient capture and accurate discrimination of circulating tumor cells using aptamer conjugated magnetic beads and surface-enhanced Raman scattering imaging. *Anal. Bioanal. Chem.* **2015**, *407*, 8883–8892. [CrossRef] [PubMed]

116. Yoon, J.; Choi, N.; Ko, J.; Kim, K.; Lee, S.; Choo, J. Highly sensitive detection of thrombin using SERS-based magnetic aptasensors. *Biosens. Bioelectron.* **2013**, *47*, 62–67. [CrossRef] [PubMed]

117. Teller, C.; Halámek, J.; Makower, A.; Scheller, F.W. A set of piezoelectric biosensors using cholinesterases. In *Biosensors and Biodetection*; Springer: New York, NY, USA, 2009; pp. 3–22.

118. Skládal, P. Piezoelectric biosensors. *TRAC Trends Anal. Chem.* **2016**, *79*, 127–133. [CrossRef]

119. Song, W.; Zhu, Z.; Mao, Y.; Zhang, S. A sensitive quartz crystal microbalance assay of adenosine triphosphate via DNAzyme-activated and aptamer-based target-triggering circular amplification. *Biosens. Bioelectron.* **2014**, *53*, 288–294. [CrossRef] [PubMed]

120. Feng, J.; Dai, Z.; Tian, X.; Jiang, X. Detection of Listeria monocytogenes based on combined aptamers magnetic capture and loop-mediated isothermal amplification. *Food Control* **2018**, *85*, 443–452. [CrossRef]

Article

Characterizations of Anti-Alpha-Fetoprotein-Conjugated Magnetic Nanoparticles Associated with Alpha-Fetoprotein for Biomedical Applications

Shu-Hsien Liao [1],*, Han-Sheng Huang [1], Jen-Jie Chieh [1], Yu-Kai Su [1], Yuan-Fu Tong [1] and Kai-Wen Huang [2,3],*

[1] Institute of Electro-Optical Science and Technology, National Taiwan Normal University, Taipei 116, Taiwan; hansheng9527@gmail.com (H.-S.H); jjchieh@ntnu.edu.tw (J.-J.C.); a20296111@gmail.com (Y.-K.S.); 60548011s@ntnu.edu.tw (Y.-F.T.)

[2] Department of Surgery and Hepatitis Research Center, National Taiwan University Hospital, Taipei 100, Taiwan

[3] Graduate Institute of Clinical Medicine, National Taiwan University, Taipei 100, Taiwan

* Correspondence: shliao@ntnu.edu.tw (S.-H.L.); skywing@ntuh.gov.tw (K.-W.H.);
Tel.: +886-2-7734-6743 (S.-H.L.); +886-2-2312-3456 (ext. 66144) (K.-W.H.)

Received: 7 August 2017; Accepted: 1 September 2017; Published: 3 September 2017

Abstract: In this work, we report characterizations of biofunctionalized magnetic nanoparticles (BMNPs) associated with alpha-fetoprotein (AFP) for biomedical applications. The example BMNP in this study is anti-alpha-fetoprotein (anti-AFP) conjugated onto dextran-coated Fe_3O_4 labeled as Fe_3O_4-anti-AFP, and the target is AFP. We characterize magnetic properties, such as increments of magnetization ΔM_H and effective relaxation time $\Delta\tau_{eff}$ in the reaction process. It is found that both ΔM_H and $\Delta\tau_{eff}$ are enhanced when the concentration of AFP, Φ_{AFP}, increases. The enhancements are due to magnetic interactions among BMNPs in magnetic clusters, which contribute extra M_H after the association with M_H and in turn enhance τ_{eff}. The screening of patients carrying hepatocellular carcinoma (HCC) is verified via $\Delta M_H/M_H$. The proposed method can be applied to detect a wide variety of analytes. The scaling characteristics of $\Delta M_H/M_H$ show the potential to develop a vibrating sample magnetometer system with low field strength for clinic applications.

Keywords: magnetic immunoassay; biofunctionalized magnetic nanoparticles; biomarker; alpha-fetoprotein; hepatocellular carcinoma; magnetization enhancement

1. Introduction

Immunoassays are biochemical tests used to detect or quantify a specific substance, such as analytes in samples of blood or bodily fluid, using immunological reactions. Immunoassay methods include the enzyme-linked immunosorbent assay (ELISA) [1], radioimmunoassay (RIA) [2], real-time polymerase chain reaction (real-time PCR) [3], immunonephelometry [4], etc. Some immunoassays, such as ELISA, require two antigens and separation of the unbound antigens, which can be tedious and time-consuming. On the other hand, magnetic immunoassay (MIA) is a novel type of diagnostic technology using magnetic nanoparticles (MNPs) as labels to replace conventional ELISA, RIA, real-time PCR, etc. MNPs are coated with dextran so that they are encapsulated or glued together with polymers in sizes of nanometers or even micrometers. In immunomagnetic tests, MNPs are first biofunctionalized against antibodies to target antigens. Reagents consisting of biofunctionalized magnetic nanoparticles (BMNPs) are then mixed with samples. Due to the molecular interactions among BMNPs and biomarkers, magnetic clusters are conjugated in the reaction process and their magnetic properties change after the association. The magnetic signal due to the changes of magnetic properties is analyzed in order to determine the unknown amount of biomarkers. Magnetic properties (magnetic relaxation [5,6], remanent magnetization [7], Brownian relaxation [8], saturation magnetization [9], spin-spin relaxation of NMR [10], and alternative-current (AC) susceptibility reduction [11–15], etc.) have been developed recently. Magnetic immunoassays can be carried out simply by mixing reagents and tested samples together and taking physical measurements. Additionally, the background noise of magnetic detection is negligible; hence, high detection sensitivity can be achieved.

Based on the increment of saturation magnetization, ΔM_S, Chieh et al. [16] recently reported another assay method that used a vibrating sample magnetometer (VSM) to label tumor biomarkers of alpha-fetoprotein (AFP) in clinical studies via the $\Delta M_S/M_S$-versus-Φ_{AFP} curve at the saturation field H_S, where Φ_{AFP} was the concentration of AFP. The authors demonstrated that VSM can be used to screen patients carrying hepatocellular carcinoma (HCC) with sensitivity better than the criterion set in clinics (0.02 μg/mL). It would be interesting to see whether we can screen HCC patients with high detection sensitivity at low magnetic fields (H). Therefore, in this work, we propose a detection method based on the scaling characteristic of the normalized increment of magnetization at low magnetic fields. It is found that M_{AFP} and τ_{eff} are enhanced when Φ_{AFP} increases, where M_{AFP} is the magnetization of the reagent and τ_{eff} is the effective relaxation time. We attribute those results to the molecular interactions among BMNPs in the associated magnetic clusters, which contribute extra magnetization and in turn enhance τ_{eff}. The scaling characteristic of $(\Delta M_{AFP}/M_{AFP,0})$-versus-$\Phi_{AFP}$ curves at low magnetic fields is demonstrated, and the screening of HCC patients via the scaling characteristic is verified in clinical studies.

2. Experiments

The MNPs in this study were dextran-coated Fe_3O_4 (MF-DEX-0060, MagQu Co., Ltd., New Taipei City, Taiwan) with a mean core diameter of ~35 nm, as detected by x-ray diffraction (D-500, Siemens). The BMNPs were Fe_3O_4-anti-AFP (MF-AFP-0060, MagQu Co. Ltd., New Taipei City, Taiwan), and the biotarget was AFP, which is a biomarker for diagnosing HCC. When the AFP level is abnormally high before surgery or other therapy, it is expected to fall to normal levels following the successful removal of all cancer cells.

In performing the AFP tests, the BMNPs consisting of Fe_3O_4-anti-AFP were first mixed with AFP. The changes of magnetic properties after the reaction process were then characterized using a VSM (Model Hystermag, MagQu Co., Taiwan) and AC susceptometer. The data of the normalized increments of magnetization $\Delta M/M$ were analyzed for a magnetic immunoassay. The AC susceptibility was measured by a highly balanced AC susceptometer in order to monitor the real-time reaction process. The AC susceptibility $\chi_{ac}(\omega)$ can be expressed as follows:

$$\chi_{ac} = \chi' + i\chi'' \tag{1}$$

where $i = (-1)^{1/2}$, $\chi''/\chi' = \tan\theta = \omega\tau_{eff}(t)$, and θ is the phase lag of the time-varying magnetization $M(t)$ with respect to the applied AC magnetic field $H(t)$.

Figure 1a shows the detection schematic of the VSM used for characterizing M after the BMNPs had conjugated with AFP. In the measurement of M, the sample vibrated with a frequency of ~30 Hz. The magnetic signal was detected with a second-order gradient coil. An electromagnet provided a magnetic field of up to 1.0 Tesla, so that the M–H curves of reagents were characterized. In assaying AFP, a reagent composed of 40 μL Fe_3O_4-anti-AFP was mixed with 60 μL AFP. We measured the M–H curves and analyzed the magnetization enhancement (ΔM) at low external fields (H) to establish the relationship between $\Delta M/M$ and the concentrations of AFP (Φ_{AFP}). Figure 1b shows the high-T_C SQUID-based AC susceptometer for characterizing the AC magnetic susceptibility. The excitation frequency is ~16 kHz. The magnetic signal of BMNPs is picked up by a gradient coil that is coupled to a high-T_C SQUID via a flux transformer. The detailed design of the pickup coil, gradient coil, and compensation coil in a homemade AC susceptometer that did not use a high-T_c SQUID was reported in [17,18].

(a)

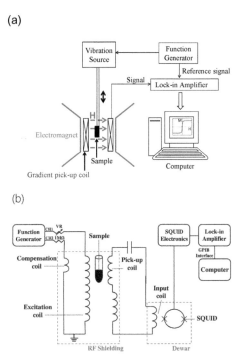

(b)

Figure 1. Detection scheme of (**a**) vibrating sample magnetometer; (**b**) high-Tc SQUID-based AC susceptometer.

The reagent was composed of anti-AFP-conjugated Fe_3O_4 labeled as Fe_3O_4-anti-AFP. The bio-target was AFP. Figure 2 depicts Fe_3O_4-anti-AFP, AFP, and a magnetic cluster composed of Fe_3O_4-anti-AFP-AFP.

(a) Bio-functionalized Fe_3O_4-antiAFP

(b) AFP

(a) + (b) → (c)

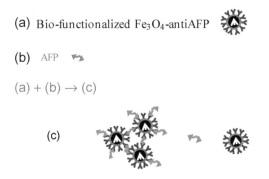

(c)

A magnetic cluster composed of Fe_3O_4-antiAFP-AFP along with biomarker and functionalized Fe_3O_4-antiAFP

Figure 2. Pictures showing (**a**) biofunctionalized Fe_3O_4-anti-AFP; (**b**) AFPs; (**c**) magnetic cluster composed of Fe_3O_4-anti-AFP-AFP.

3. Results and Discussion

This section addresses and discusses the results from the characterization of magnetic properties when biofunctionalized Fe_3O_4-anti-AFPs are associated with AFP. Additionally, we present the results from the real-time association of Fe_3O_4-anti-AFP with AFP via the time-dependency studies of $\tau_{eff}(t)$ in the reaction process using the technique of AC susceptibility. We also briefly summarize the findings. Finally, we present the clinical research on screening HCC patients via normalized increments of magnetization and address and discuss advances in sensitive bio-sensing.

Figure 3 shows ΔM_H as a function of Φ_{AFP} at μ_0H = 0.02 T, 0.06 T, and 0.16 T and ΔM_H = $M_H(\Phi_{AFP}) - M_H(\Phi_{AFP} = 0)$. For a fixed magnetic field at μ_0H = 0.02 T, ΔM_H = 0.015 emu/g when Φ_{AFP} = 0.01 μg/mL, and ΔM_H increases to $\Delta M_{\mu0H = 0.02\,T}$ = 0.13 emu/g when Φ_{AFP} = 10 μg/mL. For μ_0H = 0.16 T, $\Delta M_{\mu0H = 0.16\,T}$ = 0.03 emu/g when Φ_{AFP} = 0.01 μg/mL, and ΔM_H increases to $\Delta M_{\mu0H = 0.16\,T}$ = 0.23 emu/g when Φ_{AFP} = 10 μg/mL. Hence, we have demonstrated an enhancement of ΔM_H when Φ_{AFP} increases at a fixed magnetic field. We attribute those enhancements to the fact that more magnetic clusters are associated and stronger magnetic interactions among BMNPs are present.

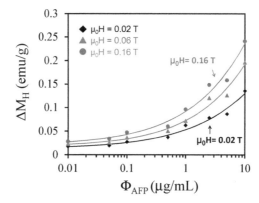

Figure 3. The increments of magnetization ΔM_H as a function of Φ_{AFP} at low magnetic fields at μ_0H = 0.02 T, 0.06 T, 0.16 T.

Figure 4 shows the normalized increment of magnetization, $\Delta M_{AFP}/M_{AFP,0}$, as a function of Φ_{AFP} at μ_0H = 0.02 T, 0.06 T, and 0.16 T, where ΔM_{AFP} = $M(\Phi_{AFP}) - M(\Phi_{AFP} = 0)$, $M_{AFP,0}$ = $M_H(\Phi_{AFP} = 0)$. It is found that $\Delta M_{AFP}/M_{AFP,0}$ as a function of Φ_{AFP} in external magnetic fields can be scaled to a universal logistic function described by the following formula [15]:

$$\Delta M_{AFP}/M_{AFP,0} = (A - B)/\{1 + [(\Phi_{AFP})/(\Phi_0)]^\gamma\} + B \qquad (2)$$

where A and B are dimensionless quantities and Φ_0 is dimensionless. The fitting parameters are as follows: A = 0.173, B = 34.2, Φ_0 = 3410 μg/mL, and γ = 0.5. We have established a relationship between $\Delta M_{AFP}/M_{AFP,0}$ and Φ_{AFP} with Φ_{AFP} varied from 0.01 μg/mL to 10 μg/mL. Therefore, the unknown amounts of AFP can be determined via a scaling characteristic of the $(\Delta M_{AFP}/M_{AFP,0})$-versus-$\Phi_{AFP}$ curve, which is versatile and can be applied to assay other biomarkers. In assaying other biomarkers, the relationship between $\Delta M_{biomarker}/M_{biomarker,0}$ and $\Phi_{biomarker}$ is first established and then $\Delta M_{biomarker}/M_{biomarker,0}$ and the $\Phi_{biomarker}$ curve are applied to determine the unknown amount of biomarkers quantitatively.

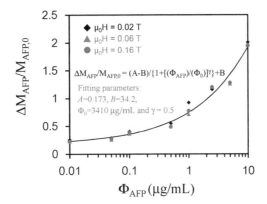

Figure 4. The normalized increment of magnetization $\Delta M_{AFP}/M_{AFP,0}$ as a function of Φ_{AFP} with data analyzed at $\mu_0 H$ = 0.02 T, 0.06 T and 0.16 T.

To observe the real-time association of τ_{eff} when Fe_3O_4-anti-AFPs are associated with AFP directly, we characterize the time-dependent τ_{eff} via the following formula: $\tan\theta = \omega\tau_{eff}$, where $\chi''/\chi' = \tan\theta$ and χ' and χ'' are the real and imaginary parts of AC susceptibility in Equation (1). Figure 5a shows $\tau_{eff}(t)$ as a function of time in the reaction process. The reagent shows τ_{eff} = ~1.3 µs, and τ_{eff} is stable to τ_{eff} = 1.3 µs at t = 7200 s. It takes approximately 6000 s for the reagent to complete the association and τ_{eff} is increased to τ_{eff} = ~1.75 µs with Φ_{AFP} = 1 µg/mL. Therefore, a detection time of 7200 s is suggested. The real-time association of Fe_3O_4-anti-AFP with AFP is verified.

The Brownian relaxation time, τ_B, is a function of the hydrodynamic volume of a magnetic particle, V_H, the viscosity of the medium, η, the Boltzmann's constant, k, and the absolute temperature, T, which is expressed as follows [19]:

$$\tau_B = 3\,V_H\eta/kT \tag{3}$$

In the reaction process, we assume that the viscosity and temperature are constant. The Brownian relaxation time is proportional to the hydrodynamic volume of the magnetic particle. The ratio of the increase in τ_{eff} after the reaction process is 1.35 with an Φ_{AFP} value of 1 µg/mL. The effective diameter of the magnetic cluster is 2.4 times larger than a single magnetic particle when Φ_{AFP} is 1 µg/mL. It presents the formation of magnetic clusters during the reaction process.

Figure 5b shows $\Delta\tau_{eff}/\tau_{eff,0}$ as a function of Φ_{AFP} with Φ_{AFP} ranging from Φ_{AFP} = 0.001 µg/mL to Φ_{AFP} = 1 µg/mL. The reagent shows τ_{eff} = 1.3 µs, and τ_{eff} is enhanced to τ_{eff} = ~1.75 µs when Φ_{AFP} = 1 µg/mL. The enhancement of τ_{eff} is due to the presence of magnetic clusters in the reaction process. The magnetic interaction among BMNPs enhances M, which in turn increases τ_{eff}. The $(\Delta\tau_{eff}/\tau_{eff,0})$-versus-$\Phi_{AFP}$ curve follows the characteristic curve [15]:

$$\Delta\tau_{eff}/\tau_{eff,0} = (A_1 - B_1)/\{1+[(\Phi_{AFP})/(\Phi_0)]^\gamma\} + B_1, \tag{4}$$

where $\Delta\tau_{eff} = \tau_{eff}(7200\ s) - \tau_{eff}(t = 0)$ and $\tau_{eff,0} = \tau_{eff}(t = 0)$. The curve is fitted to the following parameters: A_1 = −0.013 µs, B_1 = 0.56 µs, Φ_0 = 0.15 µg/mL, and γ = 0.52. Equation (4) reveals the concentration dependency of the characteristic of $\Delta\tau_{eff}/\tau_{eff,0}$ after the BMNPs have completed the association with AFP. The $(\Delta\tau_{eff}/\tau_{eff,0})$-versus-$\Phi_{AFP}$ curve shown in Figure 5b can be applied to screening patients carrying HCC. Normalized $\Delta\tau_{eff}/\tau_{eff,0}$ is analyzed instead of $\Delta\tau_{eff}$ for a magnetic immunoassay, because this enables us to eliminate minor differences in magnetic signals due to minor differences in sample amounts used from run to run, which will enhance the detection sensitivity.

Detection sensitivity can be defined by the noise level with standard deviations for the detected signal at low concentrations [20]. In this study, the detection sensitivity levels are 0.0024 µg/mL and

0.0177 µg/mL, as determined by measuring $\Delta\tau_{eff}/\tau_{eff,0}$ and $\Delta M_{AFP}/M_{AFP,0}$ respectively. The reference criterion of the AFP serum level for HCC is 0.02 µg/mL. The sensitivity of both methods reaches the criteria for a clinical AFP assay. The feasibility of AFP is demonstrated by measuring $\Delta\tau_{eff}/\tau_{eff,0}$ and $\Delta M_{AFP}/M_{AFP,0}$.

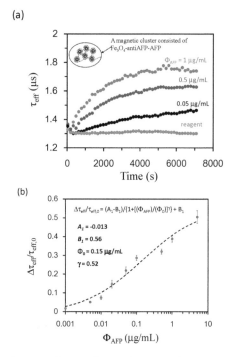

Figure 5. (a) τ_{eff} as a function of time, (b) $\Delta\tau_{eff}/\tau_{eff,0}$ as a function of Φ_{AFP} with Φ_{AFP} from $\Phi_{AFP} = 0.001$ µg/mL to $\Phi_{AFP} = 1$ µg/mL.

In this study, we characterized magnetic properties when BMNPs are associated with AFPs for biomedical applications. The findings in the characterization of magnetic properties are briefly summarized as follows. First, M and τ_{eff} are enhanced when reagents composed of BMNPs are conjugated with AFP in the reaction process. The magnetic interactions among BMNPs in magnetic clusters enhance M, which in turn increases τ_{eff}. Second, the real-time association of BMNPs with AFP was demonstrated in the time-dependent τ_{eff}. Third, bio-detection based on the $(\Delta\tau_{eff}/\tau_{eff,0})$-versus-$\Phi_{biomarkers}$ curve provided a sensitive methodology for assaying unknown amounts of AFP, and BMNPs could be applied to assay large molecules such as AFP as well as small molecules such as C-reactive protein(CRP) [21]. Finally, the proposed detection methodology based on the $(\Delta\tau_{eff}/\tau_{eff,0})$-versus-$\Phi_{biomarkers}$ curve was versatile, and the $(\Delta M_{AFP}/M_{AFP,0})$-versus-$\Phi_{AFP}$ curves shown in Figure 4 were scaled to a characteristic function described by Equation (2). The results confirm that both changes in $\Delta M_{AFP}/M_{AFP,0}$ and $\Delta\tau_{eff}/\tau_{eff,0}$ are caused by the formation of magnetic clusters and can be applied to sense a wide variety of biomarkers.

The sensitivity levels of $\Delta\tau_{eff}/\tau_{eff,0}$ and $\Delta M_{AFP}/M_{AFP,0}$ reach the criteria for a clinical AFP assay. The cost of a high-T_C SQUID-based AC susceptometer is much higher than that of a VSM with a low-strength magnet. The low-strength VSM has high potential for commercial and clinical applications. Therefore, the screening of HCC patients can be addressed by measuring $\Delta M_H/M_{H,0}$. Since the data shown in Figure 4 are scaled to a characteristic function described by Equation (2), it would be interesting to verify whether we can also obtain high detection sensitivity at low magnetic fields via Equation (2). Hence, we can apply Equation (2) at a low magnetic field, say $\mu_0H = 0.065$ T, to analyze AFP levels in clinical studies. To verify this, we show in Figure 6a ($\Delta M_{AFP}/M_{AFP,0}$)-versus-Φ_{AFP} with data analyzed at $\mu_0H = 0.065$ T, where $\Delta M_{AFP} = M_H(\Phi_{AFP}) - M_H(\Phi_{AFP} = 0)$ and $M_{AFP,0} = M_H(\Phi_{AFP} = 0)$. The background magnetic signal of serum from healthy persons in $\Delta M_{AFP}/M_{AFP,0}$ is deducted in the data analysis. To screen patients carrying HCC and healthy persons, we mixed 40 μL 0.1 emu/g of reagent with 60 μL of serum. The data for establishing the standard curve are marked with a solid dot (•). AFP levels in serum for HCC patients are marked with an open triangle (△), while AFP levels for healthy persons are marked with an open square (□). The reference criterion of the AFP serum level for HCC is 0.02 μg/mL. We found that the average AFP levels for patients carrying HCC were higher than ~0.2 μg/mL, which is significantly higher than the criterion set in clinics (0.02 μg/mL). The average AFP levels for healthy persons were below ~0.02 μg/mL, except for one healthy person who showed a false positive (AFP level = ~0.03 μg/mL). Figure 6b shows $\Delta M_{AFP}/M_{AFP,0}$ as a function of Φ_{AFP} with data analyzed at 0.16 T. HCC patients showed AFP levels higher than the clinical criterion. Healthy persons showed AFP levels of 0.001 μg/mL, except for one healthy person with a higher AFP level of ~0.4 μg/mL. The estimated values of Φ_{AFP} were different between $\mu_0H = 0.065$ T and 0.16 T. It was probably due to the magnetic clustering effect that induces background magnetic noises. Besides, the $\Delta M_{AFP}/\Delta M_{AFP,0}$ of serum tested at 0.16 T is higher than that at 0.065 T. It leads that the estimated AFP concentration at 0.16 T is higher than that at 0.065 T. The reason may be due to the larger background magnetization of serum than that of the AFP solution. The reference magnetization, $M(\Phi_{AFP} = 0)$, in the clinical test may be considered by using the averaging magnetization of healthy persons to reduce the effect in the clinical test. Thus, the feasibility of screening HCC patients by assaying AFP levels in serum was verified.

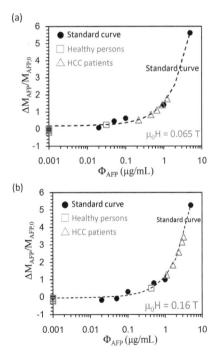

Figure 6. The normalized increment of magnetization $\Delta M_{AFP}/M_{AFP,0}$ as a function of Φ_{AFP} with data analyzed at $\mu_0H = 0.065$ T. On the standard curve, AFP levels for healthy persons and HCC patients are shown.

The AFP level in serum was recently determined via the ΔM_S-versus-Φ_{AFP} curve at the saturation field $\mu_0 H_S$ = ~0.4 T [16], where ΔM_S is the increment of the saturated magnetization. A clear demarcation between the normal group and the HCC group was verified in the test results, which indicates the feasibility of using ΔM_S-versus-Φ_{AFP} at the saturation field as the primary analysis factor for identifying the AFP risk level in patients. In this work, the screening of HCC patients was fulfilled at low magnetic fields, which makes the detection platform simple for biomedical application users.

4. Conclusions

In summary, we performed measurements of magnetization (M–H curves) and AC susceptibility when reagents consisting of Fe_3O_4-anti-AFP were conjugated with AFP. The scaling characteristic of ($\Delta M_{AFP}/M_{AFP,0}$)-versus-Φ_{AFP} curves at low magnetic fields was demonstrated, and bio-sensing using BMNPs via increments of magnetization was proposed. We showed that BMNPs can be applied to assay large as well as small molecules. The screening of HCC patients via the scaling characteristic was verified in clinical studies. The detection mechanism based on the scaling characteristic showed potential to develop a compact VSM with a low magnetic field for biomedical applications.

Acknowledgments: This work is supported by the Ministry of Science and Technology of Taiwan under grant number: 104-2112-M-003-005, 105-2112-M-003-012 and by "Aim for the Top University Plan" of the National Taiwan Normal University and the Ministry of Education, Taiwan, R.O.C under grant number 104J1A27 and 105J1A27.

Author Contributions: Shu-Hsien Liao conceived and designed the experiments, and wrote the manuscript; Yu-Kai Su and Jen-Jie Chieh performed the experiments; Yu-Kai Su, Yuan-Fu Tong and Han-Sheng Huang analyzed the data; Shu-Hsien Liao and Kai-wen Huang coordinated and supervised the work.

Conflicts of Interest: The authors declare no conflict of interest.

References

1. Lequin, R.M. Enzyme Immunoassay (EIA)/Enzyme-Linked Immunosorbent Assay (ELISA). *Clin. Chem.* **2005**, *51*, 2415–2418. [CrossRef] [PubMed]
2. Yalow, R.S.; Berson, S.A. Immunoassay of endogenous plasma insulin in man. *J. Clin. Investig.* **1960**, *39*, 1157–1175. [CrossRef] [PubMed]
3. Rajkovic, A.; El-Moualij, B.; Uyttendaele, M.; Brolet, P.; Zorzi, W.; Heinen, E.; Foubert, E.; Debevere, J. Immunoquantitative real-time PCR for detection and quantification of Staphylococcus aureus enterotoxin B in foods. *Appl. Environ. Microbiol.* **2006**, *72*, 6593–6599. [CrossRef] [PubMed]
4. Yang, S.Y.; Wu, R.M.; Chien, C.F.; Horng, H.E.; Hong, C.-Y.; Yang, H.C. One-sample measurement in laser nephelometric immunoassay using magnetic nanoparticles. *Appl. Phys. Lett.* **2006**, *89*, 244106. [CrossRef]
5. Weitschies, W.; Kötitz, R.; Bunte, T.; Trahms, L. Determination of relaxing or remanent nanoparticle magnetization provides a novel binding-specific technique for the evaluation of immunoassays. *Pharm. Pharmacol. Lett.* **1997**, *7*, 1.
6. Lee, S.K.; Myers, W.R.; Grossman, H.L.; Cho, H.-M.; Chemla, Y.R.; Clarke, J. Magnetic gradiometer based on a high-transition temperature superconducting quantum interference device for improved sensitivity of a biosensor. *Appl. Phys. Lett.* **2002**, *81*, 3094. [CrossRef]
7. Enpuku, K.; Minotani, T.; Gima, T.; Kuroki, Y.; Itoh, Y.; Yamashita, M.; Katakura, Y.; Kuhara, S. Detection of Magnetic Nanoparticles with Superconducting Quantum Interference Device (SQUID) Magnetometer and Application to Immunoassays. *Jpn. J. Appl. Phys.* **1999**, *38*, L1102. [CrossRef]
8. Enpuku, K.; Yoshida, T.; Bhyuiya, A.K.; Watanabe, H.; Asai, M. Characterization of magnetic markers and sensors for liquid phase immmunoassay using Brownian relaxation. *IEEE Trans. Magn.* **2012**, *48*, 2838–2891.
9. Horng, H.E.; Yang, S.Y.; Hong, C.-Y.; Liu, C.M.; Tsai, P.S.; Yang, H.C.; Wu, C.C. Biofunctionalized magnetic nanoparticles for high-sensitivity immunomagnetic detection of human C-reactive protein. *Appl. Phys. Lett.* **2006**, *88*, 252506. [CrossRef]
10. Lee, H.; Sun, E.; Ham, D.; Weissleder, R. Chip–NMR biosensor for detection and molecular analysis of cells. *Nat Med.* **2008**, *14*, 869–874. [CrossRef] [PubMed]

11. Hong, C.-Y.; Wu, C.C.; Chiu, Y.C.; Yang, S.Y.; Horng, H.E.; Yang, H.C. Magnetic Susceptibility Reduction Method for Magnetically Labeled Immunoassay. *Appl. Phys. Lett.* **2006**, *88*, 212512. [CrossRef]
12. Yang, S.Y.; Jian, Z.F.; Chieh, J.J.; Horng, H.E.; Yang, H.C.; Huang, I.J.; Hong, C.Y. Wash-free, antibody-assisted magnetoreduction assays of orchid viruses. *J. Virol. Methods* **2008**, *149*, 334–337. [CrossRef] [PubMed]
13. Huang, K.-W.; Chieh, J.-J.; Horng, H.-E.; Hong, C.-Y.; Yang, H.-Y. Characteristics of magnetic labeling on liver tumors with anti-alpha-fetoprotein-mediated Fe_3O_4 magnetic nanoparticles. *Int. J. Nanomed.* **2012**, *7*, 2987–2996.
14. Yang, S.Y.; Yang, C.C.; Horng, H.E.; Shin, B.Y.; Chieh, J.J.; Hong, C.Y.; Yang, H.C. Experimental study on low-detection limit for immuomagnetic reduction assays by manupulating the reagents entities. *IEEE Trans. NanobioSci.* **2013**, *12*, 65–68. [CrossRef] [PubMed]
15. Yang, C.C.; Yang, S.Y.; Chieh, J.J.; Horng, H.E.; Hong, C.Y.; Yang, H.C. Universal behavior of bio-molecule-concentration dependent reduction in AC magnetic susceptibility of bio-reagents. *IEEE Magn. Lett.* **2012**, *3*, 1500104. [CrossRef]
16. Chieh, J.-J.; Huang, K.-W.; Shi, J.-C. Sub-tesla-field magnetization of vibrated magnetic nanoreagents for screening tumor markers. *Appl. Phys. Lett.* **2015**, *106*, 073703. [CrossRef]
17. Rosensweig, R.E. Heating magnetic fluid with alternating magnetic field. *Mater* **2002**, *252*, 370–374. [CrossRef]
18. Liao, S.H.; Yang, H.C.; Horng, H.E.; Chieh, J.J.; Chen, K.L.; Chen, H.H.; Chen, J.Y.; Liu, C.I.; Liu, C.W.; Wang, L.M. Time-dependent phase lag of bio-functionalized magnetic nanoparticles conjugated with biotargets studied with alternating current magnetic susceptometor for liquid phase immunoassays. *Appl. Phys. Lett.* **2013**, *103*, 243703. [CrossRef]
19. Huang, K.-W.; Yang, S.-Y.; Hong, Y.-W.; Chieh, J.-J.; Yang, C.-C.; Horng, H.-E.; Wu, C.-C.; Hong, C.-Y.; Yang, H.-C. Feasibility studies for assaying alpha-fetoprotein using antibody-activated magnetic nanoparticles. *Int. J. Nanomed.* **2012**, *7*, 1991–1996.
20. Koh, I.; Josephson, L. Magnetic Nanoparticle Sensors. *Sensors* **2009**, *9*, 8130–8145. [CrossRef] [PubMed]
21. Chen, K.-L.; Chen, J.-H.; Liao, S.-H.; Chieh, J.-J.; Horng, H.-E.; Wang, L.-M.; Yang, H.-C. Magnetic Clustering Effect during the Association of Biofunctionalized Magnetic Nanoparticles with Biomarkers. *PLoS ONE,* **2015**, *10*, e0135290. [CrossRef] [PubMed]

Article

Studies towards hcTnI Immunodetection Using Electrochemical Approaches Based on Magnetic Microbeads

Alejandro Hernández-Albors [1,2], **Gloria Colom** [1,2], **J.-Pablo Salvador** [1,2,*] and **M.-Pilar Marco** [1,2]

[1] Nanobiotechnology for Diagnostics (Nb4D), Department of Chemical and Biomolecular Nanotechnology, Institute for Advanced Chemistry of Catalonia (IQAC) of the Spanish Council for Scientific Research (CSIC), Jordi Girona 18-26, 08034 Barcelona, Spain; ahernandezalbors@gmail.com (A.H.-A.); gloria_colom@hotmail.com (G.C.); pilar.marco@cid.csic.es (M.-P.M.)

[2] CIBER de Bioingeniería, Biomateriales y Nanomedicina (CIBER-BBN), Jordi Girona 18-26, 08034 Barcelona, Spain

* Correspondence: jpablo.salvador@iqac.csic.es; Tel.: +34-93-400-6100

Received: 19 July 2018; Accepted: 27 July 2018; Published: 29 July 2018

Abstract: Different electrochemical strategies based on the use of magnetic beads are described in this work for the detection of human cardiac troponin I (hcTnI). hcTnI is also known as the gold standard for acute myocardial infarction (AMI) diagnosis according to the different guidelines from the European Society of Cardiology (ESC) and the American College of Cardiology (ACC). Amperometric and voltamperometric sandwich magnetoimmunoassays were developed by biofunctionalization of paramagnetic beads with specific antibodies. These bioconjugates were combined with biotinylated antibodies as detection antibodies, with the aim of testing different electrochemical transduction principles. Streptavidin labeled with horseradish peroxidase was used for the amperometric magnetoimmunoassay, reaching a detectability of 0.005 ± 0.002 µg mL^{-1} in 30 min. Cadmium quantum dots-streptavidin bioconjugates were used in the case of the voltamperometric immunosensor reaching a detectability of 0.023 ± 0.014 µg mL^{-1}.

Keywords: human cardiac troponin I; magnetic beads; magnetoimmunosensor; cadmium quantum dots; Streptavidin-Horseradish Peroxidase

1. Introduction

Cardiovascular diseases (CVDs) represent one of the leading causes of death globally. According to the World Health Organization (WHO) in 2012, 17.5 million deaths were caused by these types of pathologies, and more than 80% of these deaths were due to heart attacks and strokes [1]. There are indications that the number of deaths by CVDs will keep rising, and by 2030, it is estimated that almost 23.6 million people will die because of these diseases.

The early and quick diagnosis of CVDs is very important, not only for health and patient survival, but also for cost and time efficiency in a successful diagnosis and prognosis of the illness. Traditionally, AMI diagnosis has been based on the WHO criteria, whereby patients must meet at least two out of three conditions: Ischemia symptoms such as the characteristic chest pain, significant changes in the diagnostic electrocardiogram (ECG), and elevations of the concentration in blood of the creatine kinase (CK-MB) biomarker [2]. Despite ECG being an important tool for diagnosing and monitoring the disease [3–5], it is not a good confirmatory test to diagnose CVDs because, as it is described, around half of the patients with a severe cardiopathy do not show relevant variations in their ECG. However, most of these patients show elevations on the troponin concentration, revealing cardiac muscle necrosis [6], highlighting that an ECG test is not enough to make an early and accurate diagnosis of CVDs [7]. In addition, creatine kinase lacks cardiac tissue specificity and release kinetics give less information about the dimensions of the damage in comparison with other biomarkers [8]. For all these reasons, diagnosis criteria for AMI were modified in 2000 by the European Society of Cardiology (ESC) and the American College of Cardiology (ACC), replacing CK-MB biomarker by human cardiac troponin I (hcTnI) as the gold biomarker for the assessment of cardiac damage [8,9]. After an AMI episode, free hcTnI is released to the bloodstream indicating necrosis of the cardiac muscle and increasing the mortality risk, even at concentrations below 0.06 ng mL^{-1}. For this reason, current guidelines suggest the use of the 99th percentile of troponin concentration from a healthy reference population as a cutoff for any tool focused in the diagnostic and prognostic of the cardiac illness [10–13]. In this context, different immunological methodologies have been developed to detect hcTnI in human samples. Electrochemiluminiscent immunoassays (ECLIA) [14–16], Fluoroimmunoassays [16–20], and enzyme-linked immunosorbent assay (ELISA) [21–24] have been widely used for AMI diagnosis. However, these techniques require expensive laboratory equipment, well-trained personnel and are time-consuming techniques, especially taking into account that a timely and reliable diagnosis is required for appropriate patient treatment.

In this sense, immunosensors based on electrochemical transduction have been considered an effective analytical approach, particularly because of their accuracy, high sensitivity, simplicity, low cost and short response time, reaching in some cases low limits of detection [25–29]. Specifically, electrochemical immunosensors based in the use of magnetic beads have improved the analytical performance of the immunoassays due to the high surface-area-to-volume ratio of the particles, conferring higher probability of interaction between the target analyte and the bioreceptor immobilized onto the beads surface and, consequently, achieving faster assays kinetics [30–32]. Within electrochemical immunosensors, techniques can be found based on impedance spectroscopy, which allows label-free and real-time detection measuring small changes in the surface of the electrode. Accordingly, other works are described in the literature based on the use of functionalized magnetic beads in combination with impedance spectroscopy, reaching appropriate values of limit of detection regarding clinical guidelines [31,33–37].

In this work, different approaches based on the use of the magnetic beads-antibody bioconjugates targeting hcTnI have been developed as a proof of concept. Magnetic beads were combined successfully with materials from different fields, while avoiding non-desirable interactions, and reducing, in some cases, the assay time and setting the stage for the development of multiplexed electrochemical immunosensors based on the use of the magnetic beads. Furthermore, the electrochemical immunosensors described in this text were compared in terms of assay time, background noise, limit of detection and microbead performance.

2. Materials and Methods

2.1. Chemicals and Biochemicals

Human cardiac troponin I (hcTnI) was purchased from Life Diagnostics Inc. (West Chester, PA, USA), aliquoted and stored at −80 °C until being used. Human skeletal troponin I (hsTnI) was kindly provided by Dr. Tamas Mészáros from Semmelweiss University (Budapest, Hungary), aliquoted and stored at −80 °C until used. Streptavidin-Horseradish Peroxidase bioconjugate (Sigma-Aldrich, St. Louis, MO, USA) was dissolved in 10 mM of phosphate saline buffer (PBS) solution at pH 7.5 at a final concentration of 1 mg mL^{-1}, aliquoted and stored at −20 °C until used. EZ Link sulfo-NHS-LC-LC-Biotin reagent was purchased from Pierce Biotechnology (Rockford, IL, USA). *N*-ethyl-*N*-dimethylaminopropyl-carbodiimide (EDC), *N*-hydroxysulfo-succinimide (NHS), bovine serum albumin (BSA), avidin from egg white, casein from bovine milk and 2-(4-Hydroxyphenylazo) benzoic acid (HABA) were purchased to Sigma-Aldrich. Methyl-(octaethyleneglycol) amine (mPEG-Amine) was acquired from Laysan Bio, Inc. (Arab, AL, USA). Other chemicals and biochemicals used were purchased to Sigma Chemical Co. (St. Louis, MO, USA) and all salts were provided by Merck (Darmstadt, Germany). SiMAG-Carboxyl magnetic beads (SiMAG-Carboxyl, 1 μm Ø, Prod. No. 1402) were purchased from Chemicell GmbH (Berlin, Germany). Qdot™ 585 Streptavidin Conjugate (Qdot-SAv) was purchased from Molecular probes® (Eugine, OR, USA).

2.2. Materials and Instruments

Electrochemical measurements were carried out using a portable multipotentiostat μSTAT 8000P (DropSens S.L., Llanera, Spain). Screen-printed carbon array electrodes were purchased from DropSens (SPCE, DRP-8X110, working electrode 2.56 mm Ø). Each device displayed eight 3-electrode electrochemical cells, each of them including a carbon-based working electrode, an Ag pseudo-reference electrode and a carbon counter electrode. The magnetic separation during the different washing steps involved in the immunoassay procedure was performed using a 12-tube magnetic separator rack (MagnaRackTM Cat. No. CS15000, Carlsbad, CA, USA). A polymethylmethacrylate support (PMMA support) with 8 embedded neodymium magnets, designed and manufactured by Micro-Nano Technologies Unit, of the Unique Scientific and Technical Infrastructures (U8 of the ICTS "NANBIOSIS") from Institute of Microelectronics of Barcelona (IMB-CNM, Barcelona, Spain), was used to immobilize the modified magnetic beads onto the surface of each of the eight SPCE to perform electrochemical measurements. An IKA MS 3 digital shaker (IKA®-Werke GmbH & Co. KG, Staufen, Germany) was used at 700 rpm to incubate magnetic beads at the different stages of each of the immunoassays. The pH and the conductivity of all buffers and solutions were measured with a pH-meter pH 540 GLP and a conductimeter LF 340, respectively (WTW, Weilheim, Germany). Round bottom, non-treated plates were purchased from Nirco (Barberà del Vallés, Spain). Polystyrene Immulon 2 HB™ and MaxiSorp™ microtiter plates were purchased from Nunc (Roskilde, Denmark). Washing steps were carried out using a SLY96 PW microplate washer (SLT Labinstruments GmbH, Salzburg, Austria). Dilution plates were purchased from Nirco (Barberà del Vallés, Spain). Absorbances were read on a SpectramaxPlus (Molecular Devices, Sunnyvale, CA, USA). The electrochemical data obtained was analyzed using DropView 8400 software (DropSens S.L., Llanera, Spain). The calibration curves and different fittings were analyzed using GraphPad Prims 5.03 (GraphPad Software Inc., San Diego, CA, USA).

2.3. Buffers and Solutions

PBS was 0.01 mol L^{-1} phosphate buffer, 0.14 mol L^{-1} in NaCl and 0.003 mol L^{-1} in KCl saline solution at pH 7.5. PBST was PBS with 0.05% (v/v) Tween 20. PBST-Casein was PBST with 0.15% (w/v) casein. Coating buffer was 0.05 M carbonate-bicarbonate buffer, pH 9.6. Citrate buffer was a 0.04 M solution of sodium citrate, pH 5.5. The substrate solution for optical measurement was 0.01% TMB (3,3′,5,5′-tetramethylbenzidine) and 0.004% H_2O_2 in citrate buffer. For amperometric measurements, citrate buffer-KCl was prepared with citrate buffer containing 0.1 mol L^{-1} KCl with 0.001% TMB and 0.0004% H_2O_2. For voltamperometric measurements, an acetate buffer was composed by 0.5 M sodium acetate buffer pH 5.5 and 1 μg mL^{-1} Bi (III).

2.4. Polyclonal Antibody Production against hcTnI

Antibodies As220 and As221 were raised by immunization of native human cardiac troponin I (hcTnI). Two female New Zealand white rabbits, weighing 1 to 2 kg each, were immunized following the immunizing protocol already described [38]. The evolution of the immunization was followed by titration assays by measuring the binding of a serial dilutions of the antisera to a microplate coated with hcTnI. After an acceptable antibody titer was observed, the animals were exsanguinated, and the blood was collected in vacutainer tubes provided with a serum separation gel. Antisera were obtained after centrifugation step and stored at −80 °C in the presence of 0.02% NaN₃. Polyclonal antibodies, pAb220 and pAb221 were purified for further bioconjugation procedures, first by ammonium sulphate precipitation and then by protein A affinity chromatography [39].

The production of the antibodies was performed with the support of the U2 of the ICTS "NANBIOSIS", more specifically by the Custom Antibody Service (CAbS, CIBER-BBN, IQAC-CSIC).

2.5. Preparation of Biotinylated Antibody Bioconjugates

Biotin-labeled antibodies were prepared from purified pAb220 and pAb221 and coupled to an EZ Link sulfo-NHS-LC-LC-Biotin according to the specifications provided by the supplier, but with slight modifications. Briefly, 2 mg of each antibody were dissolved in 1 mL of PBS. Then, 27 µL of 10 mM of EZ-Link sulfo NHS-LC-LC-Biotin solution (7 mg/mL of ultrapure water) were added to each antibody drop by drop under continuous stirring for 1 h. The labeled antibodies were purified by dialysis and stored freeze-dried at −80 °C. Working aliquots at 1 mg mL^{-1} were prepared in PBS and stored at −20 °C. Biotinylated antibodies pAb220-B and pAb221-B were characterized following HABA/Avidin procedure according to the supplier protocol. In all cases, the bioconjugates showed between 3 and 4 biotin moieties per antibody.

2.6. Sandwich ELISA for cTnI

The calibration curve for hcTnI using As220 and pAb221-B antibodies was developed as follows. Immulon 2 HB™ microtiter plates were coated with the capture antiserum As220 (1/16000 dilution in coating buffer, 100 µL/well), stored overnight at 4 °C and covered with adhesive plate sealers. After this time, microplates were washed four times in 300 µL/well with PBST. hcTnI standard solutions (from 0 to 200 ng mL^{-1}) were prepared in PBST-Casein and added to the microtiter plates, 100 µL/well. After 30 min, the plates were washed again and a solution of the biotinylated antibody pAb221-B (2.5 µg mL^{-1} in PBST, 100 µL/well) was added to the plates. After 30 min and another washing step, a solution of Streptavidin-Horseradish Peroxidase bioconjugate (SAv-HRP) was added to the microplates (0.17 µg mL^{-1} in PBST, 100 µL/well). After 30 min, the plates were washed again, and the substrate solution was added (100 µL/well, protected from light). Color development was stopped after 30 min by adding 4 N H$_2$SO$_4$ (50 µL/well), and the absorbances were read at 450 nm. The standard curve was fitted to a linear regression. In this case, the limit of detection (LOD) was established as the analyte concentration corresponding to the sample blank value plus three standard deviations. In the same way, the limit of quantification (LOQ) was established as the analyte concentration corresponding to the sample blank value plus ten standard deviations.

2.7. Specificity Studies

Specificity studies for both antibodies were carried out by following the sandwich ELISA procedure previously described. In this case, two different calibration curves were performed in PBST-Casein buffer, one for hcTnI and the other for the skeletal isoform, hsTnI (from 0 to 250 ng mL^{-1}, 100 µ/well). The standard curve was fitted to a linear regression, and the slopes for each of the fittings were compared to assess the specificity of the antibodies produced against hcTnI.

2.8. Preparation of the Magnetic Beads-Antibody Bioconjugates

The magnetic microbeads-antibody bioconjugate (MB-pAb220) was prepared by covalent immobilization to magnetic microbeads with a purified fraction of pAb220 immunoglobulins (described above). Briefly, 10 mg of microbeads were washed twice with 0.1 M MES pH 5.0 buffer solution and resuspended in 250 µL of a solution of 1-ethyl-3-(3′-dimethylaminopropyl) carbodiimide (EDC, 10 mg, 0.05 mmol) prepared in MES buffer, to activate carboxylic groups. The final mixture was gently shaken for 15 min at RT. After the activation step, microbeads were washed twice with MES buffer and resuspended in 250 µL of PBS solution containing 50 µg of pAb220 antibody and incubated for two hours at RT with gentle shaking to avoid microbeads aggregation. After the coupling step, the supernatant was removed and kept for further conjugation yield quantification. Then, the beads were washed three times with PBS and resuspended in 250 µL of 10 mM PBS pH 7.5 with 1% (*w/v*) of mPEG-NH$_2$ buffer and incubated at RT overnight with gentle stirring to block unreacted sites of the microparticles. Finally, after a washing step, beads were resuspended and put into storage in 250 µL of 10 mM PBS pH 7.5 with 0.5% (*w/v*) of mPEG-NH$_2$ buffer.

2.9. Magneto-ELISA (mELISA) Protocol

The calibration curve for hcTnI detection using MB-pAb220 bioconjugate was developed as follows. All steps were performed in a round-bottom 96 microplate without any surface treatment. After each incubation step, the magnetic beads were washed by placing the microplate on the magnetic rack, allowing the beads to migrate to the magnet until the liquid was clear (approximately 1 min) and then removing the supernatant. hcTnI standards solutions (from 0 to 250 ng mL^{-1} in PBST, 100 μL) were placed in the microplate wells and mixed with a solution of MB-pAb220 (0.1 mg mL^{-1} in PBST, 0.01 mg of beads/well, 100 μL/well). After 30 min of incubation at RT with gently stirring, microbeads were washed with PBST (400 μL, 3 times) and resuspended in a solution of the pAb221-B conjugate (0.5 μg mL^{-1} in PBST, 100 μL/well). After 30 min of incubation under the same conditions described before, magnetic beads were washed again and SAv-HRP was added (0.17 μg mL^{-1} in PBST, 100 μL) for 30 min at RT. Then, beads were washed, and the substrate solution was added and incubated again for 30 min at RT. The enzymatic reaction was stopped by adding 4 N H$_2$SO$_4$ (50 μL). Finally, supernatants were removed from the magnetic beads by magnetic separation and placed in other microplate for measuring absorbance at 450 nm. The standard curve was fitted to a linear regression.

2.10. Amperometric Magneto Immunosensor (AMIS) Protocol

All steps were performed in 2 mL safe-lock tubes. After each incubation step, magnetic beads were washed and supernatants were removed by magnetic separation rack until the beads had migrated to the magnet and the liquid was clear (approximately 1 min). hcTnI standards solutions (from 0 to 125 ng mL^{-1} in PBST, 100 μL) were placed in the tubes and mixed with a solution of MB-pAb220 (0.1 mg mL^{-1} in PBST, 0.01 mg of beads/tube, 100 μL/tube), previously washed 3 times with PBST. Then, a solution of pAb221-B (0.5 μg mL^{-1} in PBST, 100 μL/tube) and a solution of of SAv-HRP (2 μg mL^{-1} in PBST, 100 μL/tube) were immediately added to the tubes with the hcTnI standards and the microbeads. The mixture was incubated for 30 min at RT with gentle stirring. Afterwards, magnetic beads were washed with PBST (3 × 800 μL), resuspended in citrate-KCl buffer (100 μL) and captured onto the surface of the working electrode using a magnet located under the SPCEs. Finally, amperometric measurements were carried out at an applied potential of −0.10 V vs. Ag pseudo- reference electrode. After current stabilization, 10 μL of substrate solution prepared in citrate buffer-KCl was added and current was recorded once again after its stabilization. The standard curve was fitted to a linear regression. A schematic representation of the whole procedure is shown in Figure 1a.

2.11. Voltamperometric Magneto Immunosensor (VMIS) Protocol

All the immunochemical steps were performed in 2 mL safe-lock tubes, and all the quantities referred to in this procedure are the amount added per tube. After each incubation step, magnetic beads were washed (3 × 800 μL, PBST) by magnetic separation, placing the tubes in a magnetic rack (approximately, 1 min), as described above. hcTnI standard solutions were prepared as previously described (from 0 to 250 ng mL^{-1} in PBST, 100 μL) and mixed with a solution of the MB-pAb220 (1 mg mL^{-1} in PBST, 0.1 mg of beads/tube, 100 μL). After 30 min of incubation at RT and gently stirring, magnetic beads were washed and resuspended in a solution of pAb221-B at 8 μg mL^{-1} (100 μL, PBST). After 30 min of incubation at RT and gently stirring, the beads were washed again as previously described and resuspended in 80 μL of PBST. After a complete resuspension of the microbeads, 20 μL of a solution of 5 nM of Qdot-SAv prepared in PBST was added to each tube and incubated at RT for 30 min with gently stirring. After this time, beads were washed and Cd^{2+} ions were released by acidic digestion adding 20 μL of 1 M HCl for 30 min at RT with gentle stirring. Afterwards, each tube was placed onto the magnetic rack and supernatants were placed onto the surface of the working electrodes, each of them containing 40 μL of acetate buffer. Square Wave Voltammetry (SWV) was used to perform the electrochemical measurements using Ag as a pseudo reference electrode. A condition voltage (0.6 V, 60 s) was initially applied, followed by a deposition voltage (−1.4 V, 180 s). After a 15 s rest period, anodic stripping voltammetry was performed (from -1.20 to -0.6 V, step voltage of 10 mV, amplitude 50 mV and frequency of 20 Hz). Under these conditions, well-defined stripping current peaks at approximately -0.85 V (oxidation/stripped voltage) were obtained. A schematic representation of the whole procedure is shown in Figure 1b. The height of the peak (current), proportionally to the concentration of the hcTnI in each sample, was calculated using DropView 8400 software. Standard curves were obtained plotting the current versus the hcTnI concentration and fitting the points to the linear regression.

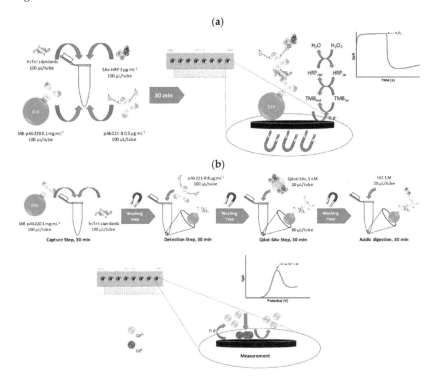

Figure 1. (a) Scheme of AMIS procedure. After all the biorecognition steps, the magnetic beads complex (MB-complex) was resuspended and immobilized onto the surface of the working electrode. Then, current intensity increased after substrate addition proportionally to the concentration of the hcTnI in the sample. (b) Scheme of the VMIS procedure. After all the biorecognition steps, the MB-complex was digested under acidic conditions provoking cadmium release from Qdot-SAv bioconjugates. Cadmium ions were placed in the electrochemical cell by magnetic rack separation. Then, deposition voltage was applied, and intensity of the peak was reached by square wave voltammetry (SWV).

3. Results and Discussion

3.1. ELISA Sandwich for hcTnI

After preparation of the pAb-Biotin bioconjugates, all antibody combinations were evaluated by sandwich ELISA using As220/As221 as the capture antibody, together with pAb221-B/pAb220-B as the detection antibody. The initial criterion followed to select the best combination was the signal-to-noise ratio (S/N ratio). The highest S/N ratio observed in all the combinations was when As220 was used as the capture probe, regardless of the pAb-biontin used. The pair formed by As220 as the capture antibody and pAb221-B as the detection antibody showed the highest signal and the lowest background noise (data not shown). This combination was selected to develop further immunoassays. According to the non-specific adsorption observed for the hcTnI protein to the surface of the microplate, further steps were carried out in order to minimize the background noise. This non-desired adsorption was because of its high isoelectric point of hcTnI (pI = 9.98) and its high hydrophilic behavior [40]. Thus, different microplates with different surface treatment were evaluated along with different proteins as additives to avoid the non-specific adsorption. Finally, Immulon 2 HB™ microplates, together with casein as an additive, showed a significant decrease of the background noise and were therefore chosen to develop the calibration curve (see Supplementary Materials, Figure S1a,b). The decrease in the background noise could be explained in terms of the decreased affinity for hydrophilic molecules shown by Immulon 2 HB™ microplates in contrast to MaxiSorp™ microplates. On the other hand, due to its small size (19 to 25 kDa, depending on the casein subunit), casein can occupy empty spaces of the microplate, blocking non-specific adsorptions more efficiently [37,38] than other typical blocking proteins, such as BSA. In this case, adding casein together with hcTnI in the same buffer significantly decreases the high background noise, avoiding an extra blocking step in the ELISA procedure. Under these conditions, calibration curves in ELISA sandwich format were developed reaching a detectability value of $0.010 \pm 0.002 \ \mu g \ mL^{-1}$ for hcTnI (Figure 2a and Table 1).

3.2. Specificity Studies

To prove the specificity of the antibodies produced against hcTnI, calibration curves for hcTnI and its skeletal isoform, skeletal human troponin I (hsTnI), were performed in ELISA. Both curves were fitted to a linear regression (See Supplementary Material, Figure S2). The slope of both immunoassays was compared in order to determine the affinity and the sensitivity of the antibodies against hsTnI and, as shown in the range of concentration evaluated, no affinity of As220 and As221 against skeletal troponin isoform was observed, suggesting that both antibodies were able to distinguish between two proteins because of their high specificity against their antigen.

3.3. Assessment of the Bioactivity of Magnetic Beads-Antibody Bioconjugates

One of the main advantages of using magnetic beads-antibody bioconjugates is their high surface-to-volume ratio, which confers a higher probability of interaction with the analyte while increasing the performance of the assay, promoting the kinetics, and the possibility of increasing the efficiency of the isolation of the analyte from the matrix due to their superparamagnetic properties [41]. First, the efficiency of the antibody coupling onto the microbeads surface was assessed by evaluation of the supernatant after the coupling step. The amount of antibody not immobilized was quantified by a Bio-Rad protein Assay, reaching in all cases a conjugation yield of $95 \pm 4\%$. Further experiments were focused in the assessment of the functionality of the antibodies by mELISA. First, the concentration of each of the immunoreagents (MB-pAb220 and pAb221-B) was optimized. The concentration of each of the antibody bioconjugates was selected according to the S/N ratio. In this sense, two different concentrations of MB-pAb220 were evaluated against two concentrations of hcTnI (0 and 1 $\mu g \ mL^{-1}$), together with three different concentrations of the pAb221-Biotin bioconjugate (2, 1 and 0.5 $\mu g \ mL^{-1}$). Concentrations of 0.1 mg mL^{-1} of MB-pAb220 and 0.5 $\mu g \ mL^{-1}$ of the biotinylated antibody were selected to perform mELISA and amperometric immunosensor given the highest S/N ratio shown (see Supplementary Material Figure S3). In this case, due to the characteristics of the beads surface, casein addition to the assay buffer was not necessary to avoid non-specific adsorption.

Finally, magnetic bead bioconjugates were characterized by developing a calibration curve followed by colorimetric transduction. The limit of detection shown by mELISA prior to the optimization was in the same order in comparison with LOD showed by ELISA, $0.023 \pm 0.001 \ \mu g \ mL^{-1}$ and $0.010 \pm 0.002 \ \mu g \ mL^{-1}$, respectively (see Figure 2a–b and Table 1).

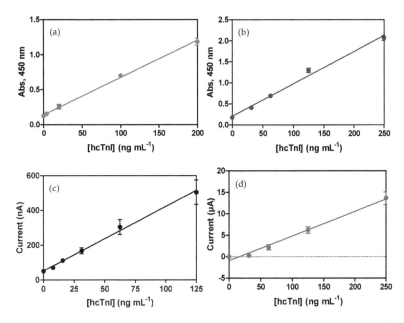

Figure 2. Calibration curves for the different immunoassays developed for the detection of hcTnI. (**a**) ELISA calibration curve for hcTnI in buffer using the immunoreagent produced. As220 was used as a capture antibody at a dilution of 1/16,000, and pAb221-B at 2.5 μg mL^{-1}. Each point was the average of at least two-well replicates, and the assay was run on two different days. (**b**) mELISA calibration curve for hcTnI in buffer using the different bioconjugates prepared as described previously. MB-pAb220 was used as a capture probe at 0.1 mg mL^{-1} and pAb221-B as a detection antibody at a concentration of 0.5 μg mL^{-1}. SAv-HRP was used at a concentration of 0.17 μg mL^{-1}. Each point was the average of at least two well replicates, and the assay was run on two different days. (**c**) AMIS calibration curve for hcTnI detection in buffer. MB-pAb220 was used as a capture probe at 0.1 mg mL^{-1} and pAb221-B as a detection antibody at a concentration of 0.5 μg mL^{-1}. SAv-HRP was used at a concentration of 2 μg mL^{-1}. Each point was the average of at least three well replicates, and the assay was run on three different days. (**d**) VMIS calibration curve for hcTnI detection in buffer using the immunoreagents produced and after different optimization steps. MB-pAb220 was used as a capture probe at 1.0 mg mL^{-1} and pAb221-B as a detection antibody at a concentration of 8 μg mL^{-1}. Qdot-SAv was used at a concentration of 5 nM. Each point was the average of at least three well replicates, and the assay was run on three different days.

Table 1. Analytical features and immunoassay characteristics for the different immunochemical approaches developed for the hcTnI detection.

	ELISA [a]	mELISA [b]	AMIS [c]	VMI [d]
Beads (mg)	-	0.01	0.01	0.1
[MBs] (mg mL^{-1})	-	0.05	0.025	0.5
Total Assay Time (min)	>120	120	30	120
Slope	5.34 ± 0.17	7.68 ± 0.311	3.71 ± 0.29	57.26 ± 3.93
Ordinate	0.14 ± 0.01	0.207 ± 0.04	52.03 ± 17.19	−0.91 ± 0.51
LOD (μg mL^{-1})	0.010 ± 0.002	0.023 ± 0.001	0.005 ± 0.002	0.023 ± 0.014
LOQ (μg mL^{-1})	0.031 ± 0.007	0.075 ± 0.044	0.020 ± 0.006	0.068 ± 0.045
R^2	0.990 ± 0.003	0.988 ± 0.01	0.989 ± 0.004	0.977 ± 0.058

[a] Enzyme-Linked Immunosorbent Assay; [b] Magnetic Enzyme-Linked Immunosorbent Assay; [c] Amperometric Magneto-Immunosensor; [d] Voltamperometric Magneto-Immunosensor.

3.4. Development of Electrochemical Immunosensors for hcTnI Detection

The two electrochemical immunosensors presented here are based on the use of magnetic beads modified with specific antibodies against hcTnI as a selective capture probe in a fast and reliable way. Two different electrochemical transductions were investigated and compared in terms of microbead features, background noise and analytical performance, among others.

3.4.1. Amperometric Magneto-Immunosensor (AMIS)

The Amperometric Magneto-Immunosensor (AMIS) is based on the registration of the current intensity produced by the enzymatic reaction catalyzed by the Horseradish-Peroxidase (HRP) present in the SAv-HRP bioconjugate when a specific substrate is used. In this case, HRP can oxidize the substrate (H_2O_2 to H_2O) leading to TMB oxidation, and subsequently reduced again by the electrode applied potential (Figure 1a). The acquired current intensity is directly proportional to the amount of the HRP immobilized onto the MB-complex and directly correlated with the concentration of hcTnI in the sample.

Using magnetic bead bioconjugates confers the possibility of reusing the surface of the electrode; due to the electrode, surface modification is not needed to perform measurements. In this regard, electrodes were evaluated after each experiment observing the background current in the absence of beads and compared with the electrochemical current of a control electrode for which no experiment was carried out. In this sense, the background current started to be significantly different after the third electrochemical measurement.

For amperometric transduction, the concentrations of all of the immunoreagents, MB-pAb220, pAb221-B and SAv-HRP, as well as their incubation times, were previously optimized by mELISA. With the main objective of significantly reducing the assay time and accomplishing the clinical guidelines recommendations, incubation time was optimized, evaluating the response of the immunoassay in terms of maximum signal and background noise when only a single step was performed, instead of a sequential-step assay. All the immunoreactants were mixed at the concentrations previously optimized by mELISA, and after 30 min of incubation time, colorimetric readout was achieved by adding the corresponding substrate (see Supplementary Materials, Figure S4a). As can be seen, the maximum response of the immunoassay in presence of hcTnI decreased significantly when the immunoassay was performed in a single step. However, further optimization of the concentration of the SAv-HRP led us to improve the S/N ratio working at 2 μg mL^{-1} (see Supplementary Materials, Figure S4b).

Finally, the calibration curve for hcTnI was performed (see Figure 2c) by mixing different standard solutions of the cardiac biomarker with MB-pAb220, pAb221-b and SAv-HRP at optimized concentrations and incubating for 30 min at RT. Electrochemical measurements of the immunocomplexes formed were obtained in the electrochemical cell after capturing the beads and the substrate addition.

Both AMIS and mELISA employed different concentration of magnetic beads according to the different immunochemical protocol. For practical purposes, the amperometric immunosensor was implemented using 2 mL safe-lock tubes. In the case of AMIS, the final concentration of beads was 0.025 mg mL^{-1}, while in the case of mELISA, the concentration of beads in the capture step was 0.05 mg mL^{-1}, and 0.1 mg mL^{-1} in the remaining steps. Despite this difference, the S/N ratio for both immunoassays at hcTnI level of 125 ng mL^{-1} was 12 $\pm$ 3 for AMIS against the 7.3 $\pm$ 0.1 for mELISA. These observations could suggest that the beads are more efficient if they are confined in a larger container than a 96-well plate microwell. On the other hand, the non-reversible formation of doublets species when MB-Ab and antigen (Ag) are incubated together, and a magnetic field is subsequently applied, has been well described [42]. In this sense, incubation of all the immunoreagents at the same time, as described in the AMIS procedure, could reduce the possibility of forming MB-Ab-Ag-Ab-MB tandems due to the presence of the detection antibody, thus improving the sensitivity of the immunoassay.

In short, according to the different analytical parameters obtained in ELISA, mELISA and AMIS, electrochemical magnetic bead-based immunosensor performance (see Table 1) not only improves the limit of detection, but also significantly reduces the total assay time.

3.4.2. Voltamperommetric Magneto-Immunosensor (VMIS)

The voltamperometric immunosensor is based on the acquisition of the characteristic oxidation potential from a specific metallic nanoprobe, such as the cadmium-containing commercial Qdot-Streptavidin bioconjugate. These Qdots are composed by a CdSe core encapsulated by a shield of ZnS coated with a commercial polymer with specific functional groups to immobilize streptavidin. After acidic digestion, the ZnS layer is dissolved and cadmium ions are released in the medium. Then, applying a deposition voltage, cadmium ions are reduced onto the surface of the working electrode and oxidized again afterwards by stripping voltammetry, showing a characteristic current peak corresponding to the oxidation potential of the cadmium (Figure 1b). The peak intensity in this case is directly related to the amount of cadmium in the Qdots immobilized by the interaction streptavidin-biotin and then is directly proportional to the hcTnI in the sample.

Different parameters were optimized before the development of the calibration curve for hcTnI detection-based Voltamperometric Magneto-Immunosensor. First, the incubation times of the Qdot-SAv bioconjugate to release the maximum amount of Cd^{2+} ions under acidic conditions were optimized. As shown in Figure 3a, no differences were observed in the current achieved after 30 min of incubation at RT and gentle stirring. Afterwards, deposition voltage was also optimized. Current signal in presence of hcTnI was significantly improved when the deposition voltage applied was changed from -1.2 to -1.4 V, with no increase in the background noise (see Figure 3b). Finally, concentrations of the different antibody bioconjugates were optimized at two different levels of hcTnI (1 and 0 $\mu g\ mL^{-1}$) to observe which combination showed the highest S/N ratio. First, MB-pAb220 concentration was optimized, evaluating a concentration range from 0.4 to 1.2 mg mL^{-1}. A saturation point was reached at a concentration of 1.0 mg mL^{-1} of magnetic beads, setting this value for further assays (see Figure 3c). Afterwards, different concentrations of the biotinylated antibody pAb221-B, ranging from 1 to 8 $\mu g\ mL^{-1}$, were assessed. As shown in Figure 3d, current values in the presence of hcTnI increased gradually with higher concentrations of pAb221-B, reaching the highest current value at 8 $\mu g\ mL^{-1}$ of the detection antibody. In all cases, the current in absence of hcTnI was nearly undetectable, revealing the total absence of nonspecific adsorptions in the immunoassay.

The calibration curve for hcTnI was performed first by mixing biomarker standard solutions (from 0 to 250 ng mL^{-1}, PBST) with the established concentration of the MB-pAb220, further steps are indicated in Figure 1b. After acidic digestion of the Qdots, the measurement of the amount of cadmium in the sample was performed in the electrochemical cell PMMA support, applying SWV procedure. As expected, the current signal recorded was, in this case, directly proportional to the concentration of hcTnI in the sample (see Supplementary Material, Figure S5). The limit of the detection achieved with this immunosensor was 0.023 $\pm$ 0.014 $\mu g\ mL^{-1}$ (see Figure 2d and Table 1).

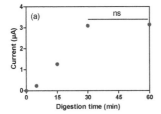 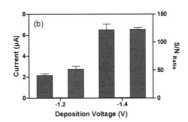

Figure 3. *Cont.*

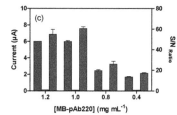 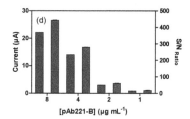

Figure 3. Optimization of the magneto bead-based voltamperometric immunosensor for hcTnI at two different levels (1 µg mL^{-1}, blue bar, 0 µg mL^{-1}, yellow bar, S/N ratio, red bar) using cadmium Qdot-SAv bioconjugates. (**a**) Digestion time in acidic conditions (1 M HCl) to release maximum amount of cadmium ions. (**b**) Deposition voltage was optimized, achieving a higher signal without any increasing of the background noise. (**c**) Concentration of beads and the effect of (**d**) detection antibody were fixed. Data are representative of two independent experiments (ns = not significant).

In principle, using cadmium quantum dots implies a signal amplification and, consequently, an improvement of the analytical performance of the immunoassay. As has been reported in previous works [43–45], the sensitivity of the immunosensor could be improved using these probes, due to the number of Cd^{2+} cations released per Qdot particle. However, the VMIS described in this work, shown similar analytical features as the mELISA immunoassays in terms of detectability. As per the number of beads and their concentration in each step, in comparison with AMIS, the number of beads used in the first capture step is significantly higher (0.1 mg in VMIS and 0.01 mg in AMIS), as is the final concentration. Following the same hypothesis stated previously (See Section 3.4.1), an increase in the concentration of the beads does not have to be directly related to an improvement of the sensitivity, but rather the contrary, as shown in this work. For these reasons, further optimization work could make it possible for VMIS to achieve the required limit of detection for hcTnI.

Nevertheless, for this magnetic bead-based immunosensor, in comparison with colorimetric immunoassays and the amperometric magneto immunosensor described here, in addition to the assay time being able to be reduced by optimization assays, could become a confirmatory tool to assess the degree of cardiovascular damage, due to the possibility of multiplexing, in combination with other metallic nanoprobes other than cadmium, as has been proved [36,46] and, consequently, detecting different biomarkers from the same patient, giving more extensive information about the magnitude of the disease.

4. Conclusions

Two electrochemical magnetic bead-based immunosensors have been developed, successfully combining a magnetic beads-antibody bioconjugate with other immunoreactants, depending on the electrochemical transduction chosen. In the case of the AMIS, the magnetic beads-antibody bioconjugate (MB-pAb220) was combined with a detection antibody produced, conjugated and characterized in the laboratory (pAb221-B), together with the rest of immunoreagents, and after a time of 30 min and the addition of the corresponding substrate, an electrochemical response directly related to the concentration of the hcTnI biomarker present in the sample was observed. In this sense, the assay time was significantly reduced in comparison with ELISA and mELISA due to the improvement of the kinetics provided by the magnetic beads, therefore fulfilling the clinical requirements in terms of time for the diagnosis of myocardial infarction. This short-time assay makes AMIS a good candidate to be a useful tool, after further optimization steps to improve detection limit in the ED department for establishing a clear diagnosis of acute myocardial infarction. Furthermore, in comparison with the rest of the techniques developed and described in this text, the detection limit achieved with this methodology was improved, approaching the 99th percentile of a healthy reference population, which is considered optimal for all techniques focused on the detection of this biomarker.

Likewise, it allows the combination of bioconjugates of a different nature, such as MB-pAb220, with the commercial bioconjugate of Qdot-Sav, as well as with the bioconjugate pAb221-B. After several optimization steps, it was possible to develop an electrochemical immunosensor for the detection of hcTnI whose potential lies in the possibility of multiplexing through the use of other metallic electrochemical nanoprobes (Pb, Zn ...) to detect different biomarkers which, along with hcTnI, can give useful information about the magnitude of the damage and establish a reliable diagnosis [47].

All the work described here reveals the high potential of magnetic beads as well as their great versatility, since they have been successfully combined with bioconjugates of different natures to develop immunochemical techniques based on different principles, such as colorimetric, amperometric and voltamperometric transduction.

Supplementary Materials: The supplementary materials are available online at http://www.mdpi.com/1424-8220/18/8/2457/s1.

Author Contributions: Conceptualization, J.-P.S. and M.-P.M.; Methodology, A.H.-A. and G.C.; Formal Analysis, J.-P.S.; Investigation, A.H.-A. and J.-P.S.; Writing-Original Draft Preparation, A.H.-A.; Writing-Review & Editing, J.-P.S. and M.-P.M.; Funding Acquisition, M.-P.M.

Funding: This work has been supported by Cajal4EU project (ENIAC-120215).

Acknowledgments: The Nb4D group (formerly Applied Molecular Receptors group, AMRg) is a consolidated research group (Grup de Recerca) of the Generalitat de Catalunya and has support from the Departament d'Universitats, Recerca i Societat de la Informació de la Generalitat de Catalunya (expedient: 2017 SGR 1441). CIBER-BBN is an initiative funded by the Spanish National Plan for Scientific and Technical Research and Innovation 2013-2016, Iniciativa Ingenio 2010, Consolider Program, CIBER Actions are financed by the Instituto de Salud Carlos III with assistance from the European Regional Development Fund. The ICTS "NANOBIOSIS", and particularly the Custom Antibody Service (CAbS, IQAC-CSIC, CIBER-BBN), is acknowledged for the assistance and support related to the immunoreagents used in this work.

Conflicts of Interest: The authors declare no conflict of interest..

References

1. World Health Organization (WHO). Cardiovascular diseases. Available online: http://www.who.int/cardiovascular_diseases/en/ (accessed on 29 July 2018).

2. Yang, Z.; Min Zhou, D. Cardiac markers and their point-of-care testing for diagnosis of acute myocardial infarction. *Clin. Biochem.* **2006**, *39*, 771–780. [CrossRef] [PubMed]

3. Anderson, J.L.; Adams, C.D.; Antman, E.M.; Bridges, C.R.; Califf, R.M.; Casey, D.E., Jr.; Chavey, W.E., II; Fesmire, F.M.; Hochman, J.S.; Levin, T.N.; et al. ACC/AHA 2007 guidelines for the management of patients with unstable angina/non-ST-Elevation myocardial infarction: A report of the American College of Cardiology/American Heart Association Task Force on Practice Guidelines (Writing Committee to Revise the 2002 Guidelines for the Management of Patients With Unstable Angina/Non-ST-Elevation Myocardial Infarction) developed in collaboration with the American College of Emergency Physicians, the Society for Cardiovascular Angiography and Interventions, and the Society of Thoracic Surgeons endorsed by the American Association of Cardiovascular and Pulmonary Rehabilitation and the Society for Academic Emergency Medicine. *J. Am. Coll. Cardiol.* **2007**, *50*, e1–e157. [CrossRef] [PubMed]

4. Savonitto, S.; Ardissino, D.; Granger, C.B.; Morando, G.; Prando, M.D.; Mafrici, A.; Cavallini, C.; Melandri, G.; Thompson, T.D.; Vahanian, A.; et al. Prognostic value of the admission electrocardiogram in acute coronary syndromes. *JAMA* **1999**, *281*, 707–713. [CrossRef] [PubMed]

5. Kaul, P.; Newby, L.K.; Fu, Y.; Hasselblad, V.; Mahaffey, K.W.; Christenson, R.H.; Harrington, R.A.; Ohman, E.M.; Topol, E.J.; Califf, R.M.; et al. Troponin T and quantitative ST-segment depression offer complementary prognostic information in the risk stratification of acute coronary syndrome patients. *J. Am. Coll. Cardiol.* **2003**, *41*, 371–380. [CrossRef]

6. Hamm, C.W.; Braunwald, E. A classification of unstable angina revisited. *Circulation* **2000**, *102*, 118–122. [CrossRef] [PubMed]

7. Kost, G.J.; Tran, N.K. Point-of-care testing and cardiac biomarkers: The standard of care and vision for chest pain centers. *Cardiol. Clin.* **2005**, *23*, 467–490. [CrossRef] [PubMed]

8. Alpert, J.; Thygesen, K.; Antman, E.; Bassand, J. Myocardial infarction redefined—A consensus document of the joint European society of cardiology/American college of cardiology committee for the redefinition of myocardial infarction. *J. Am. Coll. Cardiol.* **2000**, *36*, 959–969. [PubMed]
9. Babuin, L.; Jaffe, A.S. Troponin: The biomarker of choice for the detection of cardiac injury. *Can. Med. Assoc. J.* **2005**, *173*, 1191–1202. [CrossRef] [PubMed]
10. Zethelius, B.; Johnston, N.; Venge, P. Troponin I as a predictor of coronary heart disease and mortality in 70-year-old men: A community-based cohort study. *Circulation* **2006**, *113*, 1071–1078. [CrossRef] [PubMed]
11. Morrow, D.A.; Antman, E.M.; Tanasijevic, M.; Rifai, N.; de Lemos, J.A.; McCabe, C.H.; Cannon, C.P.; Braunwald, E. Cardiac troponin I for stratification of early outcomes and the efficacy of enoxaparin in unstable angina: A TIMI-11B substudy. *J. Am. Coll. Cardiol.* **2000**, *36*, 1812–1817. [CrossRef]
12. Hamm, C.W.; Heeschen, C.; Goldmann, B.; Vahanian, A.; Adgey, J.; Miguel, C.M.; Rutsch, W.; Berger, J.; Kootstra, J.; Simoons, M.L. Benefit of abciximab in patients with refractory unstable angina in relation to serum troponin T. levels. c7E3 fab antiplatelet therapy in unstable refractory angina (CAPTURE) study investigators. *New Engl. J. Med.* **1999**, *340*, 1623–1629. [CrossRef] [PubMed]
13. Eggers, K.M.; Jaffe, A.S.; Lind, L.; Venge, P.; Lindahl, B. Value of cardiac troponin I cutoff concentrations below the 99th percentile for clinical decision-making. *Clin. Chem.* **2009**, *55*, 85–92. [CrossRef] [PubMed]
14. He, N.-Y.; Guo, H.-S.; Gu, C.-R.; Yang, D.; Zhang, J.-N. Chemiluminescent enzyme immunoassay for cardiac troponin I detection—Optimization of experimental parameters. *Chem. J. Chin. Univ. Chin.* **2007**, *28*, 242–245.
15. Cho, I.-H.; Paek, E.-H.; Kim, Y.-K.; Kim, J.-H.; Paek, S.-H. Chemiluminometric enzyme-linked immunosorbent assays (ELISA)-on-a-chip biosensor based on cross-flow chromatography. *Anal. Chim. Acta* **2009**, *632*, 247–255. [CrossRef] [PubMed]
16. Hyytiä, H.; Järvenpää, M.-L.; Ristiniemi, N.; Lövgren, T.; Pettersson, K. A comparison of capture antibody fragments in cardiac troponin I immunoassay. *Clin. Biochem.* **2013**, *46*, 963–968. [CrossRef] [PubMed]
17. Katrukha, A.G.; Bereznikova, A.V.; Filatov, V.L.; Esakova, T.V.; Kolosova, O.V.; Pettersson, K.; Lövgren, T.; Bulargina, T.V.; Trifonov, I.R.; Gratsiansky, N.A.; et al. Degradation of cardiac troponin I: Implication for reliable immunodetection. *Clin. Chem.* **1998**, *44*, 2433–2440. [PubMed]
18. Adams, J.E.; Bodor, G.S.; Dávila-Román, V.G.; Delmez, J.A.; Apple, F.S.; Ladenson, J.H.; Jaffe, A.S. Cardiac troponin I: A marker with high specificity for cardiac injury. *Circulation* **1993**, *88*, 101–106. [CrossRef] [PubMed]
19. Eriksson, S.; Junikka, M.; Laitinen, P.; Majamaa-Voltti, K.; Alfthan, H.; Pettersson, K. Negative Interference in cardiac troponin i immunoassays from a frequently occurring serum and plasma component. *Clin. Chem.* **2003**, *49*, 1095–1104. [CrossRef] [PubMed]
20. Eriksson, S.; Halenius, H.; Pulkki, K.; Hellman, J.; Pettersson, K. Negative interference in cardiac troponin i immunoassays by circulating troponin autoantibodies. *Clin. Chem.* **2005**, *51*, 839–847. [CrossRef] [PubMed]
21. Vdovenko, M.M.; Byzova, N.A.; Zherdev, A.V.; Dzantiev, B.B.; Sakharov, I.Y. Ternary covalent conjugate (antibody-gold nanoparticle-peroxidase) for signal enhancement in enzyme immunoassay. *RSC Adv.* **2016**, *6*, 48827–48833. [CrossRef]
22. Bodor, G.S.; Porter, S.; Landt, Y.; Ladenson, J.H. Development of monoclonal antibodies for an assay of cardiac troponin-I and preliminary results in suspected cases of myocardial infarction. *Clin. Chem.* **1992**, *38*, 2203–2214. [PubMed]
23. Penttilä, K.; Penttilä, I.; Bonnell, R.; Kerth, P.; Koukkunen, H.; Rantanen, T.; Svanas, G. Comparison of the troponin T and troponin, I. ELISA tests, as measured by microplate immunoassay techniques, in diagnosing acute myocardial infarction. *Eur. J. Clin. Chem. Clin. Biochem.* **1997**, *35*, 767–774. [CrossRef] [PubMed]
24. Le Moal, E.; Giuliani, I.; Bertinchant, J.-P.; Polge, A.; Larue, C.; Villard-Saussine, S. Earlier detection of myocardial infarction by an improved cardiac TnI assay. *Clin. Biochem.* **2007**, *40*, 1065–1073. [CrossRef] [PubMed]
25. Abdorahim, M.; Rabiee, M.; Alhosseini, S.N.; Tahriri, M.; Yazdanpanah, S.; Alavi, S.H.; Tayebi, L. Nanomaterials-based electrochemical immunosensors for cardiac troponin recognition: An illustrated review. *TrAC Trends Anal. Chem.* **2016**, *82*, 337–347. [CrossRef]
26. Bhalla, V.; Carrara, S.; Sharma, P.; Nangia, Y.; Suri, C.R. Gold nanoparticles mediated label-free capacitance detection of cardiac troponin I. *Sens. Actuators B Chem.* **2012**, *161*, 761–768. [CrossRef]

27. Billah, M.M.; Hays, H.C.W.; Hodges, C.S.; Ponnambalam, S.; Vohra, R.; Millner, P.A. Mixed self-assembled monolayer (mSAM) based impedimetric immunosensors for cardiac troponin I (cTnI) and soluble lectin-like oxidized low-density lipoprotein receptor-1 (sLOX-1). *Sens. Actuators B Chem.* **2012**, *173*, 361–366. [CrossRef]

28. Tuteja, S.K.; Chen, R.; Kukkar, M.; Song, C.K.; Mutreja, R.; Singh, S.; Paul, A.K.; Lee, H.; Kim, K.-H.; Deep, A.; et al. A label-free electrochemical immunosensor for the detection of cardiac marker using graphene quantum dots (GQDs). *Biosens. Bioelectron.* **2016**, *86*, 548–556. [CrossRef] [PubMed]

29. Venge, P.; Johnston, N.; Lindahl, B.; James, S. Normal plasma levels of cardiac troponin i measured by the high-sensitivity cardiac troponin I access prototype assay and the impact on the diagnosis of myocardial ischemia. *J. Am. Coll. Cardiol.* **2009**, *54*, 1165–1172. [CrossRef] [PubMed]

30. Conzuelo, F.; Ruiz-Valdepeñas, M.V.; Campuzano, S.; Gamella, M.; Torrente-Rodriguez, R.M.; Reviejo, A.J.; Pingarrón, J.M. Rapid screening of multiple antibiotic residues in milk using disposable amperometric magnetosensors. *Anal. Chim. Acta* **2014**, *820*, 32–38. [CrossRef] [PubMed]

31. Esteban-Fernández de Ávila, B.; Escamilla-Gómez, V.; Campuzano, S.; Pedrero, M.; Salvador, J.P.; Marco, M.P.; Pingarrón, J.M. Ultrasensitive amperometric magnetoimmunosensor for human C-reactive protein quantification in serum. *Sens. Actuators B Chem.* **2013**, *188*, 212–220. [CrossRef]

32. Pinacho, D.; Sánchez-Baeza, F.; Pividori, M.-I.; Marco, M.-P. Electrochemical detection of fluoroquinolone antibiotics in milk using a magneto immunosensor. *Sensors* **2014**, *14*, 15965–15980. [CrossRef] [PubMed]

33. Bruls, D.M.; Evers, T.H.; Kahlman, J.A.; van Lankvelt, P.J.; Ovsyanko, M.; Pelssers, E.G.; Schleipen, J.J.; de Theije, F.K.; Verschuren, C.A.; van der Wijk, T.; et al. Rapid integrated biosensor for multiplexed immunoassays based on actuated magnetic nanoparticles. *Lab Chip* **2009**, *9*, 3504–3510. [CrossRef] [PubMed]

34. Dittmer, W.U.; Evers, T.H.; Hardeman, W.M.; Huijnen, W.; Kamps, R.; de Kievit, P.; Neijzen, J.H.; Nieuwenhuis, J.H.; Sijbers, M.J.; Dekkers, D.W.; et al. Rapid, high sensitivity, point-of-care test for cardiac troponin based on optomagnetic biosensor. *Clin. Chim. Acta* **2010**, *411*, 868–873. [CrossRef] [PubMed]

35. Liang, W.; Li, Y.; Zhang, B.; Zhang, Z.; Chen, A.; Qi, D.; Yi, W.; Hu, C. A novel microfluidic immunoassay system based on electrochemical immunosensors: An application for the detection of NT-proBNP in whole blood. *Biosens. Bioelectron.* **2012**, *31*, 480–485. [CrossRef] [PubMed]

36. Tang, D.; Hou, L.; Niessner, R.; Xu, M.; Gao, Z.; Knopp, D. Multiplexed electrochemical immunoassay of biomarkers using metal sulfide quantum dot nanolabels and trifunctionalized magnetic beads. *Biosens. Bioelectron.* **2013**, *46*, 37–43. [CrossRef] [PubMed]

37. Yang, Z.; Wang, H.; Guo, P.; Ding, Y.; Lei, C.; Luo, Y. A multi-region magnetoimpedance-based bio-analytical system for ultrasensitive simultaneous determination of cardiac biomarkers myoglobin and C-reactive protein. *Sensors* **2018**, *18*, 1765. [CrossRef] [PubMed]

38. Ballesteros, B.; Barceló, D.; Camps, F.; Marco, M.-P. Preparation of antisera and development of a direct enzyme-linked immunosorbent assay for the determination of the antifouling agent Irgarol 1051. *Anal. Chim. Acta* **1997**, *347*, 139–147. [CrossRef]

39. Baines, M.G.; Thorpe, R. Purification of immunoglobulin G (IgG). In *Immunochemical Protocols*; Manson, M.M., Ed.; Humana Press: Totowa, NJ, USA, 1992; pp. 79–104.

40. Filatov, V.L. Troponin: Structure, properties, and mechanism of functioning. *Biochemistry* **1999**, *64*, 969–985. [PubMed]

41. Park, H.; Hwang, M.P.; Lee, K.H. Immunomagnetic nanoparticle-based assays for detection of biomarkers. *Int. J. Nanomed.* **2013**, *8*, 4543–4552. [CrossRef]

42. Cohen-Tannoudji, L.; Bertrand, E.; Baudry, J.; Robic, C.; Goubault, C.; Pellissier, M.; Johner, A.; Thalmann, F.; Lee, N.K.; Marques, C.M.; et al. Measuring the kinetics of biomolecular recognition with magnetic colloids. *Phys. Rev. Lett.* **2008**, *100*, 108301. [CrossRef] [PubMed]

43. Wu, H.; Liu, G.; Wang, J.; Lin, Y. Quantum-dots based electrochemical immunoassay of interleukin-1α. *Electrochem. Commun.* **2007**, *9*, 1573–1577. [CrossRef]

44. Valera, E.; García-Febrero, R.; Pividori, I.; Sánchez-Baeza, F.; Marco, M.P. Coulombimetric immunosensor for paraquat based on electrochemical nanoprobes. *Sens. Actuators B Chem.* **2014**, *194*, 353–360. [CrossRef]

45. Valera, E.; Muriano, A.; Pividori, I.; Sánchez-Baeza, F.; Marco, M.P. Development of a Coulombimetric immunosensor based on specific antibodies labeled with CdS nanoparticles for sulfonamide antibiotic residues analysis and its application to honey samples. *Biosens. Bioelectron.* **2013**, *43*, 211–217. [CrossRef] [PubMed]

46. Hansen, J.A.; Wang, J.; Kawde, A.-N.; Xiang, Y.; Gothelf, K.V.; Collins, G. Quantum-dot/aptamer-based ultrasensitive multi-analyte electrochemical biosensor. *J. Am. Chem. Soc.* **2006**, *128*, 2228–2229. [CrossRef] [PubMed]

47. Zethelius, B.; Berglund, L.; Sundström, J.; Ingelsson, E.; Basu, S.; Larsson, A.; Venge, P.; Arnlöv, J. Use of multiple biomarkers to improve the prediction of death from cardiovascular causes. *N. Engl. J. Med.* **2008**, *358*, 2107–2116. [CrossRef] [PubMed]

Article

Biosensing System for Concentration Quantification of Magnetically Labeled *E. coli* in Water Samples

Anna Malec [1], Georgios Kokkinis [1], Christoph Haiden [1] and Ioanna Giouroudi [1,2,*]

[1] Institute of Sensor and Actuator Systems, TU Wien, Gusshausstrasse 27–29, 1040 Vienna, Austria; annmalec@gmail.com (A.M.); georgios.kokkinis@tuwien.ac.at (G.K.); christoph.haiden@tuwien.ac.at (C.H.)
[2] BioSense Institute, Dr Zorana Đinđića 1, 21000 Novi Sad, Serbia
* Correspondence: ioanna.giouroudi@biosense.rs or ioanna.giouroudi@gmail.com; Tel.: +381-63-823-0-113

Received: 20 June 2018; Accepted: 9 July 2018; Published: 12 July 2018

Abstract: Bacterial contamination of water sources (e.g., lakes, rivers and springs) from waterborne bacteria is a crucial water safety issue and its prevention is of the utmost significance since it threatens the health and well-being of wildlife, livestock, and human populations and can lead to serious illness and even death. Rapid and multiplexed measurement of such waterborne pathogens is vital and the challenge is to instantly detect in these liquid samples different types of pathogens with high sensitivity and specificity. In this work, we propose a biosensing system in which the bacteria are labelled with streptavidin coated magnetic markers (MPs—magnetic particles) forming compounds (MLBs—magnetically labelled bacteria). Video microscopy in combination with a particle tracking software are used for their detection and quantification. When the liquid containing the MLBs is introduced into the developed, microfluidic platform, the MLBs are accelerated towards the outlet by means of a magnetic field gradient generated by integrated microconductors, which are sequentially switched ON and OFF by a microcontroller. The velocities of the MLBs and that of reference MPs, suspended in the same liquid in a parallel reference microfluidic channel, are calculated and compared in real time by a digital camera mounted on a conventional optical microscope in combination with a particle trajectory tracking software. The MLBs will be slower than the reference MPs due to the enhanced Stokes' drag force exerted on them, resulting from their greater volume and altered hydrodynamic shape. The results of the investigation showed that the parameters obtained from this method emerged as reliable predictors for *E. coli* concentrations.

Keywords: magnetic labeling; magnetic microparticles; magnetophoresis; bacteria quantification; biosensing

1. Introduction

Microbial pathogen detection is of utmost priority for water quality control since microbial contamination threatens the health and well-being of wildlife, livestock, and human populations and can lead to serious illness and even death. Most waterborne pathogens are introduced into water supplies by human or animal urine and faeces (enteric pathogens), but they can also exist naturally in water environments as indigenous aquatic micro-organisms and the toxic compounds they produce. Traditional water monitoring techniques are typically still based on laboratory analyses of representative field-collected samples; this necessitates considerable effort and expense, and the sample may change before analysis [1]. Furthermore, currently available equipment is so large that it cannot usually be made portable. Therefore, there is an increasing demand of robust and efficient techniques of contaminants detection and of monitoring devices that save tremendous amounts of time, reagent, and sample if it is installed at contaminated sites. However, the current state of the art often fails to address same day measurements combined with multiple detection analyses and the ability to quantify the contaminants concentration in the highly diluted water samples of these water sources [2–4]. Moreover, major challenges that limit commercialization are instrumental complexity and large data mining capabilities.

Microfluidic systems [5–10] that use magnetic fields for performing the aforementioned tasks have provided a promising alternative that can fulfill the increasing requirements of such portable robust devices [11–15]. This is because magnetic fields can be well tuned and applied either externally or from a directly integrated solution in the biosensing system. In combination with these applied magnetic fields, magnetic nanoparticles are utilized. Magnetic nanoparticles have the advantage of manipulating them inside microfluidic channels by utilizing high gradient magnetic fields and their flexibility due to functionalization by means of surface modification and specific binding [16]. Specifically, these methods involve the labeling of the biological entity with magnetic particles (MPs).

In this paper, we present the proof of concept for a biosensing system able to monitor the presence of magnetically labeled *E. coli* and provide same day information on the bacterial concentration in the water sample.

2. Design and Methods

The system aims to detect and, most importantly, to quantify the amount of *E. coli* present within a water sample. The proposed platform is designed in such a manner so as to be applied not only for the quantification of *E. coli* (which is exclusively tested in this work) but also for any other disease-causing micro-agent, if one uses the appropriate antibodies and magnetic particles with a size relative to the analyte to be detected.

The innovative aspect of the proposed device is that it utilizes a single particle tracking system that analyzes the dynamics and volumetric changes of an MP after microorganisms are bound to its functionalized surface. The attachment of the microorganisms on the surface of the MPs is ensured by an appropriate biological binding protocol. The protocol efficiency is optimized by adjusting various parameters and conditions. Next, the measurements are conducted, provided that the labeling protocol is consistent (i.e., the amount of liquid, ratios, and handling of the sample and environmental conditions remain unchanged). Only the *E. coli* concentration is altered, which, in turn, is expected to increase/decrease the number of *E. coli* attached to the functionalized MPs (see Figure 1) and affect the particle's dynamics.

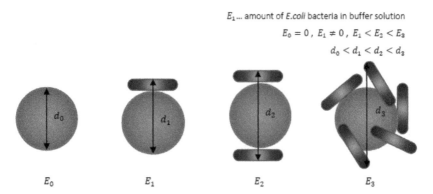

E_1... amount of *E.coli* bacteria in buffer solution

$$E_0 = 0, \ E_1 \neq 0, \ E_1 < E_2 < E_3$$
$$d_0 < d_1 < d_2 < d_3$$

Figure 1. Schematic representation of the volumetric changes of a functionalized magnetic particle (MP) with respect to the number of *E. coli* bacteria attached to its surface. The amount of the attachments depends on the amount of *E. coli* bacteria present in a buffer solution. The diameter d_0 corresponds to a reference diameter of a plain, not functionalized, unloaded MP without any *E. coli* bacteria attached. The variables d_1, d_2 and d_3 indicate the sizes of the compounds formed (magnetically labelled bacteria (MLB)) when the concentration of *E. coli* bacteria within the water solution increased.

The change in shape/volume and the resultant surface chemistry depend not only on the size/type of the attached pathogen but also on the formation of other layers/coatings arising from the procedure followed during the binding protocol. Depending on the number of bacteria attached, the behavior and the dynamics of the MLBs suspended in a liquid would change. The particle dynamics are investigated by applying an external stimulus (a magnetic field gradient) and the particle's response is analyzed. The liquid in which the reference MPs and MLBs is suspended is static. The velocity pattern of the MLBs and the reference MPs when accelerated from the inlet to the outlet by the application of the magnetic field gradient defines the concentration of *E. coli* present in the water sample.

The bacteria quantification principle is based on the decreased MLB's velocity due to inhibiting factors such as the Stokes' drag force and the altered hydrodynamic shape of the initial, functionalized MP after bacteria are attached to its surface [17–27]. A microfluidic platform with integrated microconductors (MCs) and a polydimethylsiloxane (PDMS) channel is used to manipulate the motion of the reference MPs and that of the MLBs. The reference channel is filled with the water sample containing plain, non-functionalized reference MPs and the measurement channel is filled with the water sample containing the MLBs. Experiments were conducted provided that the suspension liquid is static (no flow conditions, $u = 0$). By controlling the direct current (DC) through the microconductors (switching it ON and OFF by means of a microcontroller), a magnetic field gradient at each conductor at a time is created and the MPs and MLBs are set in motion at the same time. The measurement begins as follows; when the MPs and MLBs reach the position where the magnetic field is stronger (i.e., it is captured on the conductor where $I \neq 0$), the current is switched OFF [17,20,21]. This procedure is repeated for each adjacent MC and continues until the MPs and MLBs travel the entire distance from the inlet (MC 1) to the outlet (MC 9) of the chip as seen in Figure 2.

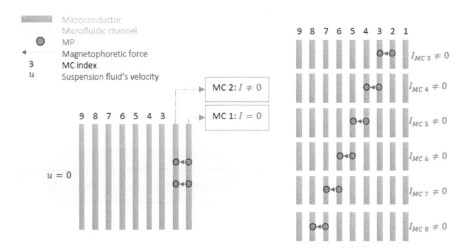

Figure 2. Schematic of the developed reference microfluidic channel with the integrated microconductors (MCs). A MP is manipulated from the right to the left by switching the current ON and OFF on adjacent MCs. The magnetic force is always collinear to the gradient of the magnetic field.

The operating principle of this microfluidic platform is described in detail in [3]. The behavior and dynamics of differently loaded MPs are observed with a video fluorescence microscope. The magnetically labeled *E. coli* bacteria are also labeled with a fluorophore for imaging purposes. The recorded video is converted to gray-scaled frames.

2.1. Particle Tracking Principle

A MATLAB script (2011b, MathWorks, Natick, MU, USA) utilizing 2D particle tracking (Crocker–Grier algorithm [28]) is used to link the exact position (centroids) of objects appearing over subsequent frames [29,30]. Based on this approach, a single frame i ($i = 1, \ldots, N$) is processed to detect multiple bright spots (representing particles) over a dark background (here the variability in pixel intensity is used to 'find' a particle) (see Figure 3b). This procedure is repeated (for all the frames; N is the total number of all frames) giving x_i- and y_i-coordinates (of each individual particle's centroid for each frame), which are then linked together to form trajectories [31].

Once particles are detected and their positions are known (for an entire sequence of video images), their locations are matched with successive and proceeding frames (i.e., it is determined which particle in a given frame most likely corresponds to the particle in the preceding frame). In order to avoid linking errors, parameters such as threshold, mask, noise length, object length, linking distance, minimum trajectory length (see Section below) must be properly adapted (i.e., when changing the magnification of an optical system or when using differently sized magnetic micromarkers these parameters must be checked again).

Figure 3. Magnetic micromarkers appear as bright spots on a dark background image. (**a**) multiple microparticles captured on a single frame i and (**b**) close-up for two particles with centroids assigned to each particle. For every frame, the coordinates x_i and y_i of the particles' centroids are calculated.

During recording of a particle's dynamics, an important parameter is the camera frame rate. It defines a time step Δt between successive frames. This parameter is given in frames per seconds (fps). Next, after selecting the desired output frames for further analysis (a combination of frames converted from multiple videos is also possible), a preview is provided (see Figure 4) to enable adjusting of parameters for a detection procedure. Proper adjustment of the parameters highlighted in Figure 4 in the graphical user interface (GUI) box on the left is crucial when recognizing an MP-*E. coli* complex as one, single complex (see Figure 5B) and not as two separate particles (see Figure 5A). At the same time, the parameters should be chosen in such a manner that two neighboring micromarkers are recognized as separate (see Figure 5C,D).

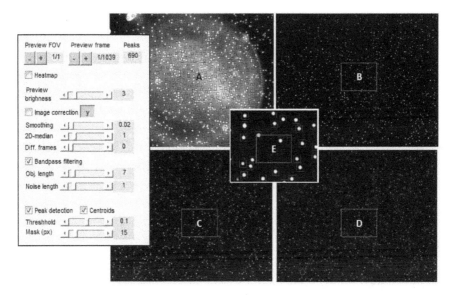

Figure 4. Graphical user interface (GUI) enabling preview of a selected frame for proper configuration of the detection parameters. This particular frame was taken from a video that recorded the Brownian motion of unloaded Dynabeads^{TM} M-270 Carboxylic Acid. There are 690 particles detected on this frame; (**A**) is an original video frame before noise removal by applying bandpass filtering; (**B**) is an image with increased brightness; (**C**) shows a preview with detected particles (embedded in a green squared mask); (**D**) is the preview of the image with visible particles' centroids (multiple red points). Regular spherical shapes of particles can be seen on the zoomed picture (**E**).

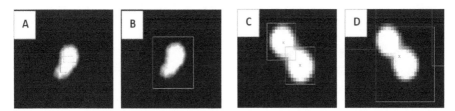

Figure 5. Particle detection and MP-*E. coli* complex recognition depend on proper adjustment of the detection parameters. Pictures (**A,B**) represent a loaded MP (MLB). Pictures (**C,D**) represent unloaded MPs: In (**A**), MLB is recognized as two separate objects (MP and *E. coli* are recognized separately) while, in picture (**B**), the complex is identified as a whole. In (**C**), two neighboring particles are distinguished as separate objects; on (**D**), they are marked as one. Proper adjustment of parameters must be done to obtain reliable results.

Apart from the detection parameters, other user-defined variables must be set to specify the linking procedure:

- Minimum and maximum size are introduced to specify the size range of particles whose trajectories are further analyzed (i.e., micromarker size range)
- Resolution—enables conversion of pixel units into real dimensions in micrometers; information about resolution must be provided while executing the additional scripts and calculations
- Time step Δt [in seconds]—is the time between two consecutive frames and is given by the camera frame rate (for 25 fps $\Delta t = 0.04$ s)

- minimum and maximum velocity—introduced to avoid the analysis of unwanted particles (e.g., foreign objects/dirt which experience different motion than the particles of interest)

To compare the dynamics of manipulated MPs, it is necessary to further process the raw data that were obtained from the tracking software routine. Additional scripts for data representation and velocity calculations were written in MATLAB. Scripts were run for each tracking attempt and motion patterns of particles were investigated and compared. Figure 6 shows exemplary displacements of one plain MP. The direction of movement is from left to right, which can be concluded from the right-side plot. There are visible characteristic points on the trajectory that correspond to the moments when an MP was captured on the MC (i.e., time intervals when the particle is not in motion $v_{MP} \approx 0$). Similar plots were drawn for loaded MPs.

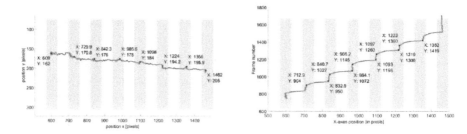

Figure 6. Method for representation of an MP's motion over a sequence of frames for a single field of view (FOV). The figure on the left shows the trajectory of an individual MP with respect to x and y coordinates in (pixels). The particle is manipulated from the most left MC to the right side. Tracking starts when the MP position is at (X: 600, Y: 162) and ends on (X: 1462, Y: 205). There are visible changes in trajectory which indicate that the MP was captured at the MC (estimated x and y positions of capture are given in data boxes). Another representation of the MPs motion is shown in the figure on the right. Here, only the motion in x-direction (i.e., the direction of the magnetic field gradient) over the sequence of succeeding frames is plotted.

2.2. Chip Design

An array of nine rectangular, parallel MCs was fabricated on a silicon wafer, below the microfluidic channels, which was wire-bonded to a printed circuit board (PCB). On top of the chip, two microfluidic channels (reference and measurement channels) made of PDMS were fabricated that were positioned perpendicularly to the MCs' array. This configuration facilitated the flow of MPs across the MCs. The current on the MC was switched ON and OFF by means of a microcontroller. The dimensions of nine parallel rectangular MCs were selected in a way to obtain a magnetic field gradient sharp enough to move the MPs and the MLBs. The height of the MC was 1 µm (500 nm of Au and 500 nm of SiO$_2$). The MCs' dimensions were: 10 µm in width separated by 8 µm gap. The simulations as well as the explanation as to why these particular values were chosen can be found in previous works [22,32]. In short, setting an upper temperature limit at 323 K, the numerical simulation indicated a maximum current density of 9 A/m^2. For the simulations in [22,32], a maximum current of 100 mA was applied in a sequential pattern and a time dependent solution was calculated and was reported in [22]. The MCs were fabricated on a 500 µm thick silicon wafer. The application of an insulation layer was essential since the chip's surface was in direct contact with the biological liquid. This layer however does not inhibit the bacterial bioadhesion to the chip (silicon-based materials are very susceptible to bacterial biofouling [33]). This problem was resolved by applying a second layer (biofilm) on the chip's surface as described below.

2.2.1. Fabrication of Microconductors

The MCs were fabricated as follows: a silicon wafer served as the bottom substrate on which an image reversal photoresist was spin-coated (AZ5214, intended for lift-off-techniques). Next, Aquatar (anti-reflecting coating) was spin-coated onto the resist film. The photoresist was exposed to ultraviolet (UV) light through the 1st mask reversal-baked. This treatment caused a reaction that resulted in cross-linking the exposed areas while unexposed areas remained photoactive. The second (flood) exposure without the mask was prepared and the photoactive areas were dissolved in the developer. Afterwards, first a Ti (titanium) adhesion layer, then the Au (gold) and lastly the Cr (Chromium) films were deposited all over the surface by thermal evaporation. The unwanted parts were removed by lifting the remaining photoresist off. On the resulting microstructure, the insulating layer SiO_2 was formed by means of plasma enhanced chemical vapor deposition (PECVD). Here, the positive photoresist AZ6624 and Aquatar were spin-coated. The structure was exposed through the designed mask. The exposed parts of the photoresist were dissolved. Afterwards, part of the passivation SiO_2 layer was removed by oxygen etching. The remaining photoresist was stripped off. As a result, the structure consisting of the insulated MCs and the uninsulated pads was obtained.

2.2.2. Microfluidic Channel

Plasma enhanced chemical vapor deposition was chosen for the microfluidic channel for its transparency, which was required for the optical microscope monitoring and software tracking. Moreover, PDMS is biocompatible and adhered to the chip's surface without slipping and in a reversible manner. Due to its elasticity and inertness, the channel's structure remained undamaged when applied on or removed from the chip. Therefore, this material could be used multiple times provided that any contaminants were cleaned after each operation. The dimensions of the microfluidic channel were selected in a way to be large enough to diminish the influence of unwanted factors (height = 110 μm, width in the middle where the measurement takes place = 500 μm while width of the channel from the inlet and towards the outlet = 90 μm, length = 50 mm). That is, to ignore the effects of interfacial flow turbulences due to the chemistry and the hydrophobicity of the channel's wall. Moreover, the channel's dimensions were adjusted to study the dynamics of multiple MPs simultaneously. The fabrication process of the microfluidic channel was as follows; on a glass wafer substrate, the negative type dry film photoresist (Ordyl SY 300 (Elga Europe, Milan, Italy)) was laminated. The mask was aligned and the light-sensitive material was exposed to the UV light. The exposed areas of the photoresist were hardened while the unexposed were dissolved in an Ordyl SY developer and removed. The obtained structure served as a mold on which liquid PDMS was slowly poured. This viscous mixture was composed of ten base units (Sylgard 184 (Sigma Aldrich, Munich, Germany)) and one unit of a curing agent and it was hardened by heating at 70 °C for 1 h on the hot plate. The obtained elastomeric PDMS channel was slowly pulled off from the mold. Lastly, the PDMS material was punched with needles in order to provide access to the channel (i.e., inlet and outlet) for the sample injection.

2.2.3. Surface Modification

Surface modification was necessary to avert unwanted bacteria (or protein) interactions (e.g., adhesion) at the chip's surface [17]. One of the strategies to avoid the adhesion of the bacteria was to apply a chemical modification on the chip with an outermost sodium alginate (SA) layer. SA exhibits a 'brush like repulsive structure' that keeps bacteria apart [34]. Polyethylene (POI) together with SA is layered over the insulation layer based on the layer-by-layer (LBL) electrostatic self-assembly (ESA). Treated with oxygen plasma, as described below, the SiO_2 passivation layer becomes negatively charged, which enables attraction of positively charged polyethyleneimine (PEI). The latter one due to the cationic character attracts the anionic SA.

Specifically, the surface was functionalized by means of plasma etching which improves adhesion properties prior to coating (i.e., oxygen plasma encourages hydroxylation [35], which allows binding of the next layer via reactive −OH groups). In order to complete the surface modification steps, the chip was:

1. Rinsed with acetone, isopropanol and deionized (DI) water to remove contaminants
2. Dried for 30 min at 150 °C at a hot plate
3. Oxygen Plasma etching (here hydroxylation of the SiO_2 passivation layer took place)
4. Dipped for 10 min in the branched, PEI 2 gL^{-1} solution that served as an adhesion promoter and rinsed with DI water
5. Dipped in sodium alginate also for 10 min 2 gL^{-1}, rinsed in DI water and dried carefully with Nitrogen

This modification applies only to the surface of the chip. The PDMS channel does not have to undergo this process since: (a) MPs are not in close contact with the microfluidic channel walls and (b) the material has an intrinsic high hydrophobicity that results in inhibition of bacterial adhesion itself.

3. Experiments

3.1. Experimental Set-Up

The experimental set-up is shown in Figure 7. The microfluidic platform, placed under the optical microscope, was connected to the electronic breadboard and a DC power supply. The videos were recorded using a Nikon Camera D5100 and stored on the PC where they were later processed using the 'particle tracking software'. In order to avoid error displacements, the camera was tightly screwed to the microscope C-mount adapter.

In these experiments, we proved the working principle, fluorescence microscopy was employed to obtain the strong image contrast between the MP-*E. coli* complex and the background and ensure the "proof of concept", but this step is not necessary for future measurements. A high-intensity light source (high-pressure mercury vapor arc-discharge lamp) was used to evoke the sufficient photon excitation from the fluorophore (Alexa Fluor from the secondary antibody). The proper lens was selected to have high adequate magnification and a long enough working distance (here, the height of the microchannel is of relevance). A close-up of the microfluidic platform placed under the microscope can be seen in Figure 7. The same objective was used for all the measurement sets: A Plan Fluor objective with 2.6−1.8 mm working distance and 60× magnification. The field of view (FOV) was adjusted in order to eliminate the tracking errors that come from the scattering of light on the channel's wall.

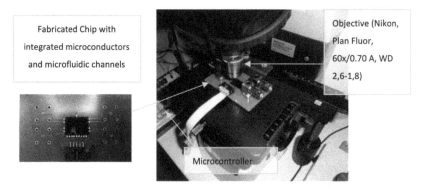

Figure 7. The microfluidic platform placed underneath the fluorescence microscope stage under 60× magnification, Plan Fluor objective with 2.6–1.8 mm working distance. After the measurement set with varying known *E. coli* concentrations was conducted, the new chip was sealed to the same PCB (printed circuit board) and measurements were repeated. WD: working distance.

The sample is applied by pipetting a drop (2 µL) of the liquid into the channel's inlet. Under-pressure was applied at the outlet in order to fill the channel and the excess was removed with swabs. The no-flow condition is checked under the microscope. After each measurement, the PDMS channel was rinsed with Acetone, Isopropanol and DI water and dried with Nitrogen.

3.2. Magnetic Markers

For these experiments magnetic markers Dynabeads™ M-280 coated with streptavidin (ThermoFisher, Waltham, MA, USA) were used to label the biological target (wild type K-12 *E. coli* strain) by means of biotinylated polyclonal antibodies (rabbit anti-*E. coli* Abcam@ ab20640). The loading of the magnetic marker was achieved by strong noncovalent bonds between the streptavidin layer of the marker and the biotin molecules attached on the surface of the antibodies (such that the biotin-streptavidin lock-and-key coupling system can only be broken under harsh conditions: pH 4, high temperature or salt concentration).

3.3. E. coli Sample Preparation

K-12 wild-type *Escherichia coli* bacteria were cultured on a plastic disposable petri dish layered with a solid plain nutrient AGAR (derived from the polysaccharide agarose) with growth temperature at 37 °C and storage temperature at 3 °C. Four bacteria concentrations, ($c = 10^3$ CFU/mL, 2×10^3 CFU/mL, 3×10^3 CFU/mL, 4×10^3 CFU/mL), were washed three times (washing; initial suspension was centrifuged for 8 min at 4.7×10^3 rpm, the supernatant was removed, the sedimented bacteria were resuspended in 1 mL 0.01M Phosphate-buffered saline (PBS)-Tween20 (0.01% v/v) and vortexed for 1 min at 10^4 rpm). In addition, 20 mL of this *E. coli* suspension was mixed with 7 µL of the original antibody concentration and incubated for 1 h on a multiple rotator (room temperature) to yield binding. Afterwards, the sample was again washed five times and re-suspended in 100 mL of 0.01 M PBS–BSA (bovine serum albumin) (0.1% w/v). Multiple washing was conducted throughout the sample preparation so as to eliminate the risk of unspecific binding. Simultaneously, 100 mL of original concentration Dynabeads™ M-280 Streptavidin (MPs) were magnetically washed and vortexed three times in 1 mL 0.01 M PBS–Tween 20 (0.01% v/v) then condensed back to 100 mL. Afterwards, the two samples were combined: 1 µL of washed MPs and 40 µL of the previously prepared complex (*E. coli*-Ab20640), the sample was left for 1 h to incubate on a multiple rotator to induce uniform biotin-streptavidin binding along the antibody-MP suspension. Next, a PBS-BSA washing buffer was added (to dilute the sample and facilitate magnetic washing) and the vial was left on a magnetic stand for approximately ≈1.5 min (during this time, MPs were attracted on the side wall towards the magnet). In the last step of the binding protocol, the supernatant was carefully discarded from the vial and the *E. coli*-loaded-MPs that were re-suspended in 100 mL PBS-BSA.

4. Results and Discussion

A DC power supply supplied the MCs with a constant DC voltage value of 12 V. The current in every individual MC was 53 mA. The peak force exerted on a particle is approximately 1 pN. The sample with plain, non-functionalized MPs was prepared and the MPs were additionally coupled with primary and secondary antibodies. Furthermore, the secondary antibodies were additionally labeled with a fluorophore for visualization purposes to obtain images with high contrast (bright spots over the dark background), which was essential for the tracking software.

Once the particles were injected in the PDMS channel and the 'no-flow' condition was satisfied, the manipulation started. After obtaining videos free of artifacts, the sample could be used in the software tracking procedure (see Section 2.1). Figure 8 represents all tracked particles in the reference channel, without *E. coli* attached to their surface, within 1800 frames (i.e., the centroids of bright spots that correspond to particles). Frames were collected at 25 fps (with high quality' and 'high sensitivity' settings) and the frames' dimensions were 1920 × 1080 pixels.

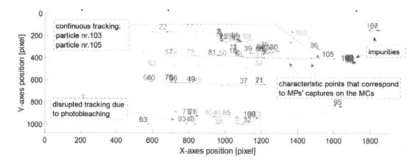

Figure 8. Positions for all detected and tracked particles, without *E. coli* attached to their surface, over the sequence of frames (1920 × 1080 pixels) collected at 25 fps. The particles were manipulated horizontally. There are some characteristic points visible on these trajectories. They correspond to the position of the MP as it was captured at a MC. There are over 105 trajectories detected for this sample.

The MPs were set in motion horizontally i.e., along the *x*-axes of the frame (for a width of 1920 pixels). The pixels were converted into real dimensions by applying the previously calculated resolution (*Res* = 7.2 pix/μm). The displacement of the tracked MPs (without *E. coli*) over a sequence of frames is presented in Figure 9.

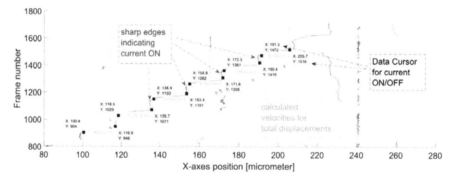

Figure 9. The displacement of the tracked MPs over a sequence of frames without any *E. coli* attached to their surface. There are two characteristic manipulation paths for: MP *nr*. 103 (green) and MP Nr. 105 (dark red).

From this graph, the estimated MPs' positions, corresponding to the beginning and the end of motion, could be extracted (i.e., these are the sharp edges on the graph that indicate the current switching). From these positions (i.e., data seen in the Data Cursors in Figure 9), the mean velocities for the displacement (i.e., between two neighboring MCs) were calculated using a MATLAB script.

The plots in Figures 8 and 9 were drawn for all of the following samples that contained particles attached to known varying concentrations of *E. coli* (i.e., first, the plot of the positions for tracked MPs in pixels along the *x*- and *y*-axis was drawn and then the displacement of the tracked MPs over the sequence of frames). All the samples were successfully manipulated and tracked. The developed MATLAB script was used to calculate the velocities for all of the samples with varying concentrations $c = 10^3$ CFU/mL, 2×10^3 CFU/mL, 3×10^3 CFU/mL, 4×10^3 CFU/mL. Figures 10 and 11 present the plots drawn for *E. coli* concentration of 2×10^3 CFU/mL.

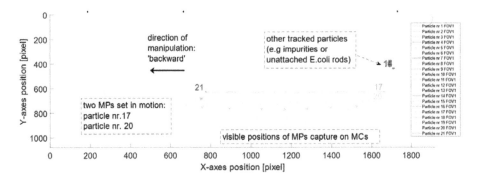

Figure 10. For MLBs with concentration of 2×10^3 CFU/mL: Positions for all the detected and tracked MLBs over a sequence of frames (1920 × 1080 pixels) recorded at 25 fps. There are some characteristic points visible on these trajectories. They correspond to the position of the MLBs as they were captured above the MCs. There are over 21 particles (including impurities) detected for this sample. MLB Nr. 17 and MLB Nr. 20 correspond to two different MLBs that were set in motion due to the magnetic field gradient.

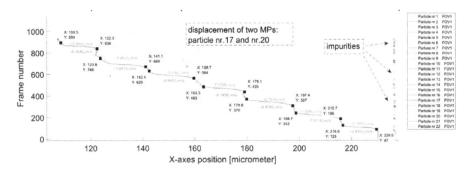

Figure 11. For MLBs with known concentration of 2×10^3 CFU/mL: the displacement of tracked particles over the sequence of frames. There are two characteristic manipulations of MLBs: MLB Nr. 17 (yellow) and MLB Nr. 20 (blue). The rest of the tracked particles are impurities (on the right) and can be clearly differentiated from MLB.

For this sample, two MLBs were present on the recorded FOV. The outputs of the tracking routine were excellent due to the uninterrupted trajectories. There were only a few impurities detected. Such a good quality of the tracking process is a result of the high fluorescence intensity of the MLBs, sufficient sample washing and setting appropriate parameters on the software (as a result no background noise was tracked). The plots of the movement along the *x*-axis for two MLBs overlap, which is accurate because both of the MLBs are loaded with the same *E. coli* amounts.

After the analysis and calculations of the velocities for all samples the data were represented in the form of a box plot (see Figure 12). All of the MPs (reference channel) and MLBs (measurement channel) were manipulated with the same current value of 53 mA. A decrease in velocity as a response to an increased *E. coli* concentration was observed.

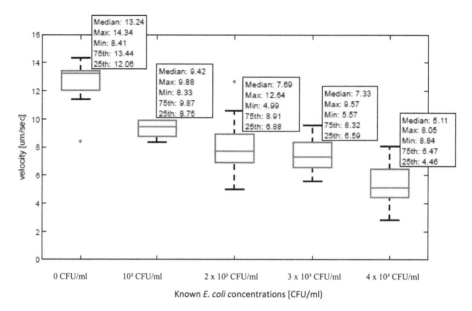

Figure 12. The velocities of the manipulated MPs and MLBs for different *E. coli* concentrations.

5. Conclusions

The results obtained indicate that the developed biosensing system is a less complex, real-time approach for the detection of biological markers and pathogens suspended in a small volume liquid sample. In our future research, we intend to conduct experiments to determine whether the proposed technique is applicable not only for *E. coli*, but also for other bacteria as this approach has the potential to be applied not only for detection of *E. coli*, but also for a variety of microorganisms if the appropriate biological binding/labeling protocol is provided as well as for obtaining information about dynamics in life cycle of single/multiple microorganism(s) in response to condition/environmental changes. As this work was a proof of principle, a lab microscope was used to verify the feasibility of the system. Portable solutions have already been presented in [30,31] and it is the aim of future work to miniaturize the system for fluorescence single particle tracking in order for it to be transportable for in-field measurements.

Author Contributions: Data curation, G.K.; Formal analysis, A.M.; Funding acquisition, I.G.; Investigation, A.M.; Methodology, C.H., G.K.; Project administration, I.G.; Software, C.H., A.M.; Supervision, G.K., I.G.; Validation, C.H.; Visualization, A.M.; Writing—original draft, I.G.

Funding: This research was partially funded by the Austrian Science Fund (FWF) Stand-Alone Project 'Magnetic Microfluidic Biosensor for the Detection and Quantification of Biomolecules' with Project No. P28544-N30. This paper is part of the ANTARES project that has received funding from the European Union's Horizon 2020 research and innovation programme under the European Union's Horizon 2020 research and innovation programme with grant agreement No. 664387.

Conflicts of Interest: The authors declare no conflict of interest.

References

1. Jang, A.; Zou, Z.; Lee, K.K.; Ahn, C.H.; Bishop, P.L. State-of-the-art lab chip sensors for environmental water monitoring. *Meas. Sci. Technol.* **2011**, *22*. [CrossRef]

2. Maas, M.B.; Perold, W.J.; Dicks, L.M.T. Biosensors for the Detection of *Escherichia coli*. *Water SA* **2017**, *43*. [CrossRef]
3. Giouroudi, I.; Kokkinis, G. Recent Advances in Magnetic Microfluidic Sensors. *Nanomaterials* **2017**, *7*, 171. [CrossRef] [PubMed]
4. Giouroudi, I.; Keplinger, F. Microfluidic Biosensing Systems using Magnetic Nanoparticles. *Int. J. Mol. Sci.* **2013**, *14*, 18535–18556. [CrossRef] [PubMed]
5. Berthier, J.; Silberzan, P. *Microfluidics for Biotechnology*; Artech House: Norwood, MA, USA, 2010; ISBN 13-978-1-59693-443-6.
6. Sackmann, E.K.; Fulton, A.L.; Beebe, D.J. The present and future role of microfluidics in biomedical research. *Nature* **2014**, *507*, 181–189. [CrossRef] [PubMed]
7. Nguyen, N.-T.; Wereley, S.T. *Fundamentals and Applications of Microfluidics*; Artech House: Norwood, MA, USA, 2002.
8. Beebe, D.J.; Mensing, G.A.; Walker, G.M. Physics and Applications of Microfluidics in Biology. *Annu. Rev. Biomed. Eng.* **2002**, *4*, 261–286. [CrossRef] [PubMed]
9. Yager, P.; Edwards, T.; Fu, E.; Helton, K.; Nelson, K.; Tam, M.R.; Weigl, B.H. Microfluidic diagnostic technologies for global public health. *Nature* **2006**, *442*, 412–418. [CrossRef] [PubMed]
10. Chin, C.D.; Laksanasopin, T.; Cheung, Y.K.; Steinmiller, D.; Linder, V.; Parsa, H.; Wang, J.; Moore, H.; Rouse, R.; Umviligihozo, G. Microfluidics-based diagnostics of infectious diseases in the developing world. *Nat. Med.* **2011**, *17*, 1015–1019. [CrossRef] [PubMed]
11. Llandro, J.; Palfreyman, J.J.; Ionescu, A.; Barnes, C.H.W. Magnetic biosensor technologies for medical applications: a review. *Med. Biol. Eng. Comput.* **2010**, *48*, 977–998. [CrossRef] [PubMed]
12. Wang, S.X.; Li, G. Advances in giant magnetoresistance biosensors with magnetic nanoparticle tags: Review and outlook. *IEEE Trans. Magn.* **2008**, *44*, 1687–1702. [CrossRef]
13. Baselt, D.R.; Lee, G.U.; Natesan, M.; Metzger, S.W.; Sheehan, P.E.; Colton, R.J. A biosensor based on magnetoresistance technology. *Biosens. Bioelectron.* **1998**, *13*, 731–739. [CrossRef]
14. Koh, I.; Josephson, L. Magnetic Nanoparticle Sensors. *Sensor* **2009**, *9*, 8130–8145. [CrossRef] [PubMed]
15. Varadan, V.K.; Chen, L.; Xie, J. *Nanomedicine: Design and Applications of Magnetic Nanomaterials, Nanosensors and Nanosystems*; John Wiley & Sons: Hoboken, NJ, USA, 2008; ISBN 978-0-470-03351-7.
16. Aivazoglou, E.; Metaxa, E.; Hristoforou, E. Microwave-assisted synthesis of iron oxide nanoparticles in biocompatible organic environment. *AIP Adv.* **2018**, *8*. [CrossRef]
17. Kokkinis, G.; Plochberger, B.; Cardoso, S.; Keplinger, F.; Giouroudi, I. Microfluidic, dual purpose sensor for in-vitro detection of Enterobacteriaceae and biotinylated antibodies. *Lab Chip* **2016**, *16*, 1261–1271. [CrossRef] [PubMed]
18. Jamalieh, M.; Kokkinis, G.; Haiden, C.; Berris, T.; Keplinger, F.; Giouroudi, I. Microfluidic Platform for Pathogen Load Monitoring. *Microelectron. Eng.* **2016**, *158*, 91–94. [CrossRef]
19. Devkota, J.; Kokkinis, G.; Berris, T.; Jamalieh, M.; Cardoso, S.; Cardoso, F.A.; Srikanth, H.; Phan, M.; Giouroudi, I. A novel approach for detection and quantification of magnetic nanomarkers using a spin valve GMR-integrated microfluidic sensor. *RSC Adv.* **2015**, *5*, 51169–51175. [CrossRef]
20. Kokkinis, G.; Jamalieh, M.; Cardoso, S.; Cardoso, F.A.; Keplinger, F.; Giouroudi, I. Magnetic based Biomolecule Detection using GMR Sensors. *J. Appl. Phys.* **2015**, *117*. [CrossRef]
21. Kokkinis, G.; Cardoso, S.; Cardoso, F.A.; Giouroudi, I. Microfluidics for the Rapid Detection of Pathogens using Giant Magnetoresistance Sensors. *IEEE Trans. Magn.* **2014**, *50*. [CrossRef]
22. Kokkinis, G.; Keplinger, F.; Giouroudi, I. On-Chip Microfluidic Biosensor using Superparamagnetic Microparticles. *Biomicrofluidics* **2013**, *7*, 7. [CrossRef] [PubMed]
23. Gooneratne, C.; Yassine, O.; Giouroudi, I.; Kosel, J. Selective Manipulation of Superparamagnetic Beads by a Magnetic Microchip. *IEEE Trans. Magn.* **2013**, *49*, 49. [CrossRef]
24. Gooneratne, C.; Giouroudi, I.; Kosel, J. A Planar Conducting Micro-Loop Structure for Transportation of Magnetic Beads: An Approach towards Rapid Sensing and Quantification of Biological Entities. *Sens. Lett.* **2012**, *210*, 769–773. [CrossRef]
25. Li, F.; Giouroudi, I.; Kosel, J. A biodetection method using magnetic particles and microtraps. *Virtual J. Biol. Phys. Res.* **2012**, *23*, 6.
26. Li, F.; Giouroudi, I.; Kosel, J. A biodetection method using magnetic particles and microtraps. *J. Appl. Phys.* **2012**, *111*. [CrossRef]

27. Gooneratne, C.; Liang, C.; Giouroudi, I.; Kosel, J. An integrated micro-chip for rapid detection of magnetic particles. *J. Appl. Phys.* **2012**, *111*. [CrossRef]

28. Crocker, J.C.; Grier, D.C. Methods of Digital Video Microscopy for Colloidal Studies. *J. Colloid Interface Sci.* **1996**, *179*, 298–310. [CrossRef]

29. Haiden, C.; Thomas, W.; Martin, J.; Franz, K.; Michael, J.V. Sizing of Metallic Nanoparticles Confined to a Microfluidic Film Applying Dark-Field Particle Tracking. *Langmuir* **2014**, *30*, 9607–9615. [CrossRef] [PubMed]

30. Haiden, C.; Thomas, W.; Martin, J.; Franz, K.; Michael, J.V. A Microfluidic Chip and Dark-Field Imaging System for Size Measurement of Metal Wear Particles in Oil. *IEEE Sens. J.* **2016**, *16*, 1182–1189. [CrossRef]

31. Haiden, C.; Thomas, W.; Martin, J.; Franz, K.; Michael, J.V. Concurrent Particle Diffusion and Sedimentation Measurements Using Two-Dimensional Tracking in a Vertical Sample Arrangement. *Appl. Phys. Lett.* **2016**, *108*. [CrossRef]

32. Dangl, A. Biosensing Based on Magnetically Induced Motion of Magnetic Microparticles. Ph.D. Thesis, Technische Universität Wien, Vienna, Austria, 2013.

33. Zhang, X.; DaShan, B.; Valerie, H.; Hongmei, H. A Brief Review of Recent Developments in the Designs That Prevent Bio-Fouling on Silicon and Silicon-Based Materials. *Chem. Cent. J.* **2017**, *11*. [CrossRef] [PubMed]

34. Israelachvili, J.N. *Intermolecular and Surface Forces*; Academic Press: Cambridge, MA, USA, 2011; p. 402.

35. García Núñez, C.; Sachsenhauser, M.; Blashcke, B.; García Marín, A.; Garrido, J.A.; Pau, J.L. Effects of Hydroxylation and Silanization on the Surface Properties of ZnO Nanowires. *ACS Appl. Mater. Interfaces* **2015**, *7*, 5331–5337. [CrossRef] [PubMed]

Article

A Multi-Region Magnetoimpedance-Based Bio-Analytical System for Ultrasensitive Simultaneous Determination of Cardiac Biomarkers Myoglobin and C-Reactive Protein

Zhen Yang [1,2,*], Huanhuan Wang [1,2], Pengfei Guo [1,2], Yuanyuan Ding [1,2], Chong Lei [3] and Yongsong Luo [1,2]

[1] School of Physics and Electronic Engineering, Xinyang Normal University, Xinyang 464000, China; 13323978535@163.com (H.W.); guopengfei2010@126.com (P.G.); m15037680221@16.com (Y.D.); eysluo@163.com (Y.L.)
[2] Key Laboratory of Microelectronics and Energy of Henan Province, Xinyang Normal University, Xinyang 464000, China
[3] Department of Micro/Nano Electronics, School of electronic information and electrical engineering, Shanghai Jiao Tong University, Dongchuan Road 800, Shanghai 200240, China; leiqhd@sjtu.edu.cn
* Correspondence: zhc025@alumni.sjtu.edu.cn; Tel.: +86-376-6391731

Received: 12 April 2018; Accepted: 28 May 2018; Published: 1 June 2018

Abstract: Cardiac biomarkers (CBs) are substances that appear in the blood when the heart is damaged or stressed. Measurements of the level of CBs can be used in course of diagnostics or monitoring the state of the health of group risk persons. A multi-region bio-analytical system (MRBAS) based on magnetoimpedance (MI) changes was proposed for ultrasensitive simultaneous detection of CBs myoglobin (Mb) and C-reactive protein (CRP). The microfluidic device was designed and developed using standard microfabrication techniques for their usage in different regions, which were pre-modified with specific antibody for specified detection. Mb and CRP antigens labels attached to commercial Dynabeads with selected concentrations were trapped in different detection regions. The MI response of the triple sensitive element was carefully evaluated in initial state and in the presence of biomarkers. The results showed that the MI-based bio-sensing system had high selectivity and sensitivity for detection of CBs. Compared with the control region, ultrasensitive detections of CRP and Mb were accomplished with the detection limits of 1.0 pg/mL and 0.1 pg/mL, respectively. The linear detection range contained low concentration detection area and high concentration detection area, which were 1 pg/mL–10 ng/mL, 10–100 ng/mL for CRP, and 0.1 pg/mL–1 ng/mL, 1 n/mL–80 ng/mL for Mb. The measurement technique presented here provides a new methodology for multi-target biomolecules rapid testing.

Keywords: myoglobin; C-reactive protein; Dynabeads; magnetoimpedance; microfluidic device

1. Introduction

Measurements of the level of cardiac biomarkers (CBs) is important when the heart is damaged or stressed, especially in the case of cardiovascular disease (CVD) risk [1,2]. Among various kinds of cardiac biomarkers, C-reactive protein (CRP) is a most important biomarker for CVD risk. The level of CRP can rise from a normal level of less than 5 mg/L to above 100 mg/L after acute inflammatory stimulus [3]. Myoglobin (Mb) is one of the early biomarker levels which increases sharply from 90 pg/mL to 250 ng/mL within 90 min after acute myocardial infarction (AMI), playing a major role in urgent diagnosis of CVD [4–6]. More and more researchers have attempted to apply different types of biosensors to detect the levels of two biomarkers [7–11]. Fluorescence-linked immunosorbent assay was widely used for detection of CBs; however, it is frequently used in in vitro diagnosis and complicated operation steps limit its application [12,13]. Immunosensors based on electrochemical impedance spectroscopy (EIS) usually use an immobilized recognition element (probe) to bind the target/analyte molecule selectively. However, the surface of the biosensors suffer from perturbations on the sensor surface that are influenced by different pH, ionic strength and co-existing molecules in biological fluids. So, the sensitivity, specificity, and relevant range are affected [14,15]. Magnetic bio-detection based on giant magnetoimpedance was proposed long ago [16,17]. Magnetoimpedance (MI) is the change of total impedance of ferromagnetic-conducting sensitive elements under application of an external magnetic field [18–20]. Recently, different MI-based bio-sensing systems for micro-sized/nano-sized magnetic particles and single biomarker detection were designed and tested [21–28].

Despite the progress in medical diagnostics, CVD is considered to be the most common disease which occurs in middle age and elderly people, especially those over 50 years old [29]. The symptoms are complex and associated with more than one biomarker. Detection of a single cardiac biomarker is usually not sufficient for precise CVD diagnostics, due to the limited specificity [30]. Therefore, there is an urgent need for development of a simple, rapid, highly sensitive and inexpensive system for simultaneous detection of several cardiac biomarkers. The advantage of a microfluidic device (MFD) is that different biomarkers can be tested in various detection regions pre-modified with different specified antibodies, i.e. the development of multi-analyte immunoassay combined with a multi-region bio-analytical system (MRBAS) is a very important task [31–35]. The studies in recent years indicated that a higher MI effect can be reached in a multilayered structure and in thin films shaped as meanders [36–40]. Both high MI response and high sensitivity with respect to an applied magnetic field are the main factors for MI bio-sensing applications.

In this paper, we designed and tested the ultrasensitive bio-analytical MI-based system for simultaneous detection of CRP and Mb, based on optimization of the structural parameters of the MI element. Immune reaction was performed in the MRBAS based on MI changes. The MI response of the sensitive element was measured in its initial state and in the presence of biomarkers. Compared with the control region, ultrasensitive and combined detection of CRP and Mb was accomplished in the different detection regions. The methodology presented here provides a vital basis for multi-analyte bio-magnetic detection like Troponin, Creatine Kinase-MB, alpha-fetoprotein (AFP) and carcinoembryonic antigen (CEA).

2. Materials and Methods

2.1. Reagents and Instruments

All the biological and chemical reagents used for the present study are listed in Table 1. For all experiments described here, deionized water was used. The MI multilayered sensitive element and Multi-region microfluidic devices (MR-MFD) were prepared by widely employed micro electromechanical system technology (MEMS) in National Key Laboratory (China). MI measurements were made using a special system based on a Hewlett-Packard 4194A Impedance analyzer (Agilent Technologies Inc., Palo Alto, CA, USA). The programmable syringe pump (PHD 4400 HPSI, Harvard Apparatus, Holliston, MA, USA) was connected with the MFD and adopted for injecting test samples. All measurements were made at room temperature.

Table 1. Biological and chemical reagents.

Solution	Detail	Company
Mercaptopropionic acid	Concentration 20 mmol/L	Aladdin Chemistry Co. Ltd (Beijing, China)
EDC	Concentration 0.2 mol/L	Aladdin Chemistry Co. Ltd (Beijingi, China).
NHS	Concentration 0.05 mol/L	Shanghai Medpep Co. Ltd (Shanghai, China)
Mouse Mb &CRP 1st antibody	Concentration 1 mg/mL	Linc-Bio Science Co. Ltd. (Shanghai, China)
BSA	1% BSA, 0.2% tween 20	Via-gene pro bio Technologies Co. Ltd. (Shanghai, China)
Human Mb &CRP antigen	0.1–100 pg/mL, 1–100 ng/mL	Linc-Bio Science Co. Ltd. (Shanghai, China)
Mb &CRP 2 nd antibody	Concentration 1 mg/mL	Linc-Bio Science Co. Ltd. (Shanghai, China)
Dynabeads® C1	Concentration 10 μg/mL	Invitrogen Co. Ltd. (Shanghai, China)
NaOH	Concentration 1 mol/L	Pinghu Chemical Reagent (Pinghu, China)
HCL	Concentration 1 mol/L	Sinpharm Chemical Reagent Co. Ltd. (Shanghai, China)
PBS	PH = 7.4	Medicago AB (Uppsala, Sweden)
C_3H_6O C_2H_5OH	AR	LingFeng Chemical Reagent Co. Ltd. (Shanghai, China)

2.2. Microfabrication of MI Element and MR-MFD

The meander-shaped MI element was designed for superior structural parameters, and the manufacturing process was presented in Figure S1, following our previous experiments described elsewhere [41]. The electrodeposited NiFe layer had good soft magnetic properties, which were similar to those previously studied in in our works [42,43]. Figure 1A shows the design diagram of MR-MFD. Two detection regions (CRP detection region 1 & Mb detection region 2), one blank control region and two inlets (sample and region buffer) were designed. The distance between any of two adjacent detection regions are 20 mm in order to minimize the magnetic interference between them. Each detection region contained a single rectangular gold film unit and possessed an area of 3×5 mm^2, and the control region possessed the same area without the gold film unit. The MR-MFD were fabricated based on SU-8 and polydimethylsiloxane (PDMS) materials by MEMS technology. The general view of the fabricated MI element is shown in Figure 1C. The MR-MFD containing the fluid reservoir and microfluidic pipeline were designed for sandwich immunoassay. Following is the step-by-step description of MR-MFD preparation:

1. Fabrication of gold layer: First a chromium adhesion layer (~60 nm) was deposited on the glass wafer with a thickness of 1 mm at a rate of 1 Å s^{-1}, followed by ~240 nm of gold at a rate of 2–3 Å s^{-1} by a radio frequency sputtering system (LH-Z550, Shanghai, China).

2. Patterning of the gold film: A photoresist layer with a thickness of 10 μm was spun onto the Au layer and afterwards patterned to several small MFD units through the mask.

3. Deleting the uncovered part of the gold layer: the uncovered part of the Au layer was removed by wet etching in the KI, I$_2$ and H$_2$O mixed solution for 45 s.

4. Deleting the photoresist: The whole glass substrate with Au film was immersed in acetone solution for 25 s.

5. SU-8 layers preparation: The SU-8 photoresist with a thickness of 500 μm was spin coated on Au film, soft baked, patterned with a mask, and developed with an SU-8 developer. Then, the same thickness of the SU-8 photoresist was spun again, finally a 5×3 mm^2 rectangular microcavity with a depth of 1 mm was achieved. Figure 1B showed the SEM of gold nanofilm.

6. PDMS casting: The pre-polymer and the curing agent of PDMS were mixed in a 10:1 ratio by weight. After thermal coagulation, the PDMS can be obtained.

7. Bonding of the SU-8 with the PDMS: The surface-treated PDMS was tightly bound to the Su-8 surface and then opened the inlets and waste chamber. Figure 1A (inset) shows the fabricated microfluidic device without PDMS.

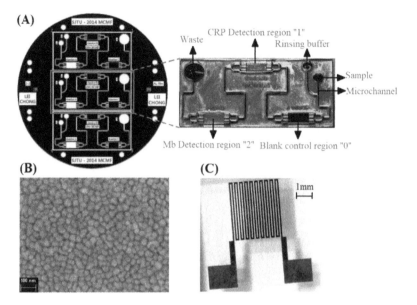

Figure 1. (**A**) The designed diagram of the microfluidic device, and the inset showing the photograph of the fabricated microfluidic device. (**B**) The Scanning Electron Microscopy image of the thin gold film. (**C**) The fabricated multilayer MI element.

2.3. Sandwich Immunoassay for Capturing CBs

The detection regions were pre-modified through self-assembly and activation, prior to PDMS bonding. To achieve selectivity and improve the efficiency of detection, the detection regions 1 and 2 were surface modified with a mouse CRP monoclonal antibody or mouse Mb monoclonal antibody, respectively. The details for the self-assembling, activation and modification process can be found elsewhere [27]. After the surface modification, the detection regions were carefully rinsed with phosphate-buffered solution (PBS) prepared with deionized water. Afterwards, the detection regions were sealed with 100 µL BSA solution (including 1% BSA, 0.2% tween 20) at 4 °C for 2 h and washed twice with PBS solution (PH = 7.4). The CRP and Mb antigen complexes with different concentrations (1 pg/mL–100 ng/mL for CRP and 0.1 pg/mL–80 ng/mL for Mb) were placed into the MR-MFD by a simple syringe pump. The complexes flowed through different regions in turn. Immunoassay time was as long as 20 min, aiming to ensure an effective combination of antigen-antibody. Finally, The PBS solution was used for rinsing. Then, biotinylated CRP & Mb polyclonal antibody (10 µL, 1 mg/mL) mixed solution was injected into the MR-MFD and washed five times. Finally, Dynabeads suspension (40 µL, 10 µg/mL) was injected. Dynabeads are polystyrene/iron oxide nanomaterials (superparamagnetic spheres) widely used for magnetic separation and biosensing [44–46]. After cultivating, washing and drying the whole double antibody immunoassay process was completed as shown in Figure 2B.

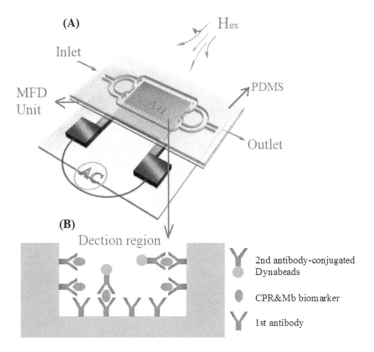

Figure 2. The principle diagram of detection for cardiac biomarkers using an MI sensitive element in the shape of the meander (**A**); general illustration of the biochemical part of the detection part (**B**).

2.4. Determination of CBs Using MI-Based Bio-Analytic System

The CRP and Mb antigen were mixed with the same concentration and volume (10 μL) in all cases under consideration. Different detection regions modified by different antibodies would capture different CBs. First of all, the MI response of the thin film element located below the blank control region was measured in its initial state. The MI response of the thin film element located below the detection regions 1 and 2 were measured in the presence of CBs. The fundamental principle for detection of Dynabead-labeled CBs is the detection of the stray fields (H_{stray}) of the Dynabeads. This is a similar principle as it was previously discussed in many cases of different magnetic biosensors [17,37–39]. When the MI element was moved to the bottom of the detection region of MR-MFD, the Dynabeads trapped in the MFD unit, located above the MI element as shown in Figure 2A, were magnetized by an external magnetic field (H_{ex}: 0–120 Oe) created by a pair of Helmholtz coils and they emitted a detectable H_{stray}. The H_{stray} modified the partially overlapping magnetic field near the MI element and resulted in changes of transverse permeability and skin penetration depth; this lead to the altering of MI. The different concentrations of CBs flowing through the MR-MFD corresponds to the changes of MI. The advantages of this MI-based MRBAS is that the different MI responses in the varying regions reflect the content for CRP and Mb, simultaneously. The effect of H_{stray} on MI is affected by the distance of the MR-MFD to the MI element; therefore, the thinnest glass wafer is selected to use as a substrate for MR-MFD fabrication. The MI ratios were calculated as follows:

$$MI\ ratio = \frac{\Delta Z}{Z} = 100 \times \frac{Z(H) - Z(H_{max})}{Z(H_{max})} \qquad (1)$$

where H_{max} = 120 Oe for orientation of the external magnetic field along the long side of the meander thin film elements.

3. Results and Discussion

The maximum MI sensitivity of the fabricated sensitive element of about 22% Oe^{-1} was obtained at 1.4 MHz and 8.8 Oe field. Different antigen molecules were captured in different detection regions. The antigen molecules were combined with a certain number of magnetic beads. The beads captured in different detection regions were magnetized in varying degrees and behaved as a magnetic dipole producing stray fields H_{stray}, disturbing the external magnetic field. Therefore, the original transverse permeability μ_T of the MI sensitive element experiences a different resultant magnetic field $H_R = H_{ex} + h_{AC} + H_{stray}$ (h_{AC} was AC current magnetic field) and achieves a different value of field superposition giving rise to a different value in the MI ratio. The MI response quantitatively reflects the presence, content, or the absence of biomarkers.

Figure 3 shows the relationship between the MI responses and CRP concentrations (C_{CRP}). The field dependences of the MI ratio for detecting different concentration of CRP are shown in Figures 2 and 3. Evidently, the MI ratios have been enhanced by distinct values on account of the CRP antigen conjugated with magnetic beads with different concentrations in the detection region 1, from 1 pg/mL to 10 ng/mL. It was worthwhile to note that the MI ratio increased from 195% (without CRP antigen) to 212% with increasing the concentration of the CRP antigen. For each concentration, independent measurement was performed 10 times under the same testing conditions. The relative standard deviation (RSD) is reasonable, as can be seen in Figure 3A; this indicates good precision.

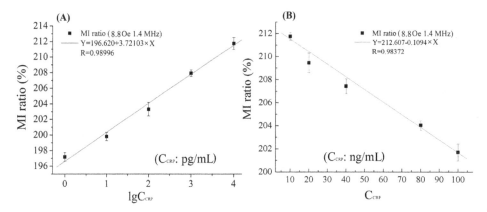

Figure 3. The relationship of MI ratio versus (**A**) low CRP concentrations and (**B**) high CRP concentrations at f = 1.4 MHz and H_{ex} = 8.8 Oe.

The linearity between the MI response signals and CRP concentrations with the range of 1 pg/mL–10 ng/mL, is represented under 8.8 Oe and 1.4 MHz by the fitting form in base-10 logarithms. As can be seen clearly from Figure 3A, 5 points locate near the curve and R = 0.98996 is very close to 1; we can say that the MI response has a good linear relationship with the CRP concentrations. However, the MI ratio began to fall down from 218 to 202% with the change of concentration from 10 ng/mL to 100 ng/mL, as can be seen in Figure S3. The linearity was also tested in this concentration range, displaying a statistical curve fitted by the linear regression. Five points locate near the curve and R = 0.98372 is very close to 1, as shown in Figure 3B. Therefore, lower and upper limitations of detection were 1 pg/mL and 100 ng/mL, respectively.

In our previous work [25], we found that Dynabeads (conjugated 10 ng/mL biomarkers) on the surface of the element were nearly magnetically saturated. The high concentration of Dynabeads had caused the high-density clusters of Dynabeads, the interaction between beads was obvious, the whole H_{stray} attenuated, and the added value of the MI was reduced. A similar result on Mb concentration (C_{Mb}) detection was obtained in the detection region 2, as shown in Figure 4. However, the linear detection ranges were 0.1 pg/mL–1 ng/mL and 1 n/mL–80 ng/mL for Mb. From Figures 3 and 4, we observed the overlap MI value in low concentration and high concentration for CRP and Mb. Therefore, dual-measurement for the same CBs sample would be better by using the designed MI-based bio-sensing system. Firstly, we can measure the sample with original concentration, and then measure the sample with decreased concentration by mixing with a buffer (e.g., in a ratio of 1:1). Then the region of concentration of the sample can be pinpointed depending on the trend of the MI ratio.

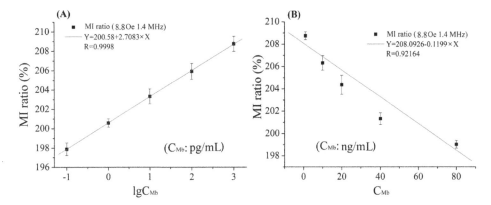

Figure 4. The relationship of MI ratio versus (**A**) low Mb concentrations and (**B**) high Mb concentrations at f = 1.4 MHz and H_{ex} = 8.8 Oe.

Sensitivity and specificity are two major factors to assess the practicality of immunosensors. In addition to the above sensitivity analysis, the specificity test of the MI-based bio-analytical system was performed by interfering antigen CEA. The single interfering antigen solution (0.1 pg/mL CEA) and four mixture antigen solutions (0.1 pg/mL CEA + 0.1 pg/mL Mb, 0.1 pg/mL CEA + 100 ng/mL CRP, 0.1 pg/mL Mb + 100 ng/mL CRP and 0.1 pg/mL CEA + 0.1 pg/mL Mb + 100 ng/mL CRP) were infused into the MR-MFD by syringe pump, respectively. The MI response was observed in controlled region 0 and detection regions 1 and 2, with a specific antigen-antibody reaction. The MI responses of the multilayered element under varying regions are shown in Figure 5. When 0.1 pg/mL CEA antigen solution was injected into the MR-MFD, there was no sharp difference for the MI ratio between control region 0 and detection region 1&2. An exclusive significant increase in the MI ratio was discovered in detection region 2 (with injecting the CEA + Mb antigens mixture solution) or in the detection region 1 (with injecting the CEA + CRP antigens mixture solution). However, significant variation of MI appeared simultaneously in detection region 1 and detection region 2 for CRP + Mb and CEA + Mb + CRP. The results indicated the selective binding of CRP in detection region 1 to the CRP antibody and Mb in detection region 2 to the Mb antibody. Thus, it illustrates the good specificity of the MI-based magnetic immunoassay for the detection of Mb and CRP. In view of the results of detection limit, we infused the mixed antigen (0.1 pg/mL CRP + 0.1 pg/mL Mb or 100 ng/mL CRP + 100 ng/mL Mb) into the MR-MFD, the same result was almost obtained like that in Figure 5 (Mb single antigen and CRP single antigen). So, in theory the detection limit for biomarkers is logical. The relative standard deviation (RSD) of 0.34% is obtained by performing 6 independent measurements (one time every ten minutes) on 0.1 pg/mL Mb under the same testing conditions as shown in Figure 6, indicating an acceptable reproducibility of the magnetic immunoassay. The stability capability test was repeated in different time (The corresponding time is 1st, 5th, 10th, 15th, 20th days, respectively). The high stability of the MI sensor contributes to the reliability of measurement results. The same reproducibility and stability test was performed on 100 ng/mL CRP. The RSD of the 6 independent assay results corresponded approximately to 0.648% and indicated the satisfactory reproducibility of the presented method. The results of the stability capability test confirm the stability of the MI element.

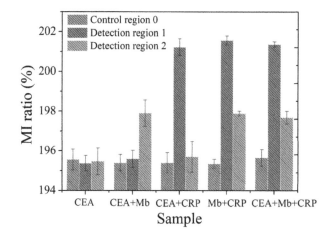

Figure 5. MI responses for the specificity test on CRP and Mb; single CEA antigen solution and four antigens mixture solution (CEA + Mb; CEA + CRP; CRP + Mb, CEA + Mb + CRP) were injected into the MR-MFD in turn.

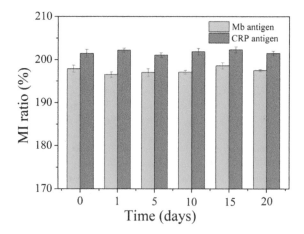

Figure 6. Stability test of the MI-based bio-analytic system in 0.1 pg/mL Mb and 100 ng/mL CRP.

The detection limit of 1 ng/mL for CRP and 0.5 ng/mL for Mb was achieved by using an integrated ribbon-based MI biosensing system in our previous work [47,48]. In this work, the fabricated multilayered thin film MI sensitive element possesses a lower minimum detectable concentration of 1 pg/mL for CRP detection and 0.1 pg/mL for the Mb detection case. This can be attributed to the higher MI sensitivity of the multilayered thin film element. Although a lower detection limit of 0.029 pg/mL for CRP can be reached by applying a NH_2-Ni-MOF electrocatalysts method [49] and detection limit is 0.01 pg/mL for Mb using an Au-WS_2 nanohybrid based SERS aptasensor [50], utilization of the MRBAS results in ultrasensitive combined detection of CBs, Mb and CRP. Bio-magnetic detection has recently become a focus of interest for researchers because of its high sensitivity, versatile diagnostic methods, convenient processes, high accuracy and low cost.

The detection limit with 0.01 pg/mL was considered more adequate for the quantification of Mb in clinical diagnostics. So, we will attempt to design more complex MR-MFD for ultrasensitive simultaneous detection of two, three or four biomarkers using the bio-magnetic measurement system in the future. Firstly, we are required to fabricate the higher sensitive MI element and MI element array by MEMS technology. Then, patterned microfluidic channel and magnetic nanoparticles labels will be considered for detection of CBs. Finally, the integrated MI-based biosensing system may be adopted.

4. Conclusions

In summary, ultrasensitive simultaneous detection of cardiac biomarkers Mb and CRP was achieved by the design and development of the MR MI-based bio-analytic system. The magnetic labels and double antibody sandwiched immunoassays were used for the measurements. The multilayered thin film MI element and MFD were manufactured by microfabrication techniques. The lower detection limits with 1 pg/mL for CRP and 0.1 pg/mL for Mb were obtained. The linear detection range contained a low concentration detection area and a high concentration detection area, which were 1 pg/mL–10 ng/mL, 10–100 ng/mL for CRP, and 0.1 pg/mL–1 ng/mL, 1 n/mL–80 ng/mL for Mb, respectively. The methodology presented here is promising for multi-analyte bio-magnetic detection of biocomponents (Troponin, CK-MB, AFP, CEA) and other biomarkers using MI-based biosensing system.

Supplementary Materials: The following are available online at http://www.mdpi.com/1424-8220/18/6/1765/ s1, Figure S1: Fabrication steps of the MI element, Figure S2: Field dependence of MI responses under the low concentration of CRP antigen (a) full view (b) partial enlargement, Figure S3: Field dependence of MI responses under the high concentration of CRP antigen (a) full view (b) partial enlargement, Figure S4: SEM characterizations CRP-conjugated Dynabeads at the concentration of 10 ng/mL (A) and SEM characterizations Mb-conjugated Dynabeads at the concentration of 10 ng/mL (B).

Author Contributions: Zhen Yang and Yongsong Luo conceived and designed the experiments; Pengfei Guo and Chong Lei contributed to the fabrication of the sensors; Huanhuan Wang and Yuanyuan Ding performed the biological test and analyzed the data; Zhen Yang performed some experiments and wrote this paper.

Funding: This research was funded by the National Natural Science Foundation of China (51602276, 61273065), Natural Science Foundation of Henan Province (162300410233), Nanhu Scholars Program for Young Scholars of XYNU, Major Pre-research Program of XYNU (2016-ZDYY-132, 2017-ZDYY-178), Start-up research grant for new faculty of XYNU (No. 16055).

Acknowledgments: The authors are grateful to the Key Laboratory of Microelectronics and Energy of Henan Province of Xinyang Normal University of China, Henan Administration of Foreign Experts Affairs, the Analytical and Testing Center in Shanghai Jiao Tong University, the Center for Advanced Electronic Materials and Devices in Shanghai Jiao Tong University.

Conflicts of Interest: The authors declare no conflict of interest.

References

1. Lee, W.B.; Chen, Y.H.; Lin, H.I.; Shiesh, S.C.; Lee, G.B. An integrated microfluidic system for fast, automatic detection of C-reactive protein. *Biosens. Bioelectron.* **2009**, *24*, 3091–3096. [CrossRef]

2. Kushner, I.; Sehgal, A.R. Is high-sensitivity C-reactive protein an effective screening test for cardiovascular risk? *Arch. Intern. Med.* **2002**, *162*, 867–869. [CrossRef] [PubMed]

3. Black, S.; Kushner, I.; Samols, D. C-reactive protein. *J. Biol. Chem.* **2004**, *47*, 48487–48490. [CrossRef] [PubMed]

4. Qureshi, A.; Gurbuz, Y.; Niazi, J.H. Biosensors for cardiac biomarkers detection: A review. *Sens. Actuators B Chem.* **2012**, *171*, 62–76. [CrossRef]

5. Wang, Q.; Liu, F.; Yang, X.H.; Wang, K.M.; Wang, H.; Deng, X. Sensitive point-of-care monitoring of cardiac biomarker myoglobin using aptamer and ubiquitous personµal glucose meter. *Biosens. Bioelectron.* **2015**, *64*, 161–164. [CrossRef] [PubMed]

6. Sallach, S.M.; Nowak, R.; Hudson, M.P.; Tokarski, G.; Khoury, N.; Tomlanovich, M.C.; Jacobsen, G.; de Lemos, J.A.; McCord, J. A Change in Serum Myoglobin to Detect Acute Myocardial Infarction in Patients with Normal Troponin I Levels. *Am. J. Cardiol.* **2004**, *94*, 864–867. [CrossRef] [PubMed]

7.   Meyer, M.H.F.; Hartmann, M.; Krgoldse, H.J.; Blankenstein, G.; Mueller-Chorus, B.; Oster, J.; Miethe, P.; Keusgen, M. CRP determination based on a novel magnetic biosensor. *Biosens. Bioelectron.* **2007**, *22*, 973–979. [CrossRef] [PubMed]

8.   Baselt, D.R.; Lee, G.U.; Natesan, M.; Metzger, S.W.; Sheehan, P.E.; Colton, R.J. A biosensor based on magnetoresistance technology. *Biosens. Bioelectron.* **1998**, *13*, 731–739. [CrossRef]

9.   Gnedenko, O.V.; Mezentsev, Y.V.; Molnar, A.A.; Lisitsa, A.V.; Ivanov, A.S.; Archakov, A.I. Highly sensitive detection of human cardiac myoglobin using a reverse sandwich immunoassay with a gold nanoparticle-enhanced surface plasmon resonance biosensor. *Anal. Chim. Acta* **2013**, *759*, 105–109. [CrossRef] [PubMed]

10. Shorie, M.; Kumar, V.; Sabherwal, P.; Ganguli, A.K. Carbon quantum dots-mediated direct fluorescence assay for the detection of cardiac marker myoglobin. *Curr. Sci.* **2015**, *108*, 1595–1596.

11. Yang, Y.N.; Lin, H.I.; Wang, J.H.; Shiesh, S.C.; Lee, G.B. An integrated microfluidic system for C-reactive protein measurement. *Biosens. Bioelectron.* **2009**, *24*, 3091–3096. [CrossRef] [PubMed]

12. Rifai, N.; Tracy, R.P.; Ridker, P.M. Clinical efficacy of an automated high-sensitivity C-reactive protein assay. *Clin. Chem.* **1999**, *45*, 2136–2141. [PubMed]

13. Xie, M.J.; Huang, H.; Hang, J.F.; Dong, Z.N.; Xiao, D.Y.; Xu, P.; Zhu, C.Y.; Xu, W.W. Evaluation of the Analytical and Clinical Performances of Time-resolved Fluoroimmunoassay for Detecting Carcinoma Antigen 50. *J. Immunoass. Immunochem.* **2015**, *3*, 265–283. [CrossRef] [PubMed]

14. Bedatty Fernandes, F.C.; Patil, A.V.; Bueno, P.R.; Davis, J.J. Optimized Diagnostic Assays Based on Redox Tagged Bioreceptive Interfaces. *Anal. Chem.* **2015**, *87*, 12137–12144. [CrossRef] [PubMed]

15. Ren, X.H.; Zhang, Y.; Sun, Y.Q.; Gao, L.L. Development of Electrochemical Impedance Immunosensor for Sensitive Determination of Myoglobin. *Int. J. Electrochem. Sci.* **2017**, *12*, 7765–7776. [CrossRef]

16. Chiriac, H.; Herea, D.D.; Corodeanu, S. Microwire array for giant magnetoimpedance detection of magnetic particles for biosensor prototype. *J. Magn. Magn. Mater.* **2007**, *311*, 425–428. [CrossRef]

17. Kurlyandskaya, G.V.; Levit, V.I. Advanced materials for drug delivery and biosensors based on magnetic label detection. *Mater. Sci. Eng. C* **2007**, *27*, 495–503. [CrossRef]

18. Beach, R.S.; Berkowitz, A.E. Sensitive field-and frequency-dependent impedance spectra of amorphous FeCoSiB wire and ribbon. *J. Appl. Phys.* **1994**, *76*, 6209–6213. [CrossRef]

19. Kraus, L. GMI modelling and material optimization. *Sens. Actuators A* **2003**, *106*, 187–194. [CrossRef]

20. Nishibe, Y.; Ohta, N.; Tsukada, K.; Yamadera, H.; Nomomura, Y.; Mohri, K.; Uchiyama, T. Sensing of passing vehicles using a lane marker on road with built-in thin film MI sensor and power source. *IEEE Trans. Veh. Technol.* **2004**, *53*, 1827–1834. [CrossRef]

21. Wang, T.; Yang, Z.; Lei, C.; Lei, J.; Zhou, Y. An integrated giant magnetoimpedance biosensor for detection of biomarker. *Biosens. Bioelectron.* **2014**, *58*, 338–344. [CrossRef] [PubMed]

22. Blyakhman, F.A.; Safronov, A.P.; Zubarev, A.Y.; Shklyar, T.F.; Makeyev, O.G.; Makarova, E.B.; Melekhin, V.V.; Larrañaga, A.; Kurlyandskaya, G.V. Polyacrylamide ferrogels with embedded maghemite nanoparticles for biomedical engineering. *Results Phys.* **2017**, *7*, 3624–3633. [CrossRef]

23. Blyakhman, F.A.; Buznikov, N.A.; Sklyar, T.F.; Safronov, A.P.; Golubeva, E.V.; Svalov, A.V.; Sokolov, S.Y.; Melnikov, G.Y.; Orue, I.; Kurlyandskaya, G.V. Mechanical, Electrical and Magnetic Properties of Ferrogels with Embedded Iron Oxide Nanoparticles Obtained by Laser Target Evaporation: Focus on Multifunctional Biosensor Applications. *Sensors* **2018**, *18*, 872. [CrossRef] [PubMed]

24. Kurlyandskaya, G.V.; Fernandez, E.; Safronov, A.P.; Blyakhman, F.A.; Svalov, A.V.; Burgoa Beitia, A.; Beketov, I.V. Magnetoimpedance biosensor prototype for ferrogel detection. *J. Magn. Magn. Mater.* **2017**, *441*, 650–655. [CrossRef]

25. Wang, T.; Zhou, Y.; Lei, C.; Lei, J.; Yang, Z. Development of an ingenious method for determination of Dynabeads protein A based on a giant magnetoimpedance sensor. *Sens. Actuators B Chem.* **2013**, *186*, 727–733. [CrossRef]

26. Safronov, A.P.; Mikhnevich, E.A.; Lotfollahi, Z.; Blyakhman, F.A.; Sklyar, T.F.; Larrañaga Varga, A.; Medvedev, A.I.; Fernández Armas, S.; Kurlyandskaya, G.V. Polyacrylamide ferrogels with magnetite or strontium hexaferrite: Next step in the development of soft biomimetic matter for biosensor applications. *Sensors* **2018**, *18*, 257. [CrossRef] [PubMed]

27. Yang, Z.; Liu, Y.; Lei, C.; Sun, X.C.; Zhou, Y. Ultrasensitive detection and quantification of E. coli O157:H7 using a giant magnetoimpedance sensor in an open-surface microfluidic cavity covered with an antibody-modified gold surface. *Microchim. Acta* **2016**, *183*, 1831–1837. [CrossRef]

28. Wang, T.; Zhou, Y.; Lei, C.; Luo, J.; Xie, S.R.; Pu, H.Y. Magnetic impedance biosensor: A review. *Biosens. Bioelectron.* **2017**, *90*, 418–435. [CrossRef] [PubMed]

29. Pearson, T.A.; Mensah, G.A.; Alexander, R.W.; Anderson, J.L.; Cannon, R.O.; Criqui, M.; Fadl, Y.Y.; Fortmann, S.P.; Hong, Y.; Myers, G.L.; et al. Markers of Inflammation and Cardiovascular Disease. *Circulation* **2003**, *107*, 499–511. [CrossRef] [PubMed]

30. Apple, F.S.; Murakami, M.M.; Pearce, L.A.; Herzog, C.A. Multi-Biomarker Risk Stratification of N-Terminal Pro-B-Type Natriuretic Peptide, High-Sensitivity C-Reactive Protein, and Cardiac Troponin T and I in End-Stage Renal Disease for All-Cause Death. *Clin. Chem.* **2004**, *50*, 2279–2285. [CrossRef] [PubMed]

31. Yu, X.; Xia, H.S.; Sun, Z.D.; Lin, Y.; Wang, K.; Yu, J.; Tang, H.; Pang, D.W.; Zhang, Z.L. On-chip dual detection of cancer biomarkers directly in serum based on self-assembled magnetic bead patterns and quantum dots. *Biosens. Bioelectron.* **2013**, *41*, 129–136. [CrossRef] [PubMed]

32. Han, K.N.; Li, C.A.; Seong, G.H. Microfluidic Chips for Immunoassays. *Annu. Rev. Anal. Chem.* **2013**, *6*, 119–141. [CrossRef] [PubMed]

33. Maeng, J.H.; Lee, B.C.; Ko, Y.J.; Cho, W.; Ahn, Y.; Cho, N.G.; Lee, S.H.; Hwang, S.Y. A novel microfluidic biosensor based on an electrical detection system for alpha-fetoprotein. *Biosens. Bioelectron.* **2008**, *23*, 1319–1325. [CrossRef] [PubMed]

34. Wu, J.D.; Dong, M.L.; Santos, S.; Rigatto, C.; Liu, Y.; Lin, F. Lab-on-a-Chip platforms for detection of cardiovascular disease and cancer biomarkers. *Sensors* **2017**, *17*, 2934. [CrossRef] [PubMed]

35. Lian, J.; Zhou, W.W.; Shi, X.Z.; Gao, Y.H. Development of Integrated Microfluidic Magnetic Biosensor for Multi-biomarker Detection. *Chin. J. Anal. Chem.* **2013**, *9*, 1302–1307. [CrossRef]

36. Chen, L.; Zhou, Y.; Zhou, Z.M.; Ding, W. Giant magnetoimpedance effects in patterned Co-based ribbons with a meander structure. *Phys. Status Solidi (a)* **2009**, *206*, 1594–1598. [CrossRef]

37. Chen, L.; Zhou, Y.; Zhou, Z.M.; Ding, W. Enhancement of magnetoimpedance effect in Co-based amorphous ribbon with a meander structure. *Phys. Status Solidi (a)* **2010**, *207*, 448–451. [CrossRef]

38. Rivero, M.A.; Maicas, M.; Lopez, E.; Aroca, C.; Sanchez, M.C.; Sanchez, P.J. Influence of the sensor shape on permalloy/Cu/permalloy magnetoimpedance. *Magn. Magn. Mater.* **2003**, *254*, 636–640. [CrossRef]

39. Morikawa, T.; Nishibe, Y.; Yamadera, H.; Nonomura, Y.; Takeuchi, M.; Sakata, J.; Taga, Y. Enhancement of giant magneto-impedance in layered film by insulator separation. *IEEE Trans. Magn.* **1996**, *32*, 4965–4967. [CrossRef]

40. Lodewijk, K.J.; Fernandez, E.; Garcia-Arribas, A.; Kurlyandskaya, G.V.; Lepalovskij, V.N.; Safronov, A.P.; Kooi, B.J. Magnetoimpedance of thin film meander with composite coating layer containing Ni nanoparticles. *J. Appl. Phys.* **2014**, *115*, 17A323. [CrossRef]

41. Wang, T.; Lei, C.; Lei, J.; Yang, Z.; Zhou, Y. Preparation of meander thin-film microsensor and investigation the influence of structural parameters on the giant magnetoimpedance effect. *Appl. Phys. A* **2012**, *109*, 205–211. [CrossRef]

42. Park, J.Y.; Allen, M.G. Integrated electroplated micromachined magnetic devices using low temperature fabrication processes. *IEEE Trans. Electron. Packag. Manuf.* **2000**, *23*, 48–55. [CrossRef]

43. Wang, T.; Zhou, Y.; Lei, C.; Lei, J.; Yang, Z. Ultrasensitive detection of Dynabeads protein A using the giant magnetoimpedance effect. *Microchim. Acta* **2013**, *180*, 1211–1216. [CrossRef]

44. Miller, M.M.; Prinz, G.A.; Cheng, S.F.; Bounnak, S. Detection of a micron-sized magnetic sphere using a ring-shaped anisotropic magnetoresistance-based sensor: A model for a magnetoresistance-based biosensor. *Appl. Phys. Lett.* **2002**, *81*, 2211–2213. [CrossRef]

45. Ferreira, H.A.; Graham, D.L.; Freitas, P.P.; Cabral, J.M.S. Biodetection using magnetically labeled biomolecules and arrays of spin valve sensors. *J. Appl. Phys.* **2002**, *93*, 7281–7286. [CrossRef]

46. Besse, P.A.; Boero, G.; Demierre, M.; Pott, V.; Popovic, R. Detection of single magnetic microbead using a miniaturized silicon Hall sensor. *Appl. Phys. Lett.* **2002**, *80*, 4199–4201. [CrossRef]

47. Yang, Z.; Liu, Y.; Lei, C.; Sun, X.C.; Zhou, Y. A flexible giant magnetoimpedance-based biosensor for the determination of the biomarker C-reactive protein. *Microchim. Acta* **2015**, *182*, 2411–2417. [CrossRef]

48. Yang, Z.; Wang, H.H.; Dong, X.W.; Yan, H.L.; Lei, C.; Luo, Y.S. Giant magnetoimpedance based immunoassay for cardiac biomarker myoglobin. *Anal. Methods* **2017**, *9*, 3636–3642. [CrossRef]

49. Shorie, M.; Kumar, V.; Kaur, H.; Singh, K.; Tomer, V.K.; Sabherwal, P. Plasmonic DNA hotspots made from tungsten disulfide nanosheets and gold nanoparticles for ultrasensitive aptamer-based SERS detection of myoglobin. *Microchim. Acta* **2018**, *185*, 158–165. [CrossRef] [PubMed]

50. Wang, Z.; Dong, P.; Sun, Z.X.; Sun, C.; Bu, H.Y.; Han, J.; Chen, S.P.; Xie, G. NH_2-Ni-MOF electrocatalysts with tunable size/morphology for ultrasensitive C-reactive protein detection via an aptamer binding induced DNA walker–antibody sandwich assay. *J. Mater. Chem. B.* **2018**, *6*, 2426–2431. [CrossRef]

Article

Eight-Channel AC Magnetosusceptometer of Magnetic Nanoparticles for High-Throughput and Ultra-High-Sensitivity Immunoassay

Jen-Jie Chieh [1,*], Wen-Chun Wei [1], Shu-Hsien Liao [1], Hsin-Hsein Chen [2], Yen-Fu Lee [2], Feng-Chun Lin [2], Ming-Hsien Chiang [1], Ming-Jang Chiu [3,4,5,6], Herng-Er Horng [1,*] and Shieh-Yueh Yang [2]

[1] Institute of Electro-Optical Science and Technology, National Taiwan Normal University, Taipei 116, Taiwan; clara6622@gmail.com (W.-C.C.); shliao@ntnu.edu.tw (S.-H.L.); willamsobain@gmail.com (M.-H.C.)
[2] MagQu Co., Ltd., New Taipei 231, Taiwan; joseph.chen@magqu.com (H.-H.C.); yf.lee@magqu.com (Y.-F.L.); venus.lin@magqu.com (F.-C.L.); syyang@maqgu.com (S.-Y.Y.)
[3] Departments of Neurology, National Taiwan University Hospital, Taipei 100, Taiwan; mjchiu@ntu.edu.tw
[4] Institute of Brain and Mind Sciences, College of Medicine, National Taiwan University, Taipei 100, Taiwan
[5] Department of Psychology, National Taiwan University, Taipei 106, Taiwan
[6] Graduate Institute of Biomedical Engineering and Bio-informatics, National Taiwan University, Taipei 106, Taiwan
* Correspondence: jjchieh@ntnu.edu.tw (J.-J.C.); phyfv001@ntnu.edu.tw (H.-E.H.)

Received: 29 January 2018; Accepted: 26 March 2018; Published: 30 March 2018

Abstract: An alternating-current magnetosusceptometer of antibody-functionalized magnetic nanoparticles (MNPs) was developed for immunomagnetic reduction (IMR). A high-sensitivity, high-critical-temperature superconducting quantum interference device was used in the magnetosusceptometer. Minute levels of biomarkers of early-stage neurodegeneration diseases were detectable in serum, but measuring each biomarker required approximately 4 h. Hence, an eight-channel platform was developed in this study to fit minimal screening requirements for Alzheimer's disease. Two consistent results were measured for three biomarkers, namely Aβ40, Aβ42, and tau protein, per human specimen. This paper presents the instrument configuration as well as critical characteristics, such as the low noise level variations among channels, a high signal-to-noise ratio, and the coefficient of variation for the biomarkers' IMR values. The instrument's ultrahigh sensitivity levels for the three biomarkers and the substantially shorter total measurement time in comparison with the previous single- and four-channels platforms were also demonstrated in this study. Thus, the eight-channel instrument may serve as a powerful tool for clinical high-throughput screening of Alzheimer's disease.

Keywords: superconducting quantum interference device; magnetic nanoparticle; immunoassay

1. Introduction

To accurately screen for and diagnose diseases at early stages, precise immunoassays such as a single-molecule array [1], mesoscale discovery assay [2], and single-molecule count [3] have been reported to exhibit higher sensitivity than current clinical methods such as enzyme-linked immunosorbent assays (ELISAs) and radioimmunoassay. However, limitations have also been found for precise immunoassays with regard to operational complexity and interference. Immunomagnetic reduction (IMR), entailing the use of a magnetic reagent comprising a solvent and bioprobe-coated magnetic nanoparticles (MNPs) as labeling markers, was recently proposed to address the mentioned limitations; the advantages of IMR include a wash-free assay, high specificity, one-antibody utility, long lifetime, and biosafety [4]. Preparation for IMR involves mixing a magnetic reagent and liquid sample with target biomolecules. Subsequently, an IMR instrument is used to measure the variation of the alternating-current (ac) magnetic susceptibility χ_{ac} of the mixture under a bioconjugation process. During bioconjugation, target molecules connect to bioprobes on MNPs, and a large magnetic cluster forms. With regard to the ac excitation field (i.e., χ_{ac}), the magnetic cluster exhibits a substantially lower response than a single MNP. An intuitive explanation of this phenomenon is that the rotational velocity, the direction of which is influenced by the ac magnetic field, is slower for a large magnetic cluster than for a single, smaller MNP. Thus, the formation of a large magnetic cluster results in the reduction of χ_{ac}. Further reduction of χ_{ac} occurs as the concentration of target molecules increases. In addition, the bounded nontarget molecules, so-called interference materials, may spin away from the MNPs because the centrifugal force generated by the rotation of MNPs is larger than the weak binding force of a nonspecific bioconjugation process. In a previous study, an excitation frequency of approximately 20 kHz was chosen because at this frequency, the magnetic moments of single MNPs with hydrodynamic diameters of approximately 50 nm could optimally oscillate with the external ac magnetic field, and the magnetic moments of the particle clusters were almost immobile [5]. These phenomena contribute to the high specificity of IMR compared with other MNP immunoassay methods such as the relaxation time method based on the same types of bioprobe-coated MNPs [6].

In IMR, first-order pick-up coils may be utilized to measure the variation of χ_{ac} for a mixture, but signal processing is required to determine the minimum detectable concentration and, thus, immunoassay sensitivity. Electronic-type IMR instruments, comprising electronic amplifiers and filters, usually achieve immunoassay sensitivity at the sub-parts-per-billion level [7,8]. However, this level of immunoassay sensitivity is only slightly better or similar to that achievable with ELISA and other precision immunoassay methods. Because superconducting quantum interference devices (SQUIDs) are highly sensitive magnetic sensors, most magnetic immunoassay methods [9–15] utilize SQUID sensors. SQUID-type IMR instruments have unique measurement configurations and minor differences in the mechanism of switching from one channel [11] to four channels [12]; thus, such instruments have enhanced immunoassay sensitivity of up to 10^{-2} parts per trillion. At this level of sensitivity, Alzheimer's disease (AD)-related biomarkers are detectable in the plasma of patients. For patients with neurodegenerative diseases, including AD, only minute biomarker amounts are present in plasma because of the barrier between the central nervous system and blood system. Hence, IMR is the only available clinical method of plasma diagnosis [11–15]. AD is prevalent in many developed countries and creates heavy financial and care burdens for many families and societies because no drug for recovery exists. The detection of AD biomarkers before early-stage AD and mild cognitive impairment (MCI) may enable early treatment with currently available drugs and thus prevention of disease progression.

For a SQUID-type IMR instrument with multiple channels, the switching mechanism could theoretically divide the measurement time of one biomarker into that of each channel in several switch cycles. In this study, sufficient switch cycles were applied to classify the measured data points of all switch cycles according to the initial and final states of the bioconjugation process. Hence, several biomarkers could be screened simultaneously in different channels, and the cycle measurement time of each channel dominated the total time. The measurement time of each channel relied on real measurement time and automatically adjusted time for the lock and unlock states of the SQUID sensor. The switching mechanism during switching introduced the large noise to the SQUID sensor as the unlock states, and then prolonged the automatically adjusted time, processed by the software program. For example of the four-channel platform [12], this switching mechanism always introduced too large noise level because the connection wire of the input coil surrounding the SQUID sensor became unconnected like the antenna during the short switch time, and then the SQUID sensor was interfered by the environmental noise to be the unlock state. Hence, the switching mechanism described in this study was developed to enhance the SQUID-type IMR instrument with eight channels.

Sensors **2018**, *18*, 1043

2. Experiments

The eight-channel ac magnetosusceptmeter for MNPs developed in this study comprised three components: sample magnetization, flux coupling, and superconducting sensing (Figure 1).

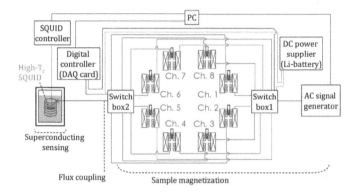

Figure 1. Configuration of the eight-channel ac magnetosusceptometer for MNPs.

The sample magnetization system consisted of eight sets of coils. Each set of coils comprised two excitation coils (referred to as excitation coils 1 and 2) and one pick-up coil. These three coils were assembled coaxially, with the pick-up coil being the innermost coil (Figure 2a). The pick-up coil was an axial gradiometer. The sample that was mixed with the reagent was located in the upper section of the pick-up coil. The eight sets of coils were arrayed as a right octagon to ensure that all sets of coils were geometrically identical. To prevent electromagnetic cross-talk among coil sets, neighboring coil sets were positioned 15 cm apart (Figure 2b).

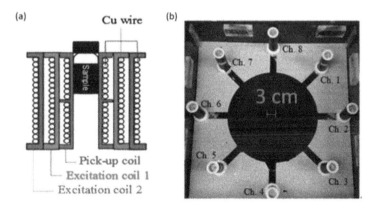

Figure 2. The eight channels of pick-up coils: (**a**) Assembly of one set of coils and (**b**) array of eight coil sets.

For each coil set, excitation coils were driven using an ac signal generator, which generated two independent ac voltages at two frequencies, f_1 and f_2, to excitation coils 1 and 2, respectively. The mixed frequency of $f_1 + 2f_2$ at approximately 20 kHz was used to suppress signals associated with no samples [5]. The ac magnetic signal that resulted from magnetized samples was detected using the pick-up coil and guided to the flux-coupling component. The eight coil sets were activated sequentially. To manipulate the activation of each coil set, electric switches were cascaded between the ac signal generator and excitation coils as well as between the pick-up coil and flux-coupling component, as indicated by switch boxes 1 and 2 (Figure 1).

The circuits of the electric switches are illustrated in Figure 3. Each coil set contained three switches, and each switch comprised one low-noise bipolar junction transistor (BJT) (C1815, UTC, New Taipei City, Taiwan) and one low-noise relay (G6K-2P-Y, Omron Corporation, Kyoto, Japan). The low noise BJT was as a buffer to isolate the DAQ noise from pickup coil and the input coil surrounding the SQUID sensor. It efficiently improved the automatically adjusted time for the unlock states of the SQUID sensor. Relays 1 and 2 (Figure 3) transferred ac voltages from the ac signal generator to excitation coils 1 and 2, respectively. Relay 3 transferred the ac signal from the sample to the flux-coupling component. Activation of the relay was controlled with the BJT, which had a power of 5 V and contained a data-acquisition (DAQ) card (PCI-6221, National Instruments, Austin, TX, USA). The DAQ card output a direct current (dc) of 5 V to the BJT, and the dc voltage generated by the Li battery was applied to the relay. The relay was activated, and either the ac voltage from the ac signal generator was transferred to the excitation coil, or the sample-induced ac signal was transferred to the flux-coupling component. Once the 5-V dc from the DAQ card was stopped, the relay was also stopped. Subsequently, no ac voltage or signal was transferred through the relay. Relays 1 and 2 were inside switch box 1, and relay 3 was inside switch box 2.

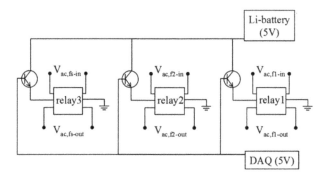

Figure 3. Circuits for electric switches used in the eight-channel AC magnetosusceptometer for MNPs.

The flux-coupling component comprised a pair of twisted wires. One end of the wires was connected to the output ports of relay 3 (i.e., Vac,fs-out in Figure 3). The other end of the wires was linked to a coil. Once Vac,fs-out was activated, an ac electric current I was induced along the wires, and an ac magnetic field B was generated at the coil terminal. The ac magnetic field B at the coil terminal was detected using a high-critical-temperature (high-T_c) SQUID magnetometer (v 5.0, JSQ GmbH), which was part of the superconducting sensing component.

The SQUID magnetometer and coil terminal were immersed in liquid nitrogen. The dewar was posited inside an electromagnetically shielded box, which exhibited a shielding factor of 100 dB at an operating frequency of approximately 20 kHz. The SQUID magnetometer was controlled with a controller (v 5.0, JSQ GmbH), the output signals of which were guided to a personal computer.

Three reagent types were used. The reagents, comprising MNPs with various bioprobes, bound specifically with β-amyloid 40 (Aβ40) (MF-AB0-0060, MagQu, Ltd., New Taipei City, Taiwan), β-amyloid 42 (Aβ42) (MF-AB2-0060, MagQu Ltd.), and tau protein (MF-TAU-0060, MagQu Ltd.). The utilized MNPs with approximately 50 nm in hydrodynamic diameter were composed of a Fe_3O_4 core and dextran in its shell [16]. For Aβ40 and a tau protein assay, an 80-μL reagent was mixed with a 40-μL sample. A 60-μL reagent was mixed with a 60-μL sample for assaying Aβ42. After the reagent was mixed with the sample, the real-time ac magnetic signal (i.e., IMR signal) was recorded to determine the reduction percentage. For each sample, IMR signal detection procedures were performed in duplicate.

3. Results and Discussion

To evaluate the specification differences among the channels, both nonsamples and a biomarker-free mixture without bioconjugation-induced variation of χ_{ac} were separately measured in all eight channels. For these measurements, the biomarker-free mixture comprised an 80-µL Aβ40 reagent and a 40-µL phosphate buffered saline solution.

The recording time was approximately 1 min for each channel, much longer than the total measurement time of tens of seconds. Figure 4 shows the recorded signal intensity of the biomarker-free mixture, represented by the gray bar, and the noise level, represented by the black bar. The observation frequency at 20 kHz was within the approximately 1-MHz bandwidth of the controller. Four critical findings validated consistency among these eight channels. First, the time-independent intensity, representative of χ_{ac} before or without the bioconjugation of target molecules (denoted as $\chi_{ac,0}$), exhibited a variance of only 7.1%. Second, the noise level ranged from 20 to 40 µV, similar to the noise level for the shorter measurement time of tens of seconds. Third, similar standard deviations were identified between the output signals and the noise level. Fourth, the signal-to-noise ratio for all channels was approximately 24.5, which is sufficiently high for reliable measurement. Notably, all reagents exhibited the same results for the biomarker-free mixture (Figure 4) because the only differences among the three reagents were the bioprobes, which were coated on the same batch of MNPs.

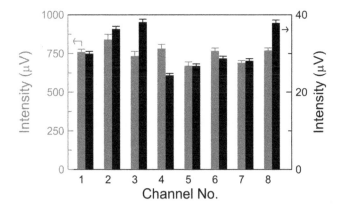

Figure 4. The consistency study of channels with the measurement of noise levels (black bars) and biomarker-free mixture (gray bars) without the bioconjugation process.

Subsequently, the assay performance of each channel was evaluated for the three biomarkers Aβ40, Aβ42, and tau protein. These biomarkers are related to AD and other neurodegenerative diseases. To achieve a high-throughput assay, the measurement guidelines were set that three biomarkers per person specimen should be measured for two consistent duplicates. Hence, each biomarker was mainly measured with two channels and supplementally with one channel, except for the tau protein—the most indicative biomarker of neurodegenerative diseases such as AD—for which two auxiliary channels were used. Channels 1 and 2 were dedicated to the tau protein; channels 3 and 4 were used for Aβ42 primarily and the tau protein secondarily; channels 7 and 8 were dedicated to Aβ40; channel 5 was reserved for the reference sample for calibration (Figure 4); and channel 6 was used as a support for the measurement of Aβ40 or Aβ42.

Immunoassay performance was evaluated using the coefficient of variation (CV) and the relations between known concentrations of target molecules and IMR values. The IMR value can be defined as follows [12–15]:

$$IMR(\%) = \frac{\chi_{ac,0} - \chi_{ac,\phi}}{\chi_{ac,0}} \times 100\%$$ (1)

where $\chi_{ac,o}$ and $\chi_{ac,\phi}$ are the χ_{ac} of the mixture before and after the bioconjugation of target molecules, respectively, the concentration of which is denoted as ϕ.

For three measurement rounds of the same biomarker concentration obtained from each main or support channel, the average and standard deviations of IMR values are plotted on the left axes in Figure 5a–c for the tau protein, Aβ42, and Aβ40, respectively. The CV, defined as the ratio of the standard deviation to the average value, was used as the stability indicator in comparison with all obtained IMR values from the various channels for each biomarker. The CV values are plotted on the right axes in Figure 5a–c for the tau protein, Aβ42, and Aβ40, respectively. All CV values for the biomarkers in concentration ranges lower than 5% validated high consistency for the IMR values obtained from the main and support channels. The biomarker ranges for the tau protein, Aβ42, and Aβ40 were 1–30,000, 0.1–1000 pg/mL, and 1–1000 pg/mL, respectively.

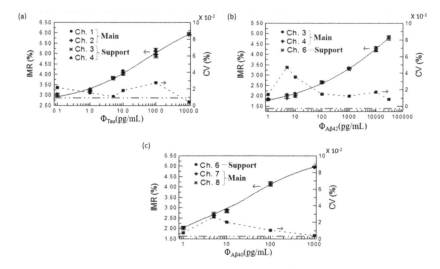

Figure 5. Relationship between the concentrations of (**a**) Aβ40; (**b**) Aβ42; and (**c**) tau protein and the measurement parameters of IMR and CV values in and among channels. IMR with a low detection limit for the various biomarkers is marked with red doted lines.

To analyze the biomarker concentration in human plasma by using the obtained IMR data from all main or support channels, the experimental relations between the average IMR values obtained from all main or support channels and the known biomarker concentrations were fitted as standard curves, marked as solid lines in Figure 5a–c. The standard curves of all biomarkers followed the same logistic function [12–15]:

$$IMR(\%) = \frac{A - B}{1 + (\frac{\phi}{\phi_0})^{\gamma}} + B$$ (2)

where A, B, ϕ_0, and γ denote fitting parameters and ϕ denotes the biomarker concentration. All fitted values of A, B, ϕ_0, and γ as well as the coefficient of determination R^2 for the biomarkers Aβ40, Aβ42, and tau protein are listed in Table 1.

Table 1. Fitted values of parameters A, B, ϕ_0, and γ from Equation (2), and coefficient of determination R^2 for the relationships between IMR signals and biomarker concentrations.

Parameter Biomarker	A	B	ϕ_0	γ	R^2
Aβ40	1.51	5.46	29.45	0.55	0.998
Aβ42	1.29	10.42	228561.4	0.23	0.999
Tau protein	2.64	6.89	53.09	0.42	0.998

The value of A in Equation (2) represents the IMR signals without biomarkers. A is the lowest level for the IMR signals. Once the biomarker concentration becomes infinite, the IMR signals approach B. Thus, B is the saturated IMR signal value at high biomarker concentrations. ϕ_0 corresponds to the biomarker concentration at which the IMR signals are $(A + B)/2$. The power γ is the slope $d[IMR(\%)]/d\phi$ where ϕ is at ϕ_0.

The critical metric of immunoassay performance, the concentration at a low detection limit, was derived from IMR of a low detection limit using Equation (2). In this study, IMR of a low detection limit for one biomarker in spite of the measurement channel was defined as $A + 3 \times CV_{max}$, and plotted with the red dotted line in Figure 5. Here, CV_{max} is the maximum CV value in the experimental range of test concentration, part of detectable concentrations. For the detection on the low concentration, the IMR value was close to the minimum IMR of no biomarkers, i.e., A value, and the definition based on three times of the IMR variation among main or support channels was strict to discrimination from the worst case of zero concentration. In addition, to validate the reliability of a low detection limit, the average intensity at a low detection limit should be larger than the noise level for each main or support channel for each biomarker. The average intensity of a low detection limit was $\chi_{ac,\phi}$, and could be obtained from $\chi_{ac,0}$, based on the biomarker-free mixture in Figure 4, by using Equation (1).

For the detection of Aβ40, first, IMR of a low detection limit was obtained as $(1.51 + 3 \times 0.025)\% = 1.585\%$ due to 0.025% of the maximum CV value occurring at 5.5 pg/mL in Figure 5c and 1.51% of the parameter A in Table 1. Second, the concentration of a low detection limit was determined to be approximately 0.03 pg/mL. Third, the average intensities at a low detection limit, relative to $\chi_{ac,\phi}$, were $770.5 \times (100\% - 1.585\%) = 758.3$ μV, $692.6 \times (100\% - 1.585\%) = 681.6$ μV, and $771.0 \times (100\% - 1.585\%) = 758.8$ μV for channels 6, 7, and 8, respectively. These values were clearly higher than the noise level plotted relative to the right axis in Figure 4. Similarly, the IMR values and concentration of low detection limits in assaying Aβ42 and tau protein were determined to be 1.39% and 2.90% and 3×10^{-4} pg/mL and 0.06 pg/mL, respectively. The IMR values of low detection-limits for the biomarkers are marked with the red doted lines in Figure 5a–c. The concentrations in the plasma of healthy people and patients with early-stage AD were found to be, respectively, (71.4 ± 30.8) pg/mL and (38.2 ± 9.2) pg/mL for the Aβ40 biomarker, (15.3 ± 1.6) pg/mL and (18.6 ± 1.5) pg/mL for the Aβ42 biomarker, and (16.2 ± 9.1) pg/mL and (53.6 ± 22.8) pg/mL for the tau protein biomarker [13–15]. Hence, the concentrations of the low detection limits were much lower than the concentrations found in healthy and early-stage AD patients. Therefore, the eight-channel ac magnetosusceptometer is sufficiently sensitive to detect and quantify these biomarkers in human plasma in screenings for early-stage AD or MCI.

The total measurement time was also a critical factor for evaluation. For each of the three biomarkers at a concentration of several picograms per milliliter, the measurement times were up to approximately 4 h when the single-channel platform was used, and they were theoretically only dependent on the conjugation process. Hence, when the multichannel platform with the switching mechanism was used, the measurement time of one biomarker, (i.e., one round) was still approximately 4 h. The total measurement time should be defined as that for the total specimens and should be compared with the ratio of the total specimens over the utilized channel number. For example, the total measurement time of seven specimens was approximately 4 h over one round for the eight-channel platform, with one channel used for calibration; 12 h over three rounds for the previously developed four-channel platform, with one channel used for calibration; and 28 h over seven rounds for the previously developed single-channel platform involving a complicated calibration process.

4. Conclusions

An eight-channel ac magnetosusceptometer for MNPs was developed for high-throughput screening. The developed magnetosusceptometer effectively measured seven biomarkers per round at a high sensitivity, detecting quantities much lower than 1 pg/mL. The instrument design involved a switching mechanism and high-T_c SQUID. High consistency was demonstrated among channels, with only 7.1% variation for the biomarker-free mixture. The developed magnetosusceptometer exhibited a signal-to-noise ratio of approximately 24.5. The instrument proved effective for the detection of minute biomarker amounts in serum, particularly for three biomarkers of AD used in this study. Results were verified in duplications of measured data.

Acknowledgments: This study was supported by the Ministry of Science and Technology, Taiwan (MOST 103-2321-B-003-006, 102-2221-E-003-008-MY2, 104-2221-E-003-016-, 104-2923-M-003-001, 103-2112-M-003-010, 103-2112-M-003-010), the Aim for the Top University Plan of National Taiwan Normal University, and the Ministry of Education, Taiwan (105J1A27).

Author Contributions: Jen-Jie Chieh conceived and designed the experiments and wrote the manuscript; Shu-Hsien Liao designed the SQUID and electronics; Hsin-Hsein Chen, Yen-Fu Lee, and Feng-Chun Lin fabricated the instrument and synthesized the reagent; Wen-Chun Wei and Ming-Hsien Chiang measured the data; Ming-Jang Chiu provided the clinical resources; Herng-Er Horng and Shieh-YuehYang coordinated and supervised the work.

Conflicts of Interest: The authors declare no conflict of interest.

References

1. Zetterberg, H.; Wilson, D.; Andreasson, U.; Minthon, L.; Blennow, K.; Randall, J.; Hansson, O. Plasma tau levels in Alzheimer's disease. *Alzheimers Res. Ther.* **2013**, *5*, 9. [CrossRef] [PubMed]
2. Cludts, I.; Meager, A.; Thorpe, R.; Wadhwa, M. Detection of neutralizing interleukin-17 antibodies in autoimmune polyendocrinopathy syndrome-1 (APS-1) patients using a novel non-cell based electrochemiluminescence assay. *Cytokine* **2010**, *50*, 129–137. [CrossRef] [PubMed]
3. Todd, J.; Freese, B.; Lu, A.; Held, D.; Morey, J.; Livingston, R.; Goix, P. Ultrasensitive flow-based immunoassays using single-molecule counting. *Clin. Chem.* **2007**, *53*, 1990–1995. [CrossRef] [PubMed]
4. Hong, C.Y.; Chen, W.H.; Chien, C.F.; Yang, S.Y.; Horng, H.E.; Yang, L.C.; Yang, H.C. Wash-free immunomagnetic detection for serum through magnetic susceptibility reduction. *Appl. Phys. Lett.* **2007**, *90*, 074105. [CrossRef]
5. Hong, C.Y.; Wu, C.C.; Chiu, Y.C.; Yang, S.Y.; Horng, H.E.; Yang, H.C. Magnetic susceptibility reduction method for magnetically labeled immunoassay. *Appl. Phys. Lett.* **2006**, *88*, 212512. [CrossRef]
6. Kötitz, R.; Weitschies, W.; Trahms, L.; Brewer, W.; Semmler, W. Determination of the binding reaction between avidin and biotin by relaxation measurements of magnetic nanoparticles. *J. Magn. Magn. Mater.* **1999**, *194*, 62–68. [CrossRef]
7. Yang, C.C.; Yang, S.Y.; Chieh, J.J.; Horng, H.E.; Hong, C.Y.; Yang, H.C. Universal Behavior of Concentration-Dependent Reduction in AC Magnetic Susceptibility of Bioreagents. *IEEE Magn. Lett.* **2012**, *3*, 1500104. [CrossRef]
8. Yang, S.Y.; Chieh, J.J.; Huang, K.W.; Yang, C.C.; Chen, T.C.; Ho, C.S.; Chang, S.F.; Chen, H.H.; Horng, H.E.; Hong, C.Y.; et al. Molecule-assisted nanoparticle clustering effect in immunomagnetic reduction assay. *J. Appl. Phys.* **2013**, *113*, 144903. [CrossRef]
9. Enpuku, K.; Minotani, T.; Gima, T.; Kuroki, Y.; Itoh, Y.; Yamashita, M.; Katakura, Y.; Kuhara, S. Detection of Magnetic Nanoparticles with Superconducting Quantum Interference Device (SQUID) Magnetometer and Application to Immunoassays. *Jpn. J. Appl. Phys.* **1999**, *38*, L1102. [CrossRef]
10. Lee, S.K.; Myers, W.R.; Grossman, H.L.; Cho, H.M.; Chemla, Y.R.; Clarke, J. Magnetic gradiometer based on a high-transition temperature superconducting quantum interference device for improved sensitivity of a biosensor. *Appl. Phys. Lett.* **2002**, *81*, 3094–3096. [CrossRef]

11. Chieh, J.J.; Yang, S.Y.; Jian, Z.F.; Wang, W.C.; Horng, H.E.; Yang, H.C.; Hong, C.Y. Hyper-high-sensitivity wash-free magnetoreduction assay on bio-molecules using high-Tc superconducting quantum interference devices. *J. Appl. Phys.* **2008**, *103*, 014703. [CrossRef]

12. Chiu, M.J.; Horng, H.E.; Chieh, J.J.; Liao, S.H.; Chen, C.H.; Shih, B.Y.; Yang, C.C.; Lee, C.L.; Chen, T.F.; Yang, S.Y.; et al. Multi-Channel SQUID-based ultra-high-sensitivity in-vitro detections for bio-markers of Alzheimer's disease via immunomagnetic reduction. *IEEE Trans. Appl. Supercond.* **2011**, *21*, 477–480. [CrossRef]

13. Yang, C.C.; Yang, S.Y.; Chieh, J.J.; Horng, H.E.; Hong, C.Y.; Yang, H.C.; Chen, K.H.; Shih, B.Y.; Chen, T.F.; Chiu, M.J. Biofunctionalized magnetic nanoparticles for specifically detecting biomarkers of Alzheimer's disease in vitro. *ACS Chem. Neurosci.* **2011**, *2*, 500–505. [CrossRef] [PubMed]

14. Chiu, M.J.; Yang, S.Y.; Chen, T.F.; Chieh, J.J.; Huang, T.Z.; Yip, P.K.; Yang, H.C.; Cheng, T.W.; Chen, Y.F.; Hua, M.S.; et al. New assay for old markers-plasma beta amyloid of mild cognitive impairment and Alzheimer's disease. *Curr. Alzheimer Res.* **2012**, *9*, 1142–1148. [CrossRef] [PubMed]

15. Chiu, M.J.; Yang, S.Y.; Horng, H.E.; Yang, C.C.; Chen, T.F.; Chieh, J.J.; Chen, H.H.; Chen, T.C.; Ho, C.S.; Chang, S.F.; et al. Combined Plasma Biomarkers for Diagnosing Mild Cognition Impairment and Alzheimer's Disease. *ACS Chem. Neurosci.* **2013**, *4*, 1530–1536. [CrossRef] [PubMed]

16. Jiang, W.; Yang, H.C.; Yang, S.Y.; Horng, H.E.; Hung, J.C.; Chen, Y.C.; Hong, C.Y. Preparation and properties of superparamagnetic nanoparticles with narrow size distribution and biocompatible. *J. Magn. Magn. Mater.* **2004**, *283*, 210–214. [CrossRef]

Article

Polyacrylamide Ferrogels with Magnetite or Strontium Hexaferrite: Next Step in the Development of Soft Biomimetic Matter for Biosensor Applications

Alexander P. Safronov [1,2], Ekaterina A. Mikhnevich [1], Zahra Lotfollahi [3,4], Felix A. Blyakhman [1,5], Tatyana F. Sklyar [1,5], Aitor Larrañaga Varga [6], Anatoly I. Medvedev [1,2], Sergio Fernández Armas [6] and Galina V. Kurlyandskaya [1,3,*]

[1] Institute of Natural Sciences and Mathematics, Ural Federal University, Ekaterinburg 620002, Russia; safronov@iep.uran.ru (A.P.S.); emikhnevich93@gmail.com (E.A.M.); Feliks.Blyakhman@urfu.ru (F.A.B.); t.f.shkliar@urfu.ru (T.F.S.); medtom@iep.uran.ru (A.I.M.)

[2] Institute of Electrophysics, Ural Division RAS, Ekaterinburg 620016, Russia

[3] Departamento de Electricidad y ElectrónicaUniversidad del País Vasco UPV/EHU, 48080 Bilbao, Spain; lotfollahi@gmail.com

[4] Deapartment of Physics, University of Birjand, Birjand 97175-615, Iran

[5] Biomedical Physics and Engineering Department, Ural State Medical University, Ekaterinburg 620028, Russia

[6] Advanced Research Facilities (SGIKER), Universidad del País Vasco UPV-EHU, 48080 Bilbao, Spain; aitor.larranaga@ehu.eus (A.L.V.); sergio.fernandez@ehu.eus (S.F.A.)

* Correspondence: galina@we.lc.ehu.es; Tel./Fax: +34-9460-13237

Received: 15 December 2017; Accepted: 15 January 2018; Published: 16 January 2018

Abstract: Magnetic biosensors are an important part of biomedical applications of magnetic materials. As the living tissue is basically a "soft matter." this study addresses the development of ferrogels (FG) with micron sized magnetic particles of magnetite and strontium hexaferrite mimicking the living tissue. The basic composition of the FG comprised the polymeric network of polyacrylamide, synthesized by free radical polymerization of monomeric acrylamide (AAm) in water solution at three levels of concentration (1.1 M, 0.85 M and 0.58 M) to provide the FG with varying elasticity. To improve FG biocompatibility and to prevent the precipitation of the particles, polysaccharide thickeners—guar gum or xanthan gum were used. The content of magnetic particles in FG varied up to 5.2 wt % depending on the FG composition. The mechanical properties of FG and their deformation in a uniform magnetic field were comparatively analyzed. FG filled with strontium hexaferrite particles have larger Young's modulus value than FG filled with magnetite particles, most likely due to the specific features of the adhesion of the network's polymeric subchains on the surface of the particles. FG networks with xanthan are stronger and have higher modulus than the FG with guar. FG based on magnetite, contract in a magnetic field 0.42 T, whereas some FG based on strontium hexaferrite swell. Weak FG with the lowest concentration of AAm shows a much stronger response to a field, as the concentration of AAm governs the Young's modulus of ferrogel. A small magnetic field magnetoimpedance sensor prototype with $Co_{68.6}Fe_{3.9}Mo_{3.0}Si_{12.0}B_{12.5}$ rapidly quenched amorphous ribbon based element was designed aiming to develop a sensor working with a disposable stripe sensitive element. The proposed protocol allowed measurements of the concentration dependence of magnetic particles in gels using magnetoimpedance responses in the presence of magnetite and strontium hexaferrite ferrogels with xanthan. We have discussed the importance of magnetic history for the detection process and demonstrated the importance of remnant magnetization in the case of the gels with large magnetic particles.

Keywords: magnetic nanoparticles; strontium hexaferrite; magnetite; ferrofluids; polyacrylamide gel; ferrogel; tissue engineering; magnetic biosensors; giant magnetoimpedance

1. Introduction

Biomedical application of magnetic materials is a rapidly extending area, which requires additional research [1–3]. Over the past decade, many efforts have been made for the development of a generation of magnetic biosensors, i.e., compact analytical devices incorporating a biological sensitive element, integrated in a physicochemical transducer employing a magnetic field [4,5]. These devices are becoming common tools used for a variety of biomedical applications [6–8]. In many cases, magnetic nanoparticles (MNPs) work as biomolecular labels [4,7] which must be provided in the form of water-based ferrofluids [2,3,9]. Apart from evaluating the concentration of the magnetic labels in a test solution, the detection of superparamagnetic nanoparticles (MNPs) after intracellular uptake was also tested [6,10]. However, one of the greatly requested applications for different cancer therapies (the detection of the MNPs incorporated into living tissues) or regenerative medicine has not yet been properly addressed. One of the fundamental reasons for this delay is the sensitivity of the existing sensing devices. Giant magnetoimpedance (MI) had attracted special attention as the phenomenon providing the basis for sensors capable of detecting picotesla magnetic fields [11,12]. The great advantage of the present day available MI-sensitive elements is their easily approached enhanced sensitivity of the order of 100%/Oe for MI ratio variation [13].

The idea of using a magnetic field sensor in combination with magnetic particles/nanoparticles working as magnetic markers for the detection of molecular recognition events was first reported in 1998 by Baselt et al. [4]. Such a device was based on giant magnetoresistance (GMR) technology and employed magnetic composite microbeads for simultaneous characterization of many biomolecular interaction events. Different geometry was proposed for a magnetoresistive biosensor prototype designed for the detection of a single micro magnetic sphere by a ring-shaped element working on anisotropic magnetoresistance effect [14]. The important disadvantage of exchange-coupled GMR sensors is the high field required for a reasonable resistance change. Microsized spin valve sensors with lower operation fields were also developed for detection of biomolecules with magnetic markers [15]. Another approach employed the Hall effect for a sensor based on standard metal-oxide-semiconductor technology for selective detection of magnetic markers [16] (Besse et al. 2002). With respect to the magnetic field sensitivity, the MI effect is the best option for the creation of magnetic biosensors: it can be mentioned that the maximum sensitivity achieved at present is ~2%/Oe for GMR materials [17].

There were attempts to develop MI-biosensors based on sensitive elements of different types: rapidly quenched wires, glass-coated microwires, amorphous ribbons and thin films [5,6,18–20]. Different MI-materials have different advantages and disadvantages, summarized in various topical reviews [12,20,21]. Although, thin films were under special focus recently due to their excellent compatibility with semiconductor electronics [5,22–24], cheap MI-biosensors with a disposable sensitive element in the form of a stripe are being developed. These disposable sensors can be used by non-skilled personnel in non-sterile environments. Amorphous Co-based rapidly quenched ribbons are excellent candidates in this case [10,13,25,26].

The development of MI biosensor to evaluate the properties of biological tissues is strongly conditioned by the availability of reliable samples. Biological materials present a wide variety of morphologies especially in the case of cancer affected tissues, characterized by the accelerated growth of irregular blood vessels [27]. In our previous works related to MI biosensors with thin film a sensitive element [5,28], we proposed to substitute biological samples at the first stage of the development of the MI biosensor prototype by the adequate model materials—synthetic ferrogels mimicking the main properties of the living tissues [29–31]. Those ferrogels were based on MNPs obtained by the electrophysical technique of laser target evaporation [32,33]. In addition, it is necessary to point out that a wide variety of morphologies of the cancer affected tissues become inevitably reflected in the corresponding wide varieties of their mechanical and magneto-electrical properties.

The choice of the MNPs was defined by an important condition for the majority of magnetic biosensing cases adapted to the magnetic label detection principle: the stray fields induced by the magnetic markers are employed as biomolecular labels providing a means for the transfer of information about the concentration of magnetic labels and therefore the biocomponent of interest [4,18]. The sensitivity limit is related to the type of MNPs—the magnetic moment of an individual particle in the external magnetic field governs the stray fields and the biodetection limit. The temptation to increase the magnetic moment of the individual magnetic label is strictly limited by the condition of superparamagnetic state [34,35] in order to avoid MNPs' agglomeration in zero field. When the magnetic particles or MNPs become incorporated into a tissue and spatially localized, their size might be much larger as a result of non-proper functioning of the living system.

We therefore propose to study various kinds of ferrogels with micron sized commercially available magnetic particles (MPs) in order to create reliable samples mimicking natural tissue and evaluate the possibility of their detection by an amorphous Co-based ribbon sensitive element. Detection of the stray fields of magnetic particles incorporated into a living system have a number of additional requests. As the living tissue is basically a "soft matter." the mechanical properties are important as well as possible deformations caused by the application of the external magnetic field. We describe our experience of the synthesis and characterization of magnetite Fe_3O_4 and strontium hexaferrite $SrFe_{12}O_{19}$ powder based ferrogels including the measurements of the change of MI of the Co-based ribbon sensitive element in the presence of ferrogels with different concentration of iron oxide magnetic particles using a specially designed MI sensor prototype as a model for biosensors.

2. Experimental

2.1. Materials

For the preparation of ferrogels we used commercial magnetic powders of oxides: magnetite Fe_3O_4 (Alfa Aesar, Ward Hill, MA, USA) and strontium hexaferrite $SrFe_{12}O_{19}$ powder mark 28PFS250 (Olkon, Kineshma, Russian Federation).

Amorphous ribbons with $Co_{68.6}Fe_{3.9}Mo_{3.0}Si_{12.0}B_{12.5}$ nominal composition (Figure 1a) were prepared by rapid quenching technique [36] and all measurements were made in their as-quenched state, i.e., without additional heat treatments or surface modifications. The saturation of magnetostriction coefficient (λs), was estimated by the inductive technique measurements of the hysteresis loops of the ribbon at 1.7 Hz under different mechanical stresses. The magnetostriction coefficient was calculated from the slope of anisotropy field vs. mechanical stress curve [37]. The requirements for the magnetoimpedance tests in magnetic biosensing conditions are: reasonably high MI effect sensitivity with respect to applied magnetic field and good corrosion stability. The composition with a molybdenum addition was selected on the basis of our previous evaluation of the MI effect and corrosion stability. We previously evaluated the resistance of the amorphous ribbons of different compositions by the weight loss in solutions of 3.0 M of orthophosphorous acid (H_3PO_3) in an ultrasonic bath [38]. Figure 1a shows a general view of the surface of $Co_{68.6}Fe_{3.9}Mo_{3.0}Si_{12.0}B_{12.5}$ amorphous ribbon. The geometry of the ribbon for magnetic and magnetoimpedance measurements was $18 \times 0.78 \times 0.026$ (mm).

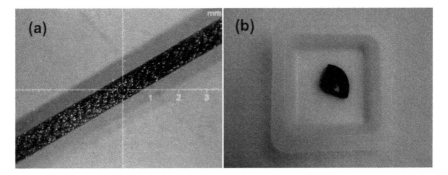

Figure 1. General view in optical microscope of the free surface of $Co_{68.6}Fe_{3.9}Mo_{3.0}Si_{12.0}B_{12.5}$ rapidly quenched amorphous ribbon based MI sensitive element (**a**); general macroscopic view of xanthan-based ferrogel FG-2a with magnetite magnetic particles in 3.19 weight % concentration (**b**).

2.2. Synthesis of Ferrogels

The basic composition of ferrogels comprised the polymeric network of polyacrylamide (PAAm), which was synthesized by free radical polymerization of monomeric acrylamide (AAm) (AppliChem, Darmstadt, Germany) in water solution at three levels of concentration (1.1 M, 0.85 M and 0.58 M) to provide gels with varying elasticity. N,N′ methylenebisacrylamide (Merck, Schtuchardt, Germany) was used as a cross-linker. Its molar concentration to monomer was 1:100. Ammonium persulfate was used as an initiator in 5 mM concentration. Powdered magnetite and strontium hexaferrite were dispersed in the reaction mixture before the synthesis. To prevent the precipitation of the particles polysaccharide thickeners were added to the reaction mixture—guar gum and xanthan gum (Sigma-Aldrich, St. Louis, MO, USA). Average molecular weight of polysaccharides was determined by viscometry using Mark-Houwink constants K = 1.7×10^{-6}, a = 1.14 for xanthan [39] and K = 3.7×10^{-4}, a = 0.74 for guar [40]. The molecular weight was found 2.2×10^6 (xanthan gum) and 1.6×10^6 (guar gum).

Thus, the synthesis of ferrogels included the following steps. First, 1% water solutions of guar or xanthan by weight were prepared by vigorous stirring, followed by equilibration for 24 h at room temperature and filtered to remove macroscopic gel fraction. Guar and xanthan gum solutions are viscoelastic at room temperature and elevated temperatures due to the formation of the physical network of polysaccharide chains. Then the weighted amount of magnetic filler was dispersed under vigorous stirring in the solution of a thickener. Due to the physical network of a polysaccharide the particles did not precipitate. Then the monomer, the cross-linker and the initiator were added to the dispersion. The reaction mixture was stirred, poured into a cylindrical polyethylene mold and placed under argon (to prevent inhibition of the reaction by oxygen) into an oven at 90 °C for 1 h until the polymerization of PAAm was completed. The total volume of the reaction mixture in each synthesis was 4.5 mL.

The amount of magnetic material was 0.5 g. The series of ferrogels filled with magnetite were denoted FG-1 if the thickener was guar gum and FG-2 if the thickener was xanthan gum. The series of ferrogels filled with strontium hexaferrite will be denoted FG-3 (thickener–guar gum) and FG-4 (thickener–xanthan gum). In each series, ferrogels varied in the concentration of the monomer–acrylamide taken in synthesis. Its concentration is placed after a mark of the series, e.g., FG-1-0.85 stands for the ferrogel based on magnetite dispersed with guar gum thickener and AAm concentration in the reaction mixture was 0.85 M. After the synthesis, ferrogels were taken out of the molds and kept in the excess of distilled water for two weeks with daily water renewal to wash out the residual monomer, salts and linear PAAm olygomers. The equilibrium swelling ratio of the gels after their stabilization was determined by gravimetry. The weight m_0 of a swollen piece of gel of ca 0.5 g was measured using an analytical balance (Mettler-Toledo MS104S, Columbus, OH, USA). Then it was dried in an oven at 90 °C down to the constant weight of the dry residue m_1. The apparent swelling ratio (α) was calculated according to the equation:

$$\alpha = \frac{m_0 - m_1}{m_1} \tag{1}$$

The value of the swelling ratio was used for the calculation of the particle content in the swollen gel applying the equation:

$$\vartheta = \frac{\gamma}{1 + \alpha} \tag{2}$$

where γ stands for the weight fraction of particles in the dry residue within the total comprised particles and cross-linked dry PAAm. The value of γ was calculated based on the proportions between the particles, the monomer, the cross-linker and the thickener taken in the synthesis. The values of the swelling ratio of ferrogels and the weight fraction of magnetic particles in the swollen gel are given in Table 1.

Table 1. Description of the synthesized gel and ferrogel samples: composition and swelling ratio.

G/FG Series	Magnetic Particles	Concentration of AAm (M)	Thickener	Swelling Ratio, α	Weight fraction of Magnetic Particles, ϑ (%)
FG-1a	Fe_3O_4	1.10	Guar	10	5.20
FG-1b	Fe_3O_4	0.85	Guar	15	4.12
FG-1c	Fe_3O_4	0.58	Guar	27	2.59
FG-2a	Fe_3O_4	1.10	Xanthan	17	3.19
FG-2b	Fe_3O_4	0.85	Xanthan	21	2.88
FG-2c	Fe_3O_4	0.58	Xanthan	93	0.77
FG-3a	$SrFe_{12}O_{19}$	1.10	Guar	15	3.70
FG-3b	$SrFe_{12}O_{19}$	0.85	Guar	51	1.24
FG-3c	$SrFe_{12}O_{19}$	0.58	Guar	54	1.31
FG-4a	$SrFe_{12}O_{19}$	1.10	Xanthan	16	3.33
FG-4b	$SrFe_{12}O_{19}$	0.85	Xanthan	52	1.21
FG-4c	$SrFe_{12}O_{19}$	0.58	Xanthan	92	0.77
G-a	-	1.10	Xanthan	31	0
G-b	-	0.85	Xanthan	90	0
G-c	-	0.58	Xanthan	303	0

Table 1 also gives a description of xanthan-based gels (without magnetic particles, with different amount of thickener) used for MI measurements. They are very important for subtraction of the corresponding signal from the ferrogel samples in the course of the evaluation of the average contribution of the stray fields of magnetic particles. Figure 1b shows general view of FG-2a ferrogel as an example. Synthesized gel samples keep their shapes constant for a very short time period. The changes of the shape for longer time periods are related to water loss and generally they can be described as shrinking the sample as a whole.

2.3. Methods

The specific surface area (S_{sp}) of powders was measured by the low-temperature sorption of nitrogen (Brunauer-Emmett-Teller physical adsorption (BET)) using Micromeritics TriStar3000 analyzer.

The X-ray diffraction (XRD) studies were performed by operating at 40 kV and 40 mA the DISCOVER D8 (Bruker, Leiderdorp, The Netherlands) diffractometer using Cu-Kα radiation (λ = 1.5418 Å), a graphite monochromator and a scintillation detector. The magnetic particles or amorphous ribbon cut in various pieces were mounted onto a zero-background silicon wafer placed in a sample holder. A fixed divergence and anti-scattering slit were used. Bruker software TOPAS-3 with Rietveld full-profile refinement was employed for the quantitative analysis of all the diffractograms.

Additionally, the average size of coherent diffraction domains (D_{XRD}) was estimated using the Scherrer approach in the case of magnetic particles [41]. The coefficient k in the Scherrer equation was taken as k = 0.9 and instrumental broadening of the peaks $FWHM_{instr}$ = 0.1.

The shape and size of the magnetic particles were studied by field-emission scanning electron microscopy (SEM) in a JEOL JSM-7000 F, equipped with a Schottky type field-emission gun. Energy-dispersive X-ray spectroscopy studies of dry ferrogels were done by ETM3000 SEM of HITACHI (Tokyo, Japan).

Magnetic measurements of the hysteresis loops (M(H)) were carried out by a vibrating sample magnetometer (VSM, Lake Shore 7404, Westerville, OH, USA) in the ±1.8 kOe field range. Although, in some cases magnetic powders were not completely saturated, for the sake of simplicity we decided to assign the saturation magnetization value to the magnetization in the field of 18 kOe. Ferrogels were measured in a polycarbonate capsule following a specially developed protocol. A vibrating sample magnetometer and a conventional inductive technique were used to study the magnetic properties of the ribbons at room temperature.

The measurements of the Young modulus of ferrogels were performed using a specially designed laboratory setup. Cylindrical gel samples ~10 mm in length and ~10 mm in diameter were used for mechanical testing. The equipment of laboratory design was built around an optical system based on a digital camera and contained the following: a bath for the gel sample, a semiconductor force transducer, an electromagnetic linear motor for applying mechanical deformations, a semiconductor optical transducer for the gel sample length measurement in dynamics. The gel sample was clamped vertically between the livers of force transducer and linear motor. To produce the "compression-decompression" cycle of gel samples, the triangular axial deformations (ε) with a constant rate 0.5 mm/s and an amplitude of up to 10% of the initial length of samples were applied. Gel tension (σ) was calculated as the recorded force normalized by the cross-sectional area of the gel sample. Gel tension (σ) was calculated as the recorded force normalized by the cross-sectional area of the gel sample. In the course of deformation, the cross-section of samples was corrected based on the assumption of a gel's constant volume under deformation (Poisson ratio is close to 0.5) [42]. The Young modulus (E) was determined as the slope of the $\sigma(\varepsilon)$ dependence:

$$tg\beta = \sigma/\varepsilon = E \tag{3}$$

The deformation of ferrogels in a uniform magnetic field was studied using the magnetic system designed and constructed at JSC "URALREDMET" (Verkhnyaya Pyshma, Russian Federation). The design of a magnetic array used for the magneto deformation studies was described early [43]. The system comprised permanent NdFeB magnets assembled with magnetic conductors and provided a vertical uniform magnetic field of 0.420 ± 0.5% T in the central zone of 1 cm^3 in volume. The magnitude of the magnetic field inside the array was determined with a Gaussmeter TKH_4 instrument (Maurer Magnetic AG Grüningen/Switzerland).

The optical cuvette with ferrogel sample was placed in the central zone of the magnetic system. The cylindrical sample of ca 5 mm both in height and in diameter was pinned onto a needle holder affixed to the bottom of the cuvette. The cuvette was filled with water and sealed with Parafilm to prevent the deswelling of the ferrogel due to evaporation. The dimension changes of the ferrogel along the field lines and across the field lines were monitored by an EVS color VEC_545_USB optical system equipped with a condensing lens.

2.4. Magnetoimpedance Measurements

The length of the ribbon sensitive element for MI measurements was 18 mm (Figure 2). The ribbon was incorporated into a "microstripe" line (Figure 2a) and a uniform external magnetic field of up to 150 Oe was created by a pair of Helmholtz coils. The MI changes were calculated from the reflection S_{11} coefficient measured by a network analyzer (Agilent E8358A). All measurements were made using an output power of 0 dB, corresponding to the amplitude of the excitation current across the sample of about 1 mA. MI measurements were made after calibration and mathematical subtraction of the test fixture contributions following the well-established protocol [17]. The longitudinal MI was measured in configuration of an alternating current flowing parallel to the external magnetic field. Total impedance (Z) was measured as a function of the external magnetic field in a frequency range of the driving current frequency of 0.1 to 100 MHz.

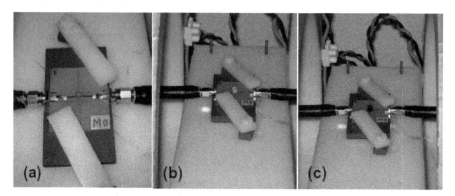

Figure 2. General view of the ribbon based MI sensitive element installed into "microstripe" line; external magnetic field is applied parallel to the long axis of the ribbon: (**a**) polymer capsule is empty (calibration measurement); (**b**) polymer capsule contains xanthan-based G-a gel without MPs; (**c**) polymer capsule contains xanthan-based FG-2a ferrogel with magnetite magnetic particles in 3.19 weight % concentration.

In our previous works for MI measurements with ferrogels, we used gel pieces of particular geometry of about 0.5 g weight [5]. The measurements were strongly conditioned by the time as gel/ferrogels were rapidly changing mass, shrinking and loosing water. In the present work, we therefore used half a polymer water resistant capsule for MI measurements. We used previously synthesized stable gels/ferrogels in order to completely fill the half of the capsule with the gel/ferrogel (Figure 2).

First, the MI responses were studied by the measurements of the amorphous ribbon sensitive element itself. For the second measurements, half of the polymer capsule was situated at the center of the ribbon element (Figure 2a). The next step was to measure MI responses with the capsule filled by gel and ferrogels of the same weight but having different concentrations of the magnetic filler. As the length of the sensitive element was 18 mm the 4 mm in diameter capsule placed in a central part provided about 22% coverage of the active surface of the sensitive element. This means that the increase of the surface coverage in the future may increase the sensitivity of the detector. At the same time, there is no reason to expect a linear dependence of the sensitivity with increase of the surface coverage. The central part of the ribbon is the most appropriate position for the detection of non-uniform magnetic fields due to the absence of the contribution of the shape anisotropy near the ends of the ribbons. Measurements with the gel were used for the subtraction of the corresponding signal from the ferrogel samples in order to evaluate the average contribution of the stray fields of magnetic particles in the spatial distribution corresponding to each ferrogel case.

3. Results and Discussion

The idea of synthesis of ferrogels as model samples mimicking natural tissues for biosensor prototype development is a multidisciplinary approach which cannot be realized as independent steps in synthesis, material characterization and sensor prototype design. Many parameters must be optimized at a time by creating synergetic combinations of different materials working in one electronic analytical device. We therefore will comparatively analyze and discuss the most important functional properties of amorphous ribbons (MI sensitive elements), magnetic particles and, based on them, gels from the point of view of such compatibility.

3.1. Characterization of the Ribbons

The XRD patterns (Figure 3) of $Co_{68.6}Fe_{3.9}Mo_{3.0}Si_{12.0}B_{12.5}$ ribbons confirmed the absence of the long-range order and the crystalline phases: a clear amorphous structure identification was possible (one very broad diffraction peak). Optical microscopy and SEM studies of the surface features show that the fabricated ribbon has a well-defined geometry (small changes of the width, Figure 1a) and a rather smooth surface with an elongated defect along the solidification direction typical for rapidly quenched ribbons (Figure 3b). The magnetostriction coefficient was very close to zero, coercivity $H_c = 0.5$ Oe and saturation magnetization $M_s = 67$ (emu/g) were defined from the shape of the hysteresis loop M(H), where M is magnetization and H is an external magnetic field applied along the ribbon axes in plane of the ribbon. H_c is defined as the field for where M = 0 and M_s is determined from the saturation of the measured magnetic moment and divided by the mass of the sample. For the sake of comparison, we include some examples of the saturation magnetization values for the amorphous ribbons of different compositions: for $Co_{68.5}Fe_{4.0}Si_{15.0}B_{12.5}$ $M_s = 76$ emu/g and for $Co_{65.9}Fe_{3.5}W_{3.1}Si_{16.5}B_{11.0}$ $M_s = 60$ emu/g [44]. The observed shape of M(H) loops indicate that the effective magnetic anisotropy is a longitudinal one but a slow saturation approach can be connected with an anisotropy distribution due to frozen-in stresses and surface anisotropy contribution [44].

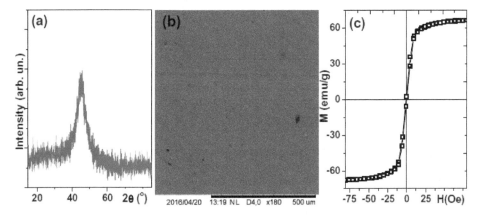

Figure 3. Selected properties of $Co_{68.6}Fe_{3.9}Mo_{3.0}Si_{12.0}B_{12.5}$ amorphous ribbons: XRD patterns (**a**); surface morphology, SEM (**b**) and VSM hysteresis loop measured at room temperature (**c**).

3.2. Characterization of the Magnetite and Hexaferrite Particles

The specific surface area measured by BET for both types of particles was 6.9 m^2/g for magnetite and 3.9 m^2/g for strontium hexaferrite samples.

Electron microscopy images of magnetic particles used as gel fillers are shown in Figure 4. Magnetite particles are polydisperse and smoothly shaped tending to be round. Their average size is below 400 nm. Strontium hexaferrite particles are very polydisperse with particle dimensions varying from 100 to 5000 nm and irregularly shaped with sharp corners in a majority of cases. As strontium hexaferrite is produced by the sintering method with consequent milling of the product, the particles had a very mixed morphology and a pronounced degree of enhanced internal stresses.

Figure 4. Scanning electron microscopy images of commercial "magnetite" (**a**) and "strontium hexaferrite" (**b**) magnetic particles. General view of dried piece of FG-2a ferrogel: secondary electrons SEM image (**c**) EDX analysis (Fe-Kα imaging) confirming iron presence and reasonably uniform distribution in magnetite-based dried FG-2a ferrogel sample with magnetite magnetic particles in 3.19 weight % concentration (**d**). The size of the image in (**c,d**) cases is 250 μm × 200 μm.

Figure 5 shows an example of XRD diffractograms of for magnetite and strontium hexaferrite $SrFe_{12}O_{19}$ powders. In all cases (see also Table 2) XRD diffractograms fitted reasonably well with the Rietveld method and crystallographic parameters were defined for all observed crystallographic phases. One can see that Fe_3O_4 is a main phase (94%) in the case of "magnetite" sample and $SrFe_{12}O_{19}$ is a main phase (87%) for "strontium hexaferrite" particles. Interestingly, the average sizes obtained by the XRD analysis appear to be rather smaller in comparison with the previous TEM estimation. It is understandable that due to the variety of shapes of the particles a precise size estimation is difficult in both cases. Even a simpler explanation can be given to these observations—the majority of large particles observed by SEM as individual particles are polycrystalline units having two or even more crystallographically different grains.

Table 2. Results of XRD analysis of the most intensive peaks for commercially available "magnetite" and "strontium hexaferrite" particles.

Particles Type	Phases	2θ (°)	FWHM$_{av}$ (°)	Intensity	Phases Content (%)	D$_{XRD}$ (nm)
"Magnetite"	Fe_3O_4	35.631	0.164	2799.33	94	130
	Fe_2O_3	33.354	0.209	27.61	1	75
	FeO(OH)	21.369	0.216	75.64	5	70
"Strontium hexaferrite"	$SrFe_{12}O_{19}$	34.349	0.136	712.19	87	230
	Fe_2O_3	33.319	0.262	166.16	13	50

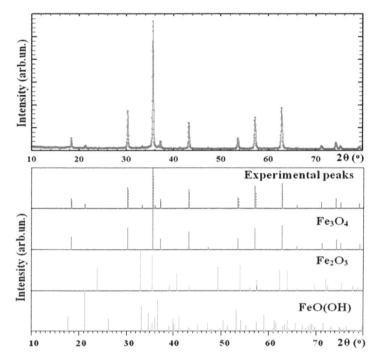

Figure 5. XRD patterns of commercial "magnetite" particles (top) and different phases identification as a result of peaks de-convolution using the databases identification for all observed phases (see also Table 2).

As the next step, the magnetic properties (M(H) hysteresis loops and primary magnetization curves) of the particles measured were studied at room temperature. In the case of magnetite MPs (Figure 6a), the material almost saturates in the field of 10 kOe showing a saturation magnetization value of about 84 emu/g, i.e., the value for bulk magnetite saturation magnetization [34,45]. In the case of strontium hexaferrite (Figure 6b), the saturation was not reached even in the field of 18 kOe which is not surprising for particles with a high degree of internal stresses and irregular shapes. At the same time, the magnetization value of about 54 emu/g is quite high, also indicating that the size of the particles is large. We observed a coercive force value of about 80 Oe for magnetite and about 1200 Oe for strontium hexaferrite samples as to be expected for multidomain MPs of both types. One can also analyze the clear difference in the change of initial magnetic permeability through the primary magnetization curves–it is certainly higher for magnetite MPs, ensuring a higher magnetic moment in a low magnetic field.

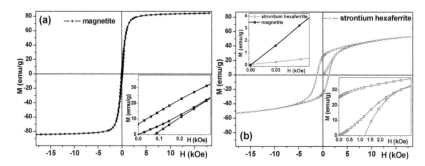

Figure 6. Hysteresis loops of magnetite (**a**) and strontium hexaferrite (**b**) particles. Insets in the right part of the main graphs show primary magnetization curves; inset in the left part of graph (**b**) shows primary magnetization curves of both types of particles in the small fields of interest for sensor applications.

3.3. Characterization of the Ferrogels

3.3.1. Ferrogels Structure

As it is seen in Table 1, the ferrogels in the study varied in three respects: (1) the chemical nature of the magnetic particles (magnetite or strontium hexaferrite), (2) the concentration of AAm in the synthesis (0.58 M, 0.85 M and 1.1 M) and (3) the chemical origin of the thickener (guar gum or xanthan gum). The main structural feature of the polymeric matrix of a ferrogel is the mesh size of the gel network, which is the average distance between neighboring cross-links in the network. The relation between the mesh size of the network and the average size of the embedded magnetic particles determines whether the particles can freely move in the network or whether they are immobilized in it [46]. The mesh size of the ferrogel network was estimated based on the values of their swelling ratio (Table 1) using the Flory-Rehner equation for the average number of monomer units in subchains between the cross-links [47]:

$$N_C = \frac{V_1(0.5\alpha^{-1} - \alpha^{-1/3})}{V_2(\ln(1 - \alpha^{-1}) + \alpha^{-1} + \chi\alpha^{-2})} \tag{4}$$

where V_1, V_2—are molar volumes of a solvent and of a polymer respectively, χ—is the Flory-Huggins parameter for a polymer–solvent mixture. We used V_1 = 18 cm^3/mol (water), V_2 = 56.2 cm^3/mol (PAAm) and χ = 0.12. The last two values were obtained by means of a quantum mechanics molecular modeling software package CAChe7.5.

As the swelling ratio diminished, if the particles were embedded in the gel matrix we took blank gels for the estimation of the upper limit for the mesh size of the polymeric network. Equation (6) gave the average number of monomer units in linear sub-chains N_C = 240, 1470 and 11,360 in respect to the polymeric network based on 1.1 M, 0.85 M and 0.58 M AAm concentration. The average distance between the cross-links, i.e., the mesh size, was then estimated using the well-known equation for the mean square end-to-end distance of the random Gaussian coil with hindered rotation [48]. It gave the upper limit for the mesh size: 5 nm, 12 nm and 33 nm for the mentioned set of AAm concentration. If we compare the mesh size to the average size of magnetite or strontium hexaferrite magnetic particles (see Figure 4), we will see that the particles are immobilized in the ferrogel network, which in this case might be considered as an elastic continuous medium for the magnetic filler. Polysaccharides, like any biopolymers, are complex multilevel systems with complex conformational and aggregative behavior, which is not fully understood even in individual solutions. For example, in our recent work one particular aspect of this behavior was discussed [49]. In addition, we should point out that, despite the chemical similarity of polysaccharides to each other, they demonstrate significant and poorly understood differences in behavior. The systems analyzed in the present work are even more complex including polyacrylamide mesh and filler. It is impossible to provide sufficient understanding of all these details at this time. We did focus on the polymer peculiarities of polysaccharides but simply used them as thickeners.

The weight fraction of magnetic particles in ferrogels was a dependent variable as it was related to the swelling ratio of the ferrogel according to Equation (2). The value of the MPs' fraction also determines the magnetic moment of the ferrogel. For magnetic biosensing, low concentrations (below 5 wt %) are usually employed in order to avoid an interaction process and interferences which would complicate the magnetization of the stray fields of the ensemble. Selected concentrations are therefore useful for comparative studies of both mechanical and magnetic properties.

3.3.2. Young Modulus of Ferrogels

As the polymeric matrix constitutes the continuous phase of a ferrogel, its mechanical strength is mostly influenced by the composition of a polymeric network. In this study, the density of the polymeric network was changed by varying the AAm concentration in the synthesis. At a low AAm monomer concentration in the synthesis the polymeric matrix contained less polymeric chains than at a high AAm concentration. Figure 7 presents the dependence of the Young modulus of the ferrogels with different thickeners on the concentration of AAm monomer taken in the synthesis. The Young modulus of ferrogels strongly depends on the AAm concentration. With a two-fold increase in concentration (from 0.58 to 1.1 M) the modulus increased ca ten-fold—from 1 kPa up to 8–12 kPa depending on the ferrogel series. In the studied concentration range of AAm, the dependence of the modulus is almost linear. It means that the influence of the density of polymeric network on the mechanical strength dominates over the influence of other composition variables (Table 1). Meanwhile, the influence of the origin of a thickener and the chemical nature of magnetic material is still noticeable.

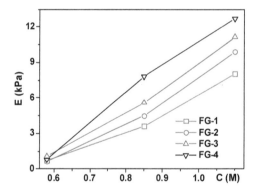

Figure 7. Young modulus of ferrogels with different concentration of acrylamide (C) taken in the synthesis of gel matrix for different FG series: FG-1 filled with magnetite if the thickener was guar gum and FG-2 if the thickener was xanthan gum, filled with strontium hexaferrite FG-3 (thickener–guar gum) and FG-4 (thickener–xanthan gum). Weight fraction ranges of magnetic particles were 2.59–5.20 wt % for FG-1, 0.77–3.19 wt % for FG-2, 1.31–3.70 wt % for FG-3 and 0.77–3.33 wt % for FG-4. Lines are for eye-guide only. C-concentrations were as high as 1.1 M, 0.85 M and 0.58 M aiming to provide gels with varying elasticity.

Concerning the origin of the thickener, xanthan gum provides higher values of the Young modulus than guar gum (compare FG-1 with FG-2 and FG-3 with FG-4 in Figure 7). It means that the addition of polysaccharide to the system is not exclusively restricted to the increase of elasticity. As mentioned above, polysaccharides form weak physical networks in solution [50]. Thus, introducing of a polysaccharide into the synthesis of a ferrogel results in the mixing of two types of polymeric networks–the chemical one provided by the cross-linked polyacrylamide and the physical one comprised of the aggregated polysaccharide macromolecules. Such systems are known as the semi-interpenetrating networks (semi-IPN) [49,50] and their properties depend on the nature of both networks.

As for the nature of the magnetic material, ferrogels FG-3 and FG-4 filled with strontium hexaferrite particles have a systematically larger value of the Young modulus than ferrogels filled with magnetite particles (compare FG-1 with FG-3 and FG-2 with FG-4 in Figure 7). Most likely, the dependence of the modulus on the magnetic material stems from the specific features of the adhesion of network polymeric subchains on the surface of the particles. Apparently, the adhesion of the gel network to the surface of strontium hexaferrite is stronger than the adhesion to the surface of magnetite. It makes the modulus of strontium hexaferrite based ferrogels (FG-3 and FG-4) higher than the modulus of magnetite based ferrogels (FG-1 and FG-2).

3.3.3. Deformation of Ferrogels in a Uniform Magnetic Field

Figure 8 shows how the volume of a ferrogel sample changes if it is placed in a uniform magnetic field of 4.2 kOe. As mentioned in the Methods section, the measured values were the diameter and the height of the cylindrical sample in the field. It was found that in all cases both the height and the diameter changed in the same way. If the gel dimension along the field (the diameter of a gel) increased so did the dimension across the field (the height of a gel). The same can be observed concerning the decrease in diameter and height. We have never observed opposite changes like the decrease in diameter but an increase in the height. Due to such similarity of the deformation along the field and across the field these data were combined to give the total volume change of a gel in field. The relative volume change was calculated by the following equation:

$$\frac{V}{V_0} = \frac{hD^2}{h_0 D_0^2} \tag{5}$$

where h_0 and D_0 are the height and the diameter of a gel without the applied field, h and D are the values if the field was applied for a certain period of time. The curves in Figure 8 give the relative increase or the decrease (in %) of the gel volume in the field. If the gel contracts in the magnetic field, the relative volume change is negative, if the gel swells–the volume change is positive.

There are two basic trends in the curves shown in Figure 8. One is the decrease of gel volume in the field (contraction), another is the opposite–the increase of the volume i.e., swelling. Both trends were reported in an earlier work for ferrogels based on the polyelectrolyte matrix of the copolymer of acrylamide and potassium acrylate [43]. The prevalence of a trend depends on the composition variables of the ferrogel. In the case of the FG-1 and FG-2 series based on magnetite, the contraction trend prevails in all compositions (Figure 8a,b). There is the obvious influence of the density of the polymeric matrix of the ferrogel. Ferrogels with a dense matrix provided by a high concentration of AAm in the synthesis, show no or a very small contraction. Weak ferrogels with the lowest concentration of AAm show a much stronger response to the field. Such a dependence is reasonable as the concentration of AAm governs the Young modulus of the ferrogels (Figure 7). At the same time, the influence of the polysaccharide does not follow the trend in the modulus. As shown in Figure 7, gels with a xanthan physical network are stronger and have higher modulus than ferrogels with a guar physical network. Despite this, the response of the FG-2 ("xanthan") series to the field is stronger than the response of the FG-1 ("guar") series. The maximal contraction obtained for magnetite based ferrogels was 30% of contraction for the gel synthesized in 0.58 M AAm with xanthan physical network.

The positive feedback of the xanthan physical network on the deformation of ferrogels in the field is also present in strontium hexaferrite based gels FG-3 and FG-4. However, the trends are more complex. Contraction, swelling and their combination are observed for these gels. Gels of FG-3 series in general swell in the magnetic field. However, in the case of FG-3-0.85 gel, small initial contractions were observed at the early stages of field application. Gels of FG-4 series do not show a uniform pattern. Gel FG-4-1.10 with a high compression modulus swells up to 10% in the field like gels of the FG-3 series. At the same time, gels FG-4-0.85 and FG-4-0.58 with a lower Young modulus contract in the field like gels of FG-1 and FG-2 series. The maximal contraction is around 30% like in the FG-2 series.

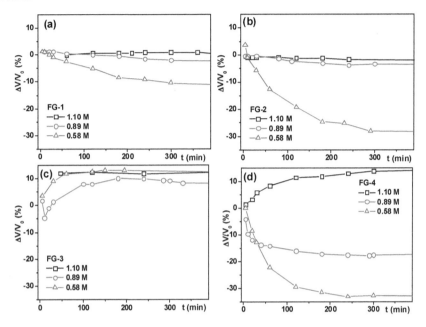

Figure 8. Time dependence of the volume deformation of ferrogels in the uniform magnetic field 0.420 T. (**a**) FG-1 series, (**b**) FG-2 series, (**c**) FG-3 series and (**d**) FG-4 series. Lines are for eye-guide only. FG-1 series filled with magnetite if the thickener was guar gum and FG-2 if the thickener was xanthan gum, filled with strontium hexaferrite FG-3 (thickener–guar gum) and FG-4 (thickener–xanthan gum). Weight fraction ranges of magnetic particles were 2.59–5.20 wt % for FG-1, 0.77–3.19 wt % for FG-2, 1.31–3.70 wt % for FG-3 and 0.77–3.33 wt % for FG-4. The legend of the plot gives the concentration of AAm in the synthesis of ferrogels, which determines the Young modulus.

We are not ready at this point to give an explanation as to why the applied external field causes changes in the ferrogel volume and the volume changes with respect to time. For now, we consider them experimental facts. In different systems both contraction and expansion were observed. In principle, the explanation of this behavior is related to the minimization of the free energy of particle interaction. It does this by changing the degree of swelling of the matrix. The degree of swelling according to modern concepts depends on at least four different thermodynamic forces, plus the forces of magnetic interaction of particles, the equations for which are also not completely defined, since they strongly depend on their mutual arrangement. All this makes the description extremely complex and uncertain at present. What is clear is that the modulus of the gel does not exceptionally govern the magnetically induced deformation. It is also clear that several structural parameters are missing from the analysis. Among them certainly are the aggregation degree of magnetic particles and the structure of aggregates. Unfortunately, these features are very difficult to characterize in such systems as a cross-linked gel. There was an experimental observation [51–53] that the linear dimension of ferrogels based on siloxane resins along the field lines increased if the magnetic particles were uniformly distributed in the polymeric matrix and it decreased if the magnetic particles in the synthesis were pre-aligned in the direction of the field. It means that the formation of aggregates of magnetic particles in the gel is crucially important for its deformation in the uniform magnetic field.

3.4. Magnetic Properties of Ferrogels

The magnetic properties of xanthan containing gels were studied in detail. Figure 9 shows hysteresis loops of magnetite-based gels and ferrogels with different concentrations of MPs. The gel matrix contributes very little to the total magnetic signal but this diamagnetic contribution can be very precisely measured by VSM and extracted from the ferrogel signal if necessary. Unfortunately, the M(H) loops for gels had a very small ferromagnetic contribution, corresponding to impurities but this contribution was very similar for gels with different amounts of thickener and it was at least 4 orders of magnitude smaller in comparison with ferromagnetic contribution of magnetic particles (Figure 9b). The values of saturation magnetization of ferrogels show a linear dependence on the magnetic filler concentration, as expected for materials with a low concentration of MPs (Figure 9d). Although the saturation magnetization M_s is a very important parameter for the characterization of magnetic materials and the understanding of their basic properties, in many technological applications and biomedicine, magnetic materials are not used in the saturated state. It is simply because the application of a high magnetic field is costly, difficult to apply (a high field) in a compact device and it is subject to special biomedical regulations [54].

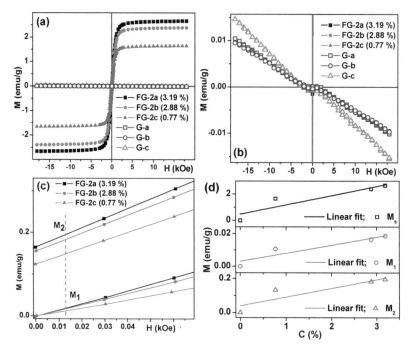

Figure 9. Magnetic properties of gels and ferrogels with magnetite particles. M(H) loops of gels/ferrogels in the field range in which the saturation tendency for ferrogels is observed (**a**). For FG-2 series the thickener was xanthan gum. M(H) loops of gels/ferrogels in low magnetic moment values scale for which the lack of the saturation tendency for gels is appreciated (**b**). Primary magnetization curves and close to remanence behavior for ferrogels: M_1 (for primary magnetization curve) and M_2 (for approaching remanence) values of the magnetic moments of ferrogels, green dashed line indicates magnetic field close to 12 Oe (**c**). Concentration dependence of M_s and M_1, M_2 parameters for FG-2 ferrogels (**d**).

Let us analyze the magnetization behavior in the positive fields starting from the demagnetized state (primary magnetization curve) and approaching zero field after saturation, i.e., approaching the state of remanence (Figure 9c). In order to create an efficient detector of the stray fields of the magnetic labels, the parameters of the two magnetic materials (amorphous ribbon and magnetic particles in the present case) must be adjusted to each other: the work interval of high sensitivity of the MI sensitive element should overlap with the field interval where magnetic particles have sufficiently high magnetic moments in order to create a sizeable magnetic field. Coming back to the M(H) hysteresis loop of the amorphous ribbon, one can see that 12 Oe value is a critical field related to an effective magnetic anisotropy in which high magnetoimpedance sensitivity can occur. Selecting two characteristic values M_1 (for primary magnetization curve) and M_2 (for approaching remanence) for a 12 Oe external magnetic field, we can plot their concentration dependence for FG-2 ferrogels, which also appears to be linear. Interestingly, the M_1 values corresponding to primary magnetization curves are of about two orders of magnitude smaller in comparison with M_s and one order of magnitude smaller in comparison with M_2 corresponding values.

Figure 9d shows the concentration dependence of M_s, M_1 and M_2 parameters for FG-2 ferrogels. The first experimental point corresponds to the case of gel with no magnetic particles included. As one can estimate from the hysteresis loop of corresponding gels, the magnetic contribution of the gel without NPs is at least two orders of magnitude smaller in comparison with contribution of ferrogels.

Figure 10 shows hysteresis loops of strontium hexaferrite-based ferrogels with different concentrations of MPs. Again, the gel matrixes contribute very little to the total magnetic signals but they can be carefully extracted from the ferrogel signals. The same as in the case of ferrogels with a magnetite filler, the values of saturation magnetization of ferrogels (Figure 10b) and two characteristic values M_1 (for primary magnetization curve) and M_2 (for approaching remanence) for a 12 Oe external magnetic field show linear dependence on the magnetic filler concentration.

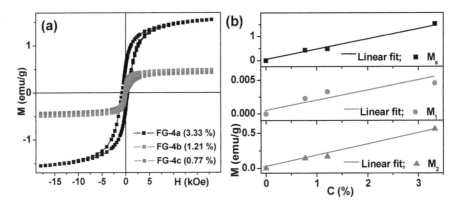

Figure 10. Magnetic properties of ferrogels with strontium hexaferrite particles (see also Table 1). M(H) loops of ferrogels in wide field range in which the saturation tendency for ferrogels is observed (**a**). For FG-4 series the thickener was xanthan gum. Concentration dependence of the saturation magnetization and M_1, M_2 parameters for FG-4 ferrogels: M_1 (for primary magnetization curve) and M_2 (for approaching remanence) values of the magnetic moments of ferrogels (**b**).

An interesting observation can be made at this point: M_2 values for strontium ferrite ferrogels of the same concentration are characterized by about twice higher magnetic moments (low field case of remanence approach) in comparison with magnetite based gels, indicating similar or better opportunities for the detection by a magnetic field sensor of strontium hexaferrite MPs based ferrogels. Coming back to the magnetic properties of the particles of magnetic fillers (Figure 6), one can notice that the magnetization value in the field of 18 kOe is much higher in the case of magnetite particles.

But the value of magnetization in the field of 12 Oe is larger for strontium hexaferrite MPs ferrogels due to magnetic hysteresis. For ferrogels with nanosized particles of a magnetic filler, due to the superparamagnetic state of the nanoparticles the simple following relation is expected: the higher the value of saturation magnetization, the higher the magnetic moment value in a small field [34]. For large scale MPs above the superparamagnetic-ferrimagnetic transition, this relation is not necessarily valid—M_s is higher for magnetite particles (and gels on their basis) but M_2 values are higher for strontium hexaferrite MPs based gels.

We would like to emphasize that the idea of using large scale MPs above the superparamagnetic-ferrimagnetic transition was not only stimulated by the fact that the magnetization signals at low field (M_1 and M_2) in the case of nanoparticles are too low for the detection of very small concentrations of them in ferrogels. In our previous work, we have demonstrated that the detection of 20 nm iron oxide nanoparticles in ferrogels in the same concentration interval was possible [5]. The main idea was to improve the detection limit following the well-known law at nanoscale–the larger the size, the higher the saturation magnetization [34]. At the same time MPs above the superparamagnetic-ferrimagnetic transition tend to aggregate and show a much more complex magnetic behavior, as we have demonstrated above. Synthesis of ferrogels with large particles can be a good solution to some extent, preventing coarse particle aggregation especially under the application of an external magnetic field. Encapsulated particles can be delivered to the point of therapy using a magnetic field for different purposes.

For example, the possibility of biomedical applications of larger nanoparticles up to 300 nm size is being discussed [27]. In solid tumors due to angiogenesis, i.e., the growth of new blood vessels, irregular gaps much bigger than the ones in healthy blood vessels are present. Nanoparticles of less than 300 nm in diameter can pass through such big gaps and accumulate in the tumors. Apart from field assisted delivery it is necessary to control the number of magnetic particles accumulated at a certain point. We propose to use a magnetic field sensor based on the magnetoimpedance effect for this purpose. In order to demonstrate the validity of the concept we have prepared ferrogel samples with large magnetic particles, systems mimicking tumor tissue with embedded particles.

3.5. Magnetoimpedance Measurements in Configuration of Biosensor Prototype

The MI ratio was calculated as follows:

$$\frac{\Delta Z}{Z} = 100 \times \frac{Z(H) - Z(H_{\max})}{Z(H_{\max})} \tag{6}$$

where $H_{\max} = 100$ Oe. For MI frequency dependence analysis, the changes of the maximum value of the total impedance $\Delta Z / Z_{\max}$ was appropriate. MI sensitivity for total impedance and its real part were denominated as $S(\Delta Z/Z)$ and $S(\Delta R/R)$ accordingly:

$$S\left(\frac{\Delta Z}{Z}\right) = \frac{\delta(\Delta Z/Z)}{\delta H} \tag{7}$$

where $\delta(\Delta Z/Z)$ is the change in the total impedance or real part GMI ratio for the magnetic field increment $\delta(H) = 0.1$ Oe. Figure 11 helps to understand the MI related definitions. Primarily the impedance modulus is calculated in Ohms from the experimentally measured voltage drop. $Z(H_{\max})$ is the impedance in the maximum field (Figure 11a). For magnetic material showing strong MI effect generally speaking $Z(H_{\max}) \neq Z(H)$. This dependence is easier to understand analyzing exact numbers for MI ratio: $Z(H_{\max}) = 0$ for the field $H = 150$ Oe but $Z(H) = 95$ Oe for the field $H = 25$ Oe (compare points for black and red arrows). The blue font part helps to represent the concept of sensitivity: let us select two values of the magnetic field (H_1 and H_2). $S\left(\frac{\Delta Z}{Z}\right) = (\Delta(Z/Z)(H_1) - \Delta(Z/Z)(H_2))(H_2 - H_1)$ corresponds to the MI sensitivity in the field interval of H_1 to H_2. Here we used a higher field increment (not $\delta(H) = 0.1$ Oe) just for simplicity.

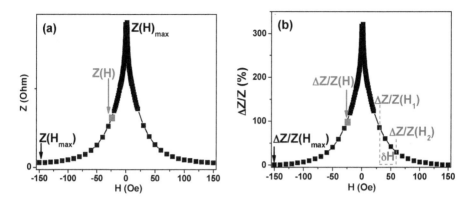

Figure 11. Magnetoimpedance related definitions for magnetoimpedance ratio $\Delta(Z/Z)(H)$ (**a**) and MI ratio sensitivity $S(\Delta(Z/Z)(H)$ calculations (**b**).

The experimental error in determining the impedance was within 1%. Direct current resistivity (R_{DC}) was also measured in all cases. MI measurements were made at room temperature.

Next, we comparatively studied MI responses of the amorphous ribbon sensitive elements in the presence of gels and FG-2 and FG-4 ferrogels (Figure 12). First of all, the MI of the sensitive element was measured in a frequency range of 1 to 100 MHz without any gel/ferrogel. The shape of $(\Delta Z/Z)_{max}$ curve is typical for Co-based amorphous ribbons [55,56] which can be described in the frame of classic electrodynamics and field dependence of transverse dynamic magnetic permeability in a condition of strong skin effect [56,57]. Figure 10a shows two MI $(\Delta Z/Z)_{max}$ curves: the "up"–$(\Delta Z/Z)_{max}$ curve was measured in the field starting from saturation in the maximum negative external magnetic field; the "down"–$(\Delta Z/Z)_{max}$ curve was measured in the field starting from saturation in the maximum positive external magnetic field. One can clearly see that both curves overlap providing high quality measurements.

Insets in Figure 12a help to understand better the definition of the $(\Delta Z/Z)_{max}$ value. The field dependence of the $\Delta Z/Z$ ratio is again typical for the MI response of amorphous ribbons with longitudinal magnetic anisotropy—one-peak shape MI response [37], as we could expect from the shape of the hysteresis loop of the ribbon (Figure 2c). However, a higher resolution representation shows a fine structure of the $\Delta Z/Z(H)$ curve showing two peaks. The existence of two peaks can be explained by the contribution of the surface anisotropy by adding a small transverse magnetic anisotropy component [37].

Here we can briefly discuss the concept of sensitivity of the magnetic field sensor with respect to the applied magnetic field. There are different field regions with close to linear $\Delta Z/Z(H)$ dependence: in a small magnetic field for 1.4 Oe < |H| < 1.9 Oe, $S\left(\frac{\Delta Z}{Z}\right) = 100\%/\text{Oe}$; and in the high magnetic field for 6 Oe < |H| < 15 Oe, $S\left(\frac{\Delta Z}{Z}\right) = 8\%/\text{Oe}$. Despite the fact of extremely high low field MI sensitivity, the linearity interval is very narrow (0.5 Oe) making it not useful for practical applications. The work interval in the high magnetic field is wide (9 Oe). We will show below that even such a modest MI sensitivity as 8%/Oe is sufficient for the measurement of ferrogel concentration dependence using this magnetic field sensor.

Figure 12b shows a frequency dependence of the MI ratio of $Co_{68.6}Fe_{3.9}Mo_{3.0}Si_{12.0}B_{12.5}$ amorphous ribbon without and with G-a gel piece (see also Figure 2b). The effect of the gel presence is clearly manifested by the increase of $(\Delta Z/Z)_{max}$ ratio. In order to quantify the observed difference, let us use the $\Lambda\Delta$ parameter defined as $\Lambda\Delta = (\Delta Z/Z)_{max}(\text{for gel}) - (\Delta Z/Z)_{max}(\text{for ribbon})$ at f = const which was useful for a rather wide frequency range of 20 to 100 MHz.

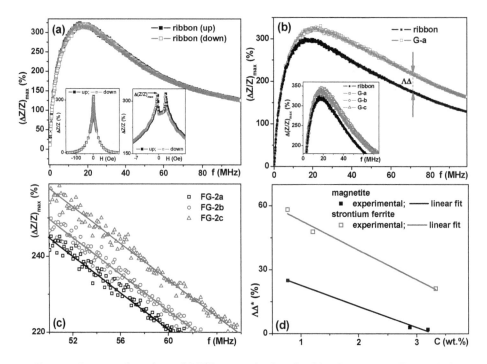

Figure 12. Frequency dependence of $(\Delta Z/Z)_{max}$ ratio for $Co_{68.6}Fe_{3.9}Mo_{3.0}Si_{12.0}B_{12.5}$ rapidly quenched amorphous ribbon. Insets show field dependence of $\Delta(Z/Z)$ ratio for 15 MHz frequency: left inset–is given for large field interval and right inset is given for low magnetic fields (**a**). Frequency dependence of $(\Delta Z/Z)_{max}$ ratio for amorphous ribbon without and with G-a gel. Green arrows explain definition of $\Lambda\Delta$ parameter. Inset shows better resolution image for $(\Delta Z/Z)_{max}$ ratio for amorphous ribbon without and with G-a, G-b and G-c gels (**b**). Selected region of frequency dependence of $(\Delta Z/Z)_{max}$ ratio for amorphous ribbon with FG-2 ferrogels, the thickener was xanthan gum (**c**). Concentration dependence of the $\Lambda\Delta^* = \Lambda\Delta(gel)-\Lambda\Delta(ferrogel)$ difference for FG-2 and FG-4 ferrogels for frequency f = 60 MHz (**d**).

Inset Figure 12b shows the frequency dependences of the MI ratio of amorphous ribbon without and with different gels. Although gel responses are close to each other, they are not exactly identical, indicating the differences in dielectric properties of the gels. On the one hand, this fact makes the extraction of the stray field responses in the case of ferrogels more complicated. On the other hand, it opens the possibility to use magnetoimpedance spectroscopy as a new technique useful for the evaluation of the properties of even non-filled gels in the future. The stray field created by the single uniformly magnetized particle can be calculated using a pure dipole model [14] for two possible geometries [58]: with magnetization of the bead either perpendicular, or parallel to the plane of the sensitive element of the sensor. In literature, there are at least four reported bead detection configurations [58–60], in which bead magnetizing field is perpendicular, parallel or under a certain angle to the sensor plane and with or without modulation across the sensitive element. In the case of ferrogels with rather large particles, simple modeling is not applicable and comparative analysis of the magnetic pre-history and MI responses become crucially important.

Figure 12c shows the frequency dependence of the MI ratio of the amorphous ribbon without and with FG-2 ferrogels (see also Figure 2b). The effect of ferrogel presence is more complex compared to the gel response. Although the corresponding gel presence leads to an increase of $(\Delta Z/Z)_{max}$ ratio, the FG response becomes close to the $(\Delta Z/Z)_{max}$ ratio of the ribbon indicating the existence of at least two contributions with different tendencies. Even so, the FG-2 responses measured directly for FG-2 with different filler concentrations (C, in weight %) show the following tendency: the higher the C value the lower the $(\Delta Z/Z)_{max}$ ratio.

Next, we recalculated the MI values measured for the samples of gels and ferrogels of two types-with magnetite (FG-2) and strontium hexaferrite (FG-4) fillers. It is worth mentioning, that all of the samples for these tests had the same mass (0.070 mg). The $D\Delta^* = D\Delta(gel) - D\Delta(ferrogel)$ parameter reflected the contribution of the stray fields of magnetic particles. Two observations are important. The first one is that in both cases $D\Delta^*$ parameter shows linear decay with the increase of the concentration of the magnetic filler. The second one is that in the case of strontium ferrite a slightly higher sensitivity for the concentration dependence was measured.

At first glance, a higher sensitivity in the case of strontium ferrite MPs seems to be contradicting our previous experience and the existing understanding of the origin of MI sensitivity, namely the disturbance of the external magnetic field by the stray field of magnetic particles. In fact, the detection of the non-single domain (not superparamagnetic particles) requires a different approach. The absolute value of the magnetic moment of superparamagnetic nanoparticles in a certain field does not depend on the magnetic history of superparamagnetic MNPs due to the absence of magnetic hysteresis. For big particles, it is not correct and the value of the magnetic moment in certain magnetic fields depends on the magnetic history and can be very different (Figures 9d and 10b). Magnetite MPs have higher M_s value than strontium hexaferrite MPs but the M_2 parameter defined above is the magnetic moment governing the value of the stray fields of large MPs with non-zero coercivity and it is higher for strontium hexaferrite MPs resulting in a slightly higher MI sensitivity to the concentration dependence.

Of course, any dependence on magnetic history can be seen as a disadvantage for the detection process but the concept is clear and the technological solution can be simple: short period application of a reasonably high external magnetic field prior to the detection process, which can be simply a field created by the permanent magnet.

4. Conclusions and Outlook

Gels and ferrogels with a polymeric network of polyacrylamide were synthesized by radical polymerization of monomeric acrylamide (AAm) in a water solution at three levels of concentration: 1.1 M, 0.85 M and 0.58 M providing gels with varying elasticity. Their mechanical and magnetic properties were comparatively analyzed. Ferrogels filled with strontium hexaferrite particles have a systematically larger value of the Young modulus than ferrogels filled with magnetite, most likely due to the specific features of the adhesion of polymeric subchains of the network on the surface of the particles. Apparently, the adhesion of the gel network to the surface of strontium hexaferrite is stronger than the adhesion to the surface of magnetite.

There are two possible trends in the deformation process of ferrogels: the decrease of gel volume in the field or the increase of the volume, i.e., swelling. The prevalence of a trend depends on the composition variables of the ferrogel. In the case of the series based on magnetite, the contraction trend prevails in all compositions: ferrogels with a dense matrix provided no or very small contraction. Weak ferrogels with the lowest concentration of AAm show a much stronger response to the field as the concentration of AAm governs the modulus of the gel. At the same time the influence of the polysaccharide does not follow the trend in modulus. Physical networks with xanthan are stronger and have higher modulus than ferrogels with guar physical networks. Despite this, the response of the "xanthan" series to the field is stronger than the response of the "guar" series. The maximum contraction obtained for magnetite based ferrogels was 30% of contraction for the gel synthesized in 0.58 M AAm with xanthan physical network. The positive feedback of the xanthan physical network on the deformation of ferrogels in the field is also present in strontium hexaferrite based gels. However, the trends are more complex. Contraction, swelling and their combination are observed in these gels. The maximum contraction is around 30% like in magnetite-based series.

Magnetic properties of magnetite and strontium hexaferrite particles and gels and "xanthan" series of ferrogels were systematically studied and comparatively analyzed. The saturation magnetizations

Sensors **2018**, *18*, 257

of the particles were in both cases quite close to the bulk values, quite in accordance with structural peculiarities and the sizes of the particles.

We have designed a small magnetoimpedance biosensor prototype with $Co_{68.6}Fe_{3.9}Mo_{3.0}Si_{12.0}B_{12.5}$ rapidly quenched amorphous ribbon based element aiming in the future to develop a sensor working with a disposable stripe sensitive element. The developed protocol allowed measurements of the concentration dependence of magnetic particles in magnetite and strontium hexaferrite ferrogels with xanthan using the MI effect. We have discussed the importance of magnetic pre-history and demonstrated importance of remnant magnetization in the detection process.

We have shown the possibility of detecting large magnetic particles using an amorphous ribbon MI sensitive element. Our main result is that we were able to create such model samples mimicking natural tissue. They can be viewed as reliable substitutes for living tissues. This study opens new possibilities for the development of biosensors for the detection of the particles in large volumes. More systematic studies are necessary in order to improve this simple prototype up to the level requested by biomedical applications and research other types of ferrogels and biocomposites but the direction seems to be promising.

Acknowledgments: This work was supported in part within the framework of the state task of the Ministry of Education and Science of Russia 3.6121.2017/8.9; RFBR grant 16-08-00609 and by the ACTIMAT grant of the Basque Country Government. Selected studies were made at SGIKER Common Services of UPV-EHU and URFU Common Services. We thank I.V. Beketov, A.A. Svalova, Burgoa Beitia, A. Amirabadizadeh, A. García-Arribas and I. Orue for their special support.

Author Contributions: A.P. Safronov, G.V. Kurlyandskaya and F.A. Blyakhman conceived and designed the experiments; A.P. Safronov, E.A. Mikhnevich, Z. Lotfollahi, T.F. Sklyar, A. Larrañaga Varga, A.I. Medvedev, S. Fernández Armas and G.V. Kurlyandskaya performed the experiments; A.P. Safronov, F.A. Blyakhman, Aitor Larrañaga Varga, Zahra Lotfollahi and G.V. Kurlyandskaya analyzed the data; A.P. Safronov, F.A. Blyakhman and G.V. Kurlyandskaya wrote the manuscript. All authors discussed the results and implications and commented on the manuscript at all stages. All authors read and approved the final version of the manuscript.

Conflicts of Interest: The authors declare no conflict of interest.

References

1. Bucak, S.; Yavuzturk, B.; Sezer, A.D. Magnetic nanoparticles: Synthesis, surface modification and application. In *Recent Advances in Novel Drug Carrier Systems*; Sezer, A.D., Ed.; InTech: Rijeka, Croatia, 2012; pp. 165–200.
2. Moroz, P.; Jones, S.K.; Gray, B.N. Status of hyperthermia in the treatment of advanced liver cancer. *J. Surg. Oncol.* **2001**, *77*, 259–269. [CrossRef] [PubMed]
3. Pavlov, A.M.; Gabriel, S.A.; Sukhorukov, G.B.; Gould, D.J. Improved and targeted delivery of bioactive molecules to cells with magnetic layer-by-layer assembled microcapsules. *Nanoscale* **2015**, *7*, 9686–9693. [CrossRef] [PubMed]
4. Baselt, D.R.; Lee, G.U.; Natesan, M.; Metzger, S.W.; Sheehan, P.E.; Colton, R.J. A biosensor based on magnetoresistance technology. *Biosens. Bioelectron.* **1998**, *13*, 731–739. [CrossRef]
5. Kurlyandskaya, G.V.; Fernandez, E.; Safronov, A.P.; Svalov, A.V.; Beketov, I.V.; Burgoa Beitia, A.; Garcıa-Arribas, A.; Blyakhman, F.A. Giant magnetoimpedance biosensor for ferrogel detection: Model system to evaluate properties of natural tissue. *Appl. Phys. Lett.* **2015**, *106*, 193702. [CrossRef]
6. Blanc-Béguin, F.; Nabily, S.; Gieraltowski, J.; Turzo, A.; Querellou, S.; Salaun, P.Y. Cytotoxicity and GMI bio-sensor detection of maghemite nanoparticles internalized into cells. *J. Magn. Magn. Mater.* **2009**, *321*, 192–197. [CrossRef]
7. Wang, T.; Guo, L.; Lei, C.; Zhou, Y. Ultrasensitive determination of carcinoembryonic antigens using a magnetoimpedance immunosensor. *RSC Adv.* **2015**, *5*, 51330–51336. [CrossRef]
8. Coisson, M.; Barrera, G.; Celegato, F.; Martino, L.; Vinai, F.; Martino, P.; Ferraro, G.; Tiberto, P. Specific absorption rate determination of magnetic nanoparticles through hyperthermia measurements in non-adiabatic conditions. *J. Magn. Magn. Mater.* **2016**, *415*, 2–7. [CrossRef]
9. Safronov, A.P.; Beketov, I.V.; Tyukova, I.S.; Medvedev, A.I.; Samatov, O.M.; Murzakaev, A.M. Magnetic nanoparticles for biophysical applications synthesized by high-power physical dispersion. *J. Magn. Magn. Mater.* **2015**, *383*, 281–287. [CrossRef]

10. Kumar, A.; Mohapatra, S.; Fal-Miyar, V.; Cerdeira, A.; Garcia, J.A.; Srikanth, H.; Gass, J.; Kurlyandskaya, G.V. Magnetoimpedance biosensor for Fe_3O_4 nanoparticle intracellular uptake evaluation. *Appl. Phys. Lett.* **2007**, *91*, 143902. [CrossRef]

11. Tomizawa, S.; Kawabata, S.; Komori, S.; Hoshino, H.; Fukuoka, Y.; Shinomiya, K. Evaluation of segmental spinal cord evoked magnetic field after sciatic nerve stimulation. *Clin. Neurophys.* **2008**, *119*, 1111–1118. [CrossRef] [PubMed]

12. Uchiyama, T.; Mohri, K.; Honkura, Y.; Panina, L.V. Recent advances of pico-Tesla resolution magneto-impedance sensor based on amorphous wire CMOS IC MI Sensor. *IEEE Trans. Magn.* **2012**, *48*, 3833–3839. [CrossRef]

13. Kurlyandskaya, G.V.; Garcia-Arribas, A.; Barandiaran, J.M.; Kisker, E. Giant magnetoimpedance stripe and coil sensors. *Sens. Actuators A* **2001**, *91*, 116–119. [CrossRef]

14. Miller, M.M.; Prinz, G.A.; Cheng, S.-F.; Bounnak, S. Detection of a micron-sized magnetic sphere using a ring-shaped anisotropic magnetoresistance-based sensor: A model for a magnetoresistance-based biosensor. *Appl. Phys. Lett.* **2002**, *81*, 2211–2213. [CrossRef]

15. Ferreira, H.A.; Graham, D.L.; Freitas, P.P.; Cabral, J.M.S. Biodetection using magnetically labeled biomolecules and arrays of spin valve sensors. *J. Appl. Phys.* **2002**, *93*, 7281–7286. [CrossRef]

16. Besse, P.-A.; Boero, G.; Demierre, M.; Pott, V.; Popovic, R. Detection of single magnetic microbead using a miniaturized silicon Hall sensor. *Appl. Phys. Lett.* **2002**, *80*, 4199–4201. [CrossRef]

17. Kurlyandskaya, G.V.; de Cos, D.; Volchkov, S.O. Magnetosensitive Transducers for Nondestructive Testing Operating on the Basis of the Giant Magnetoimpedance Effect: A Review. *Russ. J. Non-Destr. Test.* **2009**, *45*, 377–398. [CrossRef]

18. Chiriac, H.; Herea, D.D.; Corodeanu, S. Microwire array for giant magnetoimpedance detection of magnetic particles for biosensor prototype. *J. Magn. Magn. Matter.* **2007**, *311*, 425–428. [CrossRef]

19. Kurlyandskaya, G.V.; Levit, V.I. Advanced materials for drug delivery and biosensors based on magnetic label detection. *Mater. Sci. Eng. C* **2007**, *27*, 495–503. [CrossRef]

20. Kurlyandskaya, G.V. Giant magnetoimpedance for biosensing: Advantages and shortcomings. *J. Magn. Magn. Matter.* **2009**, *321*, 659–662. [CrossRef]

21. Wang, T.; Zhou, Y.; Lei, C.; Luo, J.; Xie, S.; Pu, H. Magnetic impedance biosensor: A review. *Biosens. Bioelectron.* **2017**, *90*, 418–435. [CrossRef] [PubMed]

22. Li, B.; Kavaldzhiev, M.N.; Kosel, J. Flexible magnetoimpedance sensor. *J. Magn. Magn. Matter.* **2015**, *378*, 499–505. [CrossRef]

23. Shcherbinin, S.V.; Volchkov, S.O.; Lepalovskii, V.N.; Chlenova, A.A.; Kurlyandskaya, G.V. System based on a ZVA-67 vector network analyzer for measuring high-frequency parameters of magnetic film structures. *Russ. J. Nondestruct.* **2017**, *53*, 204–212. [CrossRef]

24. Svalov, A.V.; Fernandez, E.; Arribas, A.; Alonso, J.; Fdez-Gubieda, M.L.; Kurlyandskaya, G.V. FeNi-based magnetoimpedance multilayers: Tailoring of the softness by magnetic spacers. *Appl. Phys. Lett.* **2012**, *100*, 162410. [CrossRef]

25. Beato-López, J.J.; Pérez-Landazábal, J.I.; Gómez-Polo, C. Magnetic nanoparticle detection method employing non-linear magnetoimpedance effects. *J. Appl. Phys.* **2017**, *121*, 163901. [CrossRef]

26. Beach, R.; Berkowitz, A. Sensitive field-and frequency-dependent impedance spectra of amorphous FeCoSiB wire and ribbon. *J. Appl. Phys.* **1994**, *76*, 6209–6213. [CrossRef]

27. Grossman, J.H.; McNeil, S.E. Nanotechnology in cancer medicine. *Phys. Today* **2012**, *65*, 38–42. [CrossRef]

28. Volchkov, S.O.; Dukhan, A.E.; Dukhan, E.I.; Kurlyandskaya, G.V. Computer-Aided Inspection Center for Magnetoimpedance Spectroscopy. *Russ. J. Nondestruct.* **2016**, *52*, 647–652. [CrossRef]

29. Helminger, M.; Wu, B.; Kollmann, T.; Benke, D.; Schwahn, D.; Pipich, V.; Faivre, D.; Zahn, D.; Cölfen, H. Synthesis and Characterization of Gelatin-Based Magnetic Hydrogels. *Adv. Funct. Mater.* **2014**, *24*, 3187–3196. [CrossRef] [PubMed]

30. Blyakhman, F.A.; Safronov, A.P.; Zubarev, A.Y.; Shklyar, T.F.; Makeyev, O.G.; Makarova, E.B.; Melekhin, V.V.; Larrañaga, A.; Kurlyandskaya, G.V. Polyacrylamide ferrogels with embedded maghemite nanoparticles for biomedical engineering. *Results Phys.* **2017**, *7*, 3624–3633. [CrossRef]

31. Liu, T.Y.; Hu, S.-H.; Liu, T.Y.; Liu, D.-M.; Chen, S.Y. Magnetic-sensitive behavior of intelligent ferrogels for controlled release of drug. *Langmuir* **2006**, *22*, 5974–5978. [CrossRef] [PubMed]

32. Osipov, V.V.; Platonov, V.V.; Uimin, M.A.; Podkin, A.V. Laser synthesis of magnetic iron oxide nanopowders. *Tech. Phys.* **2012**, *57*, 543–549. [CrossRef]

33. Novoselova, I.P.; Safronov, A.P.; Samatov, O.M.; Beketov, I.V.; Medvedev, A.I.; Kurlyandskaya, G.V. Water based suspensions of iron oxide obtained by laser target evaporation for biomedical applications. *J. Magn. Magn. Mater.* **2016**, *415*, 35–38. [CrossRef]

34. O'Handley, R.C. *Modern Magnetic Materials*; John Wiley & Sons: New York, NY, USA, 1972; p. 740.

35. Qu, J.; Liu, G.; Wang, Y.; Hong, R. Preparation of Fe_3O_4-chitosan nanoparticles used for hypothermia. *Adv. Powder Technol.* **2010**, *21*, 421–427. [CrossRef]

36. McHenry, M.E.; Willard, M.A.; Laughlin, D.E. Amorphous and nanocrystalline materials for applications as soft magnets. *Prog. Mater. Sci.* **1999**, *44*, 291–433. [CrossRef]

37. Spano, M.; Hathaway, K.; Savage, H. Magnetostriction and magnetic anisotropy of field annealed Metglas* 2605 alloys via dc M-H loop measurements under stress. *J. Appl. Phys.* **1982**, *53*, 2667–2669. [CrossRef]

38. Lotfollahi, Z.; García-Arribas, A.; Amirabadizadeh, A.; Orue, I.; Kurlyandskaya, G.V. Comparative study of magnetic and magnetoimpedance properties of CoFeSiB-based amorphous ribbons of the same geometry with Mo or W additions. *J. Alloy. Compd.* **2017**, *693*, 767–776. [CrossRef]

39. Brunchi, C.-E.; Morariu, S.; Bercea, M. Intrinsic viscosity and conformational parameters of xanthan inaqueous solutions: Salt addition effect. *Colloids Surf. B Biointerfaces* **2014**, *122*, 512–519. [CrossRef] [PubMed]

40. Picout, D.R.; Ross-Murphy, S.B. On the Mark–Houwink parameters for galactomannans. *Carbohydr. Polym.* **2007**, *70*, 145–148. [CrossRef]

41. Scherrer, P. Bestimmung der Grösse und der inneren Struktur von Kolloidteilchen mittels Röntgensrahlen (Determination of the size and internal structure of colloidal particles using X-rays). *Nachr. Ges. Wiss. Göttingen.* **1918**, *2*, 98–102.

42. Galicia, J.A.; Sandre, O.; Cousin, F.; Guemghar, D.; Ménager, C.; Cabuil, V. Designing magnetic composite materials using aqueous magnetic fluids. *J. Phys. Condens. Matter* **2003**, *15*, S1379–S1402.

43. Safronov, A.P.; Terziyan, T.V.; Istomina, A.S.; Beketov, I.V. Swelling and Contraction of Ferrogels Based on Polyacrylamide in a Magnetic Field. *Polym. Sci. Ser. A* **2012**, *54*, 26–33. [CrossRef]

44. Knobel, M. Giant magnetoimpedance in soft magnetic amorphous and nanocrystalline materials. *J. Phys. IV* **1998**, *8*, 212–220. [CrossRef]

45. Spizzo, F.; Sgarbossa, P.; Sieni, E.; Semenzato, A.; Dughiero, F.; Forzan, M.; Bertani, R.; Del Bianco, L. Synthesis of ferrofluids made of iron oxide nanoflowers: Interplay between carrier fluid and magnetic properties. *Nanomaterials* **2017**, *7*, 373. [CrossRef] [PubMed]

46. Galicia, J.A.; Cousin, F.; Dubois, E.; Sandre, A.; Cabuila, V.; Perzynski, R. Static and dynamic structural probing of swollen polyacrylamide ferrogels. *Soft Matter* **2009**, *5*, 2614–2624. [CrossRef]

47. Quesada-Perez, M.; Maroto-Centeno, J.A.; Forcada, J.; Hidalgo-Alvarez, R. Gel swelling theories: The classical formalism and recent approaches. *Soft Matter* **2011**, *7*, 10536–10547. [CrossRef]

48. Rubinstein, M.; Colby, R.H. *Polymer Physics*; Oxford University Press: New York, NY, USA, 2003; p. 59.

49. Safronov, A.P.; Tyukova, I.S.; Kurlyandskaya, G.V. Coil-to-helix transition of gellan in dilute solutions is a two-step process. *Food Hydrocoll.* **2018**, *74*, 108–114. [CrossRef]

50. Dickinson, E. *Food Polymers, Gels and Colloids*; Royal Society of Chemistry: Cambridge, UK, 1991.

51. Amici, E.; Clark, A.H.; Normand, V.; Johnson, N.B. Interpenetrating Network Formation in Gellan–Agarose Gel Composites. *Biomacromolecules* **2000**, *1*, 721–729. [CrossRef] [PubMed]

52. Yuan, N.; Xu, L.; Wang, H.; Fu, Y.; Zhang, Z.; Liu, L.; Wang, C.; Zhao, J.; Rong, J. Dual Physically Cross-Linked Double Network Hydrogels with High Mechanical Strength, Fatigue Resistance, Notch-Insensitivity and Self-Healing Properties. *ACS Appl. Mater. Interfaces* **2016**, *8*, 34034–34044. [CrossRef] [PubMed]

53. Filipcsei, G.; Zrinyi, M. Magnetodeformation effects and the swelling of ferrogels in a uniform magnetic field. *J. Phys. Condens. Matter* **2010**, *22*, 276001. [CrossRef] [PubMed]

54. Glaser, R. *Biophysics*; Springer: Berlin/Heidelberg, Germany; New York, NY, USA, 1999.

55. Makhotkin, V.E.; Shurukhin, B.P.; Lopatin, V.A.; Marchukov, P.Y.; Levin, Y.K. Magnetic field sensors based on amorphous ribbons. *Sens. Actuators A Phys.* **1991**, *27*, 759–762. [CrossRef]

56. Semirov, A.V.; Bukreev, D.A.; Moiseev, A.A.; Derevyanko, M.S.; Kudryavtsev, V.O. Relationship Between the Temperature Changes of the Magnetostriction Constant and the Impedance of Amorphous Elastically Deformed Soft Magnetic Cobalt-Based Ribbons. *Russ. Phys. J.* **2013**, *55*, 977–982. [CrossRef]

57. Machado, F.L.A.; de Araujo, A.E.P.; Puca, A.A.; Rodrigues, A.R.; Rezende, S.M. Surface magnetoimpedance measurements in soft-ferromagnetic materials. *Phys. Status Solidi A Appl. Res.* **1999**, *173*, 135–144. [CrossRef]
58. Megens, M.; Prins, M. Magnetic biochips: A new option for sensitive diagnostics. *J. Magn. Magn. Mat.* **2005**, *293*, 702–708. [CrossRef]
59. Chiriac, H.; Tibu, M.; Moga, A.-E.; Herea, D.D. Magnetic GMI sensor for detection of biomolecules. *J. Magn. Magn. Mater.* **2005**, *293*, 671–676. [CrossRef]
60. Li, G.; Wang, S. Analytical and micromagnetic modeling for detection of a single magnetic microbead or nanobead by spin valve sensors. *IEEE Trans. Magn.* **2003**, *39*, 3313–3315.

Article

Mechanical, Electrical and Magnetic Properties of Ferrogels with Embedded Iron Oxide Nanoparticles Obtained by Laser Target Evaporation: Focus on Multifunctional Biosensor Applications

Felix A. Blyakhman [1,2], Nikita A. Buznikov [3], Tatyana F. Sklyar [1,2], Alexander P. Safronov [2,4], Elizaveta V. Golubeva [2], Andrey V. Svalov [2], Sergey Yu. Sokolov [1,2], Grigory Yu. Melnikov [2], Iñaki Orue [5] and Galina V. Kurlyandskaya [2,6,*]

[1] Ural State Medical University, Yekaterinburg 620028, Russia; Feliks.Blyakhman@urfu.ru (F.A.B.); t.f.shkliar@urfu.ru (T.F.S.); sergey.sokolov@urfu.ru (S.Y.S.)
[2] Institute of Natural Sciences and Mathematics Ural Federal University, Yekaterinburg 620002, Russia; safronov@iep.uran.ru (A.P.S.); golubeva.elizaveta.v@gmail.com (E.V.G.); andrey.svalov@urfu.ru (A.V.S.); grisha2207@list.ru (G.Y.M.)
[3] Scientific and Research Institute of Natural Gases and Gas Technologies—Gazprom VNIIGAZ, Razvilka Leninsky District, Moscow Region 142717, Russia; n_buznikov@mail.ru
[4] Institute of Electrophysics, Ural Division RAS, Yekaterinburg 620016, Russia
[5] Advanced Research Facilities (SGIKER), Universidad del País Vasco UPV-EHU, 48080 Bilbao, Spain; inaki.orue@ehu.eus
[6] Departamento de Electricidad y Electrónica and BCMaterials, Universidad del País Vasco UPV/EHU, 48080 Bilbao, Spain
* Correspondence: galina@we.lc.ehu.es; Tel.: +34-9460-13237; Fax: +34-9460-13071

Received: 17 February 2018; Accepted: 13 March 2018; Published: 15 March 2018

Abstract: Hydrogels are biomimetic materials widely used in the area of biomedical engineering and biosensing. Ferrogels (FG) are magnetic composites capable of functioning as magnetic field sensitive transformers and field assisted drug deliverers. FG can be prepared by incorporating magnetic nanoparticles (MNPs) into chemically crosslinked hydrogels. The properties of biomimetic ferrogels for multifunctional biosensor applications can be set up by synthesis. The properties of these biomimetic ferrogels can be thoroughly controlled in a physical experiment environment which is much less demanding than biotests. Two series of ferrogels (soft and dense) based on polyacrylamide (PAAm) with different chemical network densities were synthesized by free-radical polymerization in aqueous solution with N,N'-methylene-diacrylamide as a cross-linker and maghemite Fe_2O_3 MNPs fabricated by laser target evaporation as a filler. Their mechanical, electrical and magnetic properties were comparatively analyzed. We developed a giant magnetoimpedance (MI) sensor prototype with multilayered FeNi-based sensitive elements deposited onto glass or polymer substrates adapted for FG studies. The MI measurements in the initial state and in the presence of FG with different concentrations of MNPs at a frequency range of 1–300 MHz allowed a precise characterization of the stray fields of the MNPs present in the FG. We proposed an electrodynamic model to describe the MI in multilayered film with a FG layer based on the solution of linearized Maxwell equations for the electromagnetic fields coupled with the Landau-Lifshitz equation for the magnetization dynamics.

Keywords: magnetic nanoparticles; magnetic biosensors; ferrogels; magnetic multilayers; giant magnetoimpedance

Sensors **2018**, *18*, 872

1. Introduction

Magnetic biosensors are compact analytical devices incorporating a biological or biologically derived sensitive element integrated in or associated with a physicochemical transducer employing magnetic materials and magnetic fields [1,2]. These devices are well suited to the requirements of biomedical applications where there is an increasing number of different tests and demand for simple measurement protocols [3]. They are not an alternative to complex medical equipment but in many cases they are cheap and well adapted to society's need to make medical care promptly available to everyone. The first type of magnetic transducer was introduced for magnetic permeability measurements in bioanalysis by inductance measurements with a Maxwell bridge [4]. Nowadays, different magnetic effects have been shown to be capable of creating magnetic biosensors: anisotropic magnetoresistance, giant magnetoresistance, tunneling magnetoresistance, spin-valves, magnetoelastic effect, inductive effect, the Hall effect and magnetoimpedance (MI) [5–12].

Taking the simplest classification (in accordance with the working principle) magnetic biosensors can be divided into two groups: biosensors based on magnetic labels and label-free detecting systems. The functional basis of the magnetic label detection is simple—the fringe fields of magnetic markers employed as biomolecular labels provide a means for the transfer of biological information. In the case of biomedical applications, superparamagnetic nanoparticles (MNPs) must be provided in the form of water-based ferrofluids or ferrogels [13–15]. Apart from the classic configuration of magnetic bio-detection, in which all magnetic labels are placed at a certain distance from the surface of the sensitive element as a result of molecular recognition events, more complex configurations have also been tested. MNPs can be detected inside a living cell after intracellular uptake [11,16]. However, the detection of the MNPs incorporated into living tissues, has not yet been properly addressed.

The principle reason for this delay is the sensitivity of the existing magnetic biosensors to a magnetic field. In this respect, giant MI [17,18] have attracted special attention as this phenomenon provides a basis for biosensors capable of detecting picotesla magnetic fields [19]. A superparamagnetic label (a uniformly magnetized sphere) can be roughly viewed as a magnetic dipole placed in the centre of a sphere [20]. The stray field generated by it is proportional to the $|X|^{-3}$, where X is a distance between the dipole centre and the point under consideration. In the simplest magnetic biosensor, it is the distance between the centre of the nanoparticle and the surface of the sensitive element. This means that stray field value very rapidly decreases with an increase in the distance.

In the case of MNPs incorporated into living tissues, the evaluation of their stray fields becomes a very complex problem as the MNPs are situated at different distances and the majority of them are far from the surface of the sensitive element. In living cells after intracellular uptake of MNPs, this distance becomes in the order of at least 50 μm [21]. The detection of the MNPs incorporated into living tissues should take into consideration distances of at least 1 mm order. Analysis of long-range and short-range adhesive interactions of the streptavidin-biotin bond (a typical example of the molecular recognition event widely employed in biosensing) gives energy of about 31 kT, and its effective length is approximately 9 Å [22]. This means that for the development of specified biosensors for the detection of MNPs incorporated into living tissues we must make a step from about 1 nm to 1 mm distances keeping in mind $|X|^{-3}$ law for the stray field strengths. This is why the sensitivity of magnetic biosensors with respect to the applied field becomes a crucial condition of successful label detection.

The second problem for the detection of stray fields of magnetic particles incorporated into living systems is a very high contribution of the water signal to the dielectric properties of the tissue itself [21]. This means that the magnetic signal detection by the dynamic measuring technique requires extraction of a rather small signal from a significant background.

As cells or living tissues are basically soft gels [23], the mechanical properties are important as well because deformations can be caused by the application of an external magnetic field in the case of FG. It might cause changes in interparticle distances and their configuration, which can change their dipolar interactions and magnetic response, therefore complicating the stray field interpretation.

Finally, biological samples such as cells and tissues have a number of restrictions for testing by physical methods. Living matter is a cluster of interconnected complex processes that is hard to consider separately. In order to solve this problem, mathematical and physical models are widely used in biophysical research. This strategy allows us to reproduce the selected features and properties of the living system important for each particular case. In particular, for the structural organization of cells and tissues, polyelectrolyte synthetic hydrogels can be used as a physical model [24,25]. The development of a new generation of magnetic biosensors is conditioned by the development of reliable samples mimicking some properties of the living systems. Ferrogels offer the possibility of following selected morphologies, while avoiding a huge variety of morphologies usually present in biological tissues, especially the tumor tissues affected by angiogenesis [26].

Recently we proposed to substitute biological samples at the first stage of development of the MI biosensor by a model material—synthetic hydrogel with a certain amount of MNPs. Although, successful detection and stray field evaluation has been possible in the case of thin film structures deposited onto rigid substrates [2,27], the experiments with large FG pieces confirmed the urgent need to develop a model for the description of the MI in multilayered films with a top FG layer.

In the present work, we describe our experience synthesizing and characterizing maghemite Fe_2O_3 MNPs obtained by laser target evaporation technique and ferrogels incorporating MNPs in chemically crosslinked hydrogels. The deformation of ferrogels and their electrical and magnetic properties were comparatively analyzed. FeNi/Ti and FeNi/Cu multilayered sensitive elements were deposited onto rigid or flexible substrates and magnetoimpedance sensor prototype responses were measured with and without a top FG layer. An electrodynamic model of the MI in multilayered film with a FG layer based on the solution of linearized Maxwell equations for the electromagnetic fields coupled with the Landau-Lifshitz equation for the magnetization dynamics was proposed.

2. Experimental Method

2.1. Synthesis of Iron Oxide MNPs and Preparation of Ferrofluid

Iron oxide MNPs were synthesized by laser target evaporation method (LTE) using commercial magnetite (Fe_3O_4) (Alfa Aesar, Ward Hill, MA, USA) as a precursor. Magnetite powder with specific surface area 6.9 m^2/g (average particle diameter 0.2 μm) was pressed in a pellet of 65 mm in diameter and 20 mm in height, which was then evaporated by a laser beam using a laboratory installation designed at the Institute of Electrophysics UD RAS (Yekaterinburg, Russian Federation). A Ytterbium laser with 1.07 μm wavelength operated in a pulsed regime with pulse frequency 4.85 KHz and pulse duration 60 μs was used for the evaporation. The average output power of irradiation was 262 W. The condensation of MNPs from vapor took place in a permanent flow of argon (5 L/min). The details of the synthetic method are elaborated elsewhere [28,29].

An electrostatically stabilized ferrofluid based on iron oxide MNPs was prepared in 5 mM of sodium citrate solution in distilled water. Suspension in an initial concentration 5 wt % was de-aggregated by ultrasound with permanent cooling using Cole-Parmer CPX-750 processor (Cole-Parmer Instruments Corp., Vernon Hills, IL, USA) operated at 300 W power output for 30 min. The average hydrodynamic diameter of aggregates in suspension was monitored by dynamic light scattering. After that, the suspension was centrifuged at 10,000 rpm for 5 min using Hermle Z383 centrifuge (Hermle-labortechnik, Wehingen, Germany) to remove remaining aggregates. The concentration of ferrofluid after centrifuging was 3.2 wt % of MNPs. This stock ferrofluid was then diluted with 5 mM sodium citrate solution to obtain different concentrations for the preparation of ferrogels with varying MNP content.

2.2. Synthesis of Ferrogels

Ferrogels based on the polyacrylamide (PAAm) chemical network were synthesized by free-radical polymerization in an aqueous solution with *N,N'*-methylene-diacrylamide (Merck Schuchardt,

Hohenbrunn, Germany) as a cross-linker. Two series of ferrogels (soft and dense) with different densities were obtained. The soft series was synthesized in 1.6 M solution of acrylamide (AAm) (AppliChem, Darmstadt, Germany) and cross-linker to monomer molar ratio was set at 1:100. In the dense series the concentration of AAm was 2.7 M and cross-linker to monomer molar ratio was 1:50. Further on these series are denoted as FG-I (soft) and FG-II (dense). In each series the monomer and the cross-linker were dissolved in several ferrofluids with a diminishing concentration of MNPs to obtain ferrogels with varying content of magnetic particles. Ammonium persulfate in 3 mM concentration was used as an initiator and N,N,N',N'-tetramethylethylene diamine (TEMED) (SigmaAldrich Inc., St. Louis, MO, USA) in 6 mM concentration as a catalyst. Polymerization was performed at room temperature for 1 h in polyethylene probe tubes. After the synthesis, the ferrogels were taken out from the tubes and extensively washed in distilled water for two weeks with water renewal every two days until constant weight of the gel samples was achieved.

As the ferrogel samples swelled in the consequent washing cycles after the synthesis, the final concentration of MNPs in ferrogels changed with respect to the initial concentration set up in the synthesis. The concentration of MNPs in ferrogels was determined by the gravimetric analysis. First, the equilibrium swelling ratio of the ferrogels after their equilibration was determined. The weight, m_0 of a swollen piece of gel approximately 0.5 g was measured using an analytical balance (Mettler-Toledo MS104S). Then, it was dried in an oven at 90 °C down to the constant weight of the dry residue m_1. The apparent swelling ratio (α) of the ferrogel was calculated according to the equation:

$$\alpha = \frac{m_0 - m_1}{m_1} \tag{1}$$

The value of the swelling ratio was used for the calculation of the particle content in the swollen gel applying the equation:

$$\omega = \frac{\gamma}{1 + \alpha} \tag{2}$$

where γ stands for the weight fraction of particles in the dry residue, which was measured by a thermogravimetry analysis (TGA).

The values of γ were also used for the calculation of the swelling ratio of gel matrix of ferrogels (α') which was related solely to the PAAm network excluding solid MNPs:

$$\alpha' = \frac{\alpha}{\gamma} \tag{3}$$

The values of the swelling ratio of ferrogels and the weight fraction of magnetic particles in the swollen gel are given in Table 1.

The FG samples were cut in pieces of different shapes for different studies. Figure 1 shows a general view of FG-I-1 ferrogel sample after 40 min drying in ambient conditions. The geometry 10 mm × 2 mm × 1 mm was used for MI measurements with multilayered sensitive elements in order to test the magnetic biosensor prototype.

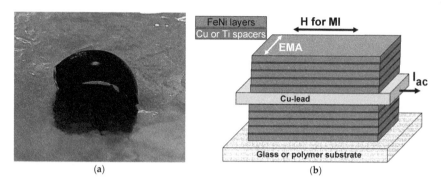

(a)   (b)

Figure 1. General view of FG-I-1 ferrogel 9 mm $\times$ 2 mm $\times$ 1 mm piece after 40 min drying in ambient conditions (**a**); Schematic description of the magnetoimpedance (MI) multilayered structures: EMA is the easy magnetization axis direction, close to the direction of a magnetic field applied during multilayered structure deposition; H is the external magnetic field direction for MI measurements, configuration of longitudinal magnetoimpedance (**b**).

2.3. Magnetic Multilayers Deposition

The magnetic MI samples were $Fe_{19}Ni_{81}$/Ti or $Fe_{19}Ni_{81}$/Cu multilayers, with a central non-magnetic copper layer deposited by dc-magnetron sputtering onto glass or polymer substrates. The optimum deposition conditions (both background pressure and a working Ar pressure) were previously discussed and proven [30,31]. A background pressure of 3×10^{-7} mbar and a working Ar pressure of 3.8×10^{-3} mbar were used in the present studies: [FeNi/Ti(6 nm)]$_x$ or [FeNi/Cu(3 nm)]$_x$ structures at the top and the bottom of the MI sandwich [32]. A transverse magnetic anisotropy was induced during the deposition process by the application of an in-plane constant magnetic field of 250 Oe for [FeNi/Ti(6 nm)]$_x$ or 100 Oe [FeNi/Cu(3 nm)]$_x$ multilayers. The shape of the MI sensitive elements was defined by the deposition conditions through the utilization of magnetic masks: 10 mm long and 0.5 mm wide rectangles. The long sides of the samples were oriented in a direction perpendicular to the direction of the applied magnetic field. Therefore, the induced magnetic anisotropy axis was formed parallel to the short side of the stripe in all of the samples. Figure 1b gives a description of the studied multilayers. As substrates we used either rigid (Corning glass) or flexible cycloolefin (ethylene-norbornene copolymer) polymer or COP, widely accepted for magnetic biosensors with microfluidic systems [30].

Table 1. Composition and swelling ratio of ferrogels.

Number	FG Series	Concentration of MNPs, C (%)	Apparent Swelling Ratio, α (Unitless)	Swelling Ratio, α' (Unitless)
1		0.00	14.4	14.4
2		0.40	13.0	13.7
3	FG-I	0.81	12.3	13.7
4		1.21	11.7	13.6
5		1.81	10.0	12.2
6		0.00	7.6	7.6
7		0.43	7.1	7.4
8	FG-II	0.94	6.4	6.8
9		1.45	6.0	6.6
10		1.90	6.0	6.7

2.4. Methods

The specific surface area (Ssp) of powdered materials was measured by the low-temperature sorption of nitrogen (Brunauer-Emmett-Teller physical adsorption (BET)) using Micromeritics

TriStar3000 analyzer (Micromeritics Instrument Corp., Norcross, GA, USA). Transmission electron microscopy (TEM) was performed using a JEOL JEM2100 microscope (JEOL Ltd., Tokyo, Japan) operated at 200 kV. The particles were spread on carbon-coated copper grids. The X-ray diffraction (XRD) studies were performed by Bruker DISCOVER D8 (Bruker, Billerica, MA, USA) diffractometer operated at 40 kV and 40 mA using Cu-Kα radiation (λ = 1.5418 Å), a graphite monochromator and a scintillation detector. The MNPs were mounted on a zero-background silicon wafer placed in a sample holder. A fixed divergence and anti-scattering slit were used. Bruker software TOPAS-3 with Rietveld full-profile refinement was employed for the quantitative analysis of all the diffractograms. The hydrodynamic diameter of MNPs/aggregates in suspension, the parameters of lognormal distribution, and zeta-potential were measured by the dynamic and electrophoretic light scattering using a particle size analyzer Brookhaven Zeta Plus (Brookhaven Instruments Corp., Holtville, NY, USA). TGA of dried ferrogels was performed using NETZSCH STA 409 thermal analyzer (NETZSCH Geratebau, Selb/Bavaria, Germany) by heating from 40 to 1000 °C at 10 K/min in an air flow of 20 mL/min.

The testing of mechanical characteristics of ferrogels in static and dynamic mode was performed using the laboratory setup. It comprised an electromagnetic linear motor for applying mechanical deformations, a semiconductor force transducer, and a semiconductor optical transducer for the measurement of the length of the ferrogel sample. A cylindrical ferrogel sample (~7 mm in length and ~90 mm^2 in cross section) placed in a bath filled with distilled water was clamped vertically between two parallel plates connected to the levers of the force transducer and the linear motor. The measurement in a quasi-static mode was performed by a step-wise application of compressive deformation to a sample with force equilibration for at least 10 s at each step. The deformation step was 1–2% of the sample height and the total deformation of the sample was up to 15%.

The testing of the electrical properties of ferrogels was performed using a laboratory installation for the electrical potential measurement in biophysical systems described elsewhere [33,34]. The measurement of the electrical potential of ferrogel was done by two identical Ag/AgCl tapered glass microelectrodes (~1 micron in tip diameter) typically used in biophysical studies for intracellular voltage measurement. The electrodes were single-pulled using a standard electrode puller, ME-3 (EMIB Ltd., Moscow, Russia) from thin-walled, single-barrel borosilicate capillary tubes, TW150F-6 (World Precision Instruments, Sarasota, FL, USA). The pulled electrodes were immersed in a 3 M KCl solution with the tip facing upward, so that the solution climbed to the tip by capillary action. One electrode was pinned into the ferrogel sample and the other was placed into outside water. The potential difference between microelectrodes was measured using an instrumental amplifier on the base of an integrated circuit, INA 129 (Burr-Brown, Dallas, TX, USA). To reduce the influence of electromagnetic interference on the potential difference measurement, special wire shields were provided around the measuring unit.

The in-plane magnetic hysteresis loops of the magnetic multilayers were measured by magneto-optical Kerr microscopy (MOKE device of laboratory design) for the whole sample size: 10 mm × 0.5 mm. The magnetization curves of ferrogels were measured by vibrating sample magnetometer (VSM: Faraday magnetometer of laboratory design) for samples of about 100 mg weight placed into a polycarbonate capsule. The capsule contribution was measured separately and carefully extracted from the M(H) data. All measurements were made at room temperature.

A thin film multilayered sensitive element was incorporated into a "microstripe" line using silver paint (Figure 2a). Figure 2b shows a general view of the ferrogel piece cut for MI measurements. A uniform external magnetic field of up to 100 Oe was created by a pair of Helmholtz coils and applied along the MI sensitive element elongated stripe, i.e., all measurements were made in longitudinal MI configuration (alternating current flowing parallel to the external magnetic field). The MI changes were calculated from the S_{11} reflection coefficient measured by a network analyzer (Agilent E8358A). An output power of 0 dB, corresponding to the amplitude of the excitation current across the sample of about 1 mA was used in all MI measurements.

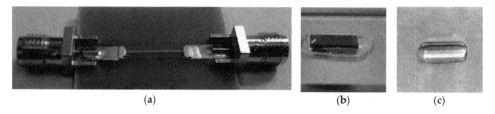

(a)   (b)   (c)

Figure 2. General view of the ribbon-based MI sensitive element installed into the "microstripe" line (**a**); general view of ferrogel (**b**) and gel; (**c**) pieces cut for MI measurements.

The calibration and mathematical subtraction of the test fixture contributions were carefully done following a previously established protocol [30]. Total impedance (Z) was measured as a function of the external magnetic field in a frequency range 0.1–300 MHz. MI ratio (4a) and MI ratio sensitivity were calculated as follows:

$$\frac{\Delta Z}{Z} = 100 \times \frac{Z(H) - Z(H_{\max})}{Z(H_{\max})} \qquad (4)$$

$$S\left(\frac{\Delta Z}{Z}\right) = \frac{\delta(\Delta Z/Z)}{\delta H} \qquad (5)$$

where $H_{\max} = 100$ Oe and $\delta\,(\Delta Z/Z)$ is the change in the total impedance MI ratio $(H) = 0.1$ Oe. For MI frequency dependence analysis, the changes of the maximum value of the total impedance $\Delta(Z/Z)_{\max}$ was provided. The experimental error in the impedance determination was within 1%. Direct current resistivity (R_{DC}) was also measured in all cases.

In our previous work, for MI measurements with ferrogels we used gel pieces of particular geometry of 1.0 to 0.5 g weight [2,35]. The measurements were strongly conditioned by the time as the gel/ferrogels changed mass rapidly, shrunk and lost water (see Figure 1a). The FG synthesized for the present study were kept in distilled water for storage. In the present work, we therefore developed a special protocol for MI measurements with gels and FG taking into account the mass loss of the piece of gel or FG of the same shape. The length was slightly shorter as about 0.1 mm from each end of the MI element was used for electrical connection. A special protocol was elaborated for MI measurements in order to minimize the experimental errors and ensure a reproducible procedure. First, the samples were cut in rectangles of 9 mm × 2 mm × 1 mm. Then each sample was taken out of the water and kept under ambient conditions for about 60 min, while its weight loss was measured by an analytical balance. The total time of the MI measurements in a decreasing field was limited to 3 min and characterized by a linear total mass loss of 10% with respect to the initial mass.

First, the MI responses were studied by measuring the multilayered sensitive element itself, i.e., without gel or ferrogel on top of it. Then, FG stripes (Figure 2b) were placed in the centre of the MI element one by one for different concentrations. Measurements with the gel were used for subtracting the corresponding signal from the ferrogel samples in order to evaluate the average contribution of the stray fields of magnetic particles in the spatial distribution corresponding to each ferrogel case.

For the modeling of the MI response in a multilayered element in the presence of a ferrogel we proposed an electrodynamic approach in order to describe the MI in a multilayered thin film structure with a layer of ferrogel on top of it. This is why very thin FG samples of a small volume were used for the MI measurements—for larger samples, calculations were limited by the available computing capacity. The theoretical approach is based on the solution of linearized Maxwell equations for the electromagnetic fields coupled with the Landau-Lifshitz equation for the magnetization dynamics [18]. The influence of the ferrogel on the permeability of the multilayered element in the configuration of the MI sandwich is described in terms of an effective stray field created by MNPs.

3. Results and Discussion

3.1. Structural Characterization of Nanoparticles and Ferrogels

Figure 3a gives TEM image of synthesized MNPs and Figure 2b gives particle size distribution (PSD) histogram which was obtained by the graphical evaluation of the diameter of 2160 particles. The particles are spherical and non-coalescent. Only a few of the particles appeared to be hexagonal or to have hexagonal corners. PSD is well fitted by the following lognormal distribution function:

$$PSD = \frac{3.38}{d} \exp\left[0.5\left(\frac{\ln(d/17.3)}{0.413}\right)^2\right] \tag{6}$$

The specific surface area of MNPs was 64 m^2/g. The surface average diameter of MNPs, calculated from this value using the equation $d_s = 6/(\varrho \times S_{sp})$ (ϱ = 4.6 g/cm^3 being iron oxide density) was 20.7 nm. This was in fair agreement with the value d_s = 24.4 nm, obtained using PSD (Equation (5)). The discrepancy between these two values apparently stems from the deviation from the spherical shape of several MNPs, which can be noticed in Figure 3. The polyhedral shape provides larger surface than sphere. Hence, the specific surface area of the ensemble of MNPs in the study is to some extent higher than that for the equivalent ensemble of exact spheres. The enlarged S_{sp} resulted in lower estimated values of d_s than that obtained from PSD.

XRD patterns of iron oxide MNPs are given in Figure 3c. The crystalline structure of MNPs corresponded to the inverse spinel lattice with a space group Fd3m. The lattice period was found, a = 0.8358 nm, which was larger than that for maghemite (γ-Fe$_2$O$_3$, a = 0.8346 nm) but lower than that for magnetite (Fe$_3$O$_4$, a = 0.8396) [36]. Based on the dependence between the lattice period of the spinel cell and the effective state of oxidation of Fe, the non-stoichiometric composition of MNPs was found Fe$_{2.75}$O$_4$.

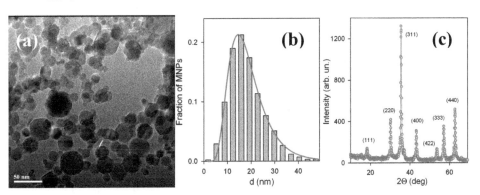

Figure 3. TEM image of iron oxide MNPs embedded in ferrogels (a). Particle size distribution obtained by the graphical analysis of 2160 particles (b); Inset (c) XRD pattern for MNPs with corresponding Miller indexes.

Ferrofluids with MNPs were stable to aggregation and sedimentation. The average hydrodynamic diameter of MNPs in suspension measured by the dynamic light scattering was 56 nm, and it did not change during a month of observation. This value corresponds to the fifth mode of PSD and is therefore larger that the value of d_s given above, which stands for the second mode. The discrepancy between them stems from the polydispersity of the ensemble of MNPs. The value of hydrodynamic diameter correlates well with such data obtained for LTE iron oxide MNPs earlier [28] and corresponds to the individual particles in ferrofluid. The stability of ferrofluids was provided by the adsorption of negatively charged citrate anions on the surface of MNPs [28,29]. It gave negative net charge to the dispersed MNPs, provided their repulsion in the suspension, and prevented aggregation. The value of zeta-potential of the suspension was −40 mV, which was higher (in absolute value) than the threshold for the electrostatic stability of water-based suspensions [37].

No visible aggregation was observed in ferrofluid in the synthesis of ferrogels, which remained clear and non-opalescent after polymerization was completed. Thus, we assumed that individual MNPs present in ferrofluid remained in the ferrogel.

The main structural feature of ferrogel is the mesh size of its network. This can be estimated based on the equilibrium swelling ratio of a gel, which is the uptake of water by the dry polymeric network. The swelling ratio for all ferrogels is given in Table 1. There is a noticeable diminishing trend in values of apparent swelling ratio α, which was calculated according to Equation (1). Meanwhile, one should take into account that values of α included the weight of embedded MNPs. It is more correct to use the value of the swelling ratio for the PAAm network, α' for the mesh size characterization. These values only slightly depended on the presence of embedded MNPs (see Table 1). We used the average value of α' to characterize the mesh size of FG-I and FG-II series. It was 13.5 ± 0.8 for the FG-I series and 7.0 ± 0.4 for the FG-II series. The obtained values were then corrected, taking into account that dry residues contained both polymer and iron oxide MNPs. The degree of swelling related solely to the polyacrylamide network in ferrogels was independent of MNPs content in the gel. That is, iron oxide MNPs do not provide extra cross-linking of the network due to the adsorption of the sub-chains on their surface.

Based on the equilibrium degree of swelling of the polymeric network (α') the average number of monomer units in linear sub-chains between cross-links (N_C) was evaluated using the Flory-Rehner equation [38]:

$$N_C = \frac{V_1(0.5\alpha^{-1} - \alpha^{-1/3})}{V_2(\ln(1 - \alpha^{-1}) + \alpha^{-1} + \chi\alpha^{-2})} \qquad (7)$$

where V_1, V_2 are molar volumes of a solvent and of a polymer respectively, χ is Flory-Huggins parameter for a polymer-solvent mixture. We used $V_1 = 18$ cm^3/mol (water), $V_2 = 56.2$ cm^3/mol (PAAm) and $\chi = 0.12$. The last two values were obtained by means of quantum mechanics molecular modeling software package CAChe7.5. Equation (7) gave the number of monomer units in linear sub-chains $N_C = 55$ for the PAAm network in the FG-I series and $N_C = 16$ for the FG-II series.

The equilibrium conformation of electrically neutral PAAm subchain in water is a random Gaussian coil with hindered rotation. The mean square end-to-end distance <R^2>, which corresponds to the distance between adjacent cross-links (in other words the mesh size of the network) can be calculated according to the equation [39]:

$$\left\langle R^2 \right\rangle = Na^2\frac{1 - \cos\theta}{1 + \cos\theta} \qquad (8)$$

where N is the number of bonds in the polymeric chain, a is the bond length, θ is the bond angle. We took $a = 0.154$ nm for the ordinary C–C bond, $\theta = 109.5°$ for the bond angle, and $N = 2N_C$ for the number of bonds. The mesh size of the network, calculated using Equation (8) is 2.4 nm for the FG-I series and 1.2 nm for the FG-II series. Both values are by an order of magnitude smaller than the average diameter of iron oxide MNPs (24.4 nm), which means that the cross-links of the polymeric network are closer to each other than the particle diameter. As a result, the MNPs cannot freely move inside the polymeric network of FG but are entrapped in it.

3.2. Deformation of Ferrogels and Their Electrical Properties

Figure 4a presents typical compression deformation plots for ferrogels of FG-I and FG-II series. Each plot is a combination of several step-wise deformation runs on the same sample. It is noticeable that deformation of soft ferrogels of the FG-I series takes place at lower values of tension than the deformation of ferrogels of the FG-II series (dense). This is a consequence of the smaller mesh size of the latter. In both series the embedding of MNPs in polymeric network substantially enlarges the tension, which is necessary to achieve a certain level of deformation. In this respect, embedded MNPs provide the enhancement of mechanical strength of a ferrogel equivalent to the additional cross-linking. Meanwhile, the swelling ratio of PAAm network (see Table 1) is not as sensitive to the presence of MNPs in the structure as mechanical strength. One common feature among plots given in Figure 4a is that there are two parts with a different slope in all of them. The initial part of the plot in the deformation range 0–7% is concave, while at higher deformations the plot is linear. Apparently, the concave part is the result of complex structural changes, which take place in the polymeric network under the applied force. The linear parts of the plots were taken for the calculation of Young modulus of ferrogels in both series.

Figure 4b shows the dependence of the Young modulus on the weight fraction of MNPs in ferrogel for FG-I and FG-II series. In FG-II (dense) series the embedding of MNPs in the polymeric network, even in a minimal concentration, resulted in a substantial increase of the Young modulus. In contrast, the addition of MNPs in a concentration larger than 1 wt % does not result in further increase of the modulus. In the FG-I (soft) series the influence of MNPs on the Young modulus of ferrogels is not as pronounced as in the FG-II series. Meanwhile, a limited increase in modulus with the weight fraction of MNPs can still be observed. There are two conclusions, which can be made from the data given in Figure 4. The first is that iron oxide MNPs in low concentration provide the increase in the Young modulus of ferrogels. The second is that this effect is enhanced by the density of the polymeric network. It is almost negligible if the mesh size of the network is large, and it becomes substantial if it decreases. This could also be attributed to the water uptake (swelling ratio) of the polymeric network. In other words, the influence of MNPs on the Young modulus becomes larger at a lower water uptake of the network. Figure 5 shows the dependence of the electrical potential of ferrogel on the weight fraction of iron oxide MNPs.

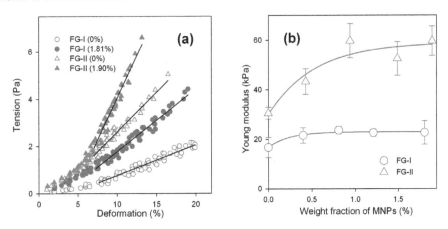

Figure 4. Typical deformation plots for ferrogels of FG-I and FG-II series. Open symbols correspond to unfilled (blank) gel, closed symbols correspond to ferrogel with the highest concentration of MNPs. Lines show the linear parts of the plots, which were used for the calculation of the Young modulus (**a**); Dependence of Young modulus of ferrogel on the weight fraction of MNPs. Lines are for an eye-guide only (**b**).

The potential in PAAm gel is negative and its value is ca. −10 mV. This value is low when compared to the typical values of the electrical potential for polyelectrolyte hydrogels such as gels of polyacrylic or polymethacrylic acid [40,41], which are larger by an order of magnitude. Certainly, it is because PAAm does not contain ionized groups in its polymeric sub-chains. In general, the electrical potential of the gel is Donnan potential [42], which is the result of the restricted ionic equilibrium on the gel boundary with supernatant. If there are ionic species, which cannot freely cross the gel boundary, it causes non-uniform distribution of free ions and eventually results in the existence of the electrical potential step on the boundary. In the case of polyelectrolyte gels, such unmovable ions are those attached to the polymeric sub-chains.

There is a question as to why PAAm still show a low negative electrical potential even though they are non-ionic. For instance, it might be the result of a very small fraction of carboxylic residue in the PAAm structure due to the hydrolysis of AAm monomeric units.

The embedding of iron oxide MNPs in gel structure results in the increase of the negative values of the potential. Most likely it is due to the net negative electrical charge of MNPs themselves. As it was noted above, the ferrofluid taken for the synthesis of ferrogels was electrostatically stabilized by sodium citrate to prevent aggregation of MNPs. The zeta potential of the ferrofluid was −40 mV. This potential is located at the interface between the dense and the diffuse part of the double electrical layer at the surface of the MNPs. If MNPs are embedded into the PAAm network, they take the double electrical layer with them. As it was shown above, the mesh size of the network is by an order of magnitude smaller than the average diameter of MNPs. Thus, negatively charged MNPs cannot move across the ferrogel boundary and such a restriction gives rise to the native Donnan potential of the ferrogel.

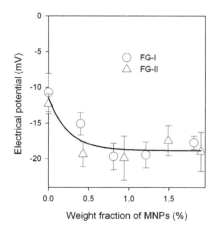

Figure 5. Dependence of the electrical potential of a ferrogel on the weight fraction of MNPs in FG-I and FG-II series. The line is for an eye-guide only.

It is noticeable in Figure 5 that the negative values of the electrical potential substantially increase at low concentration of MNPs and stay almost constant at higher concentration. There is no difference between the FG-I and FG-II series with respect to the electrical potential. It is the same for both series within the experimental error.

3.3. Magnetic Characterization of Nanoparticles and Ferrogels

Figure 6 and Table 2 summarize the results of the magnetic measurements: M(H) hysteresis loops of soft ferrogels are given in Figure 6a and of dense ferrogels in Figure 6b. Although full saturation was not reached in the field of 10 kOe for the samples with high concentration of MNPs, we used the magnetization value for this field as the saturation magnetization M_s just for simplicity. The validity of such approximation was confirmed by the fact that M(H = 10 kOe) ≈ 0.95 × M(H = 70 kOe) [28]. For practical reasons of magnetic biodetection, magnetic field H = 1 kOe was sufficient. In all cases under consideration, including measurements of the MNPs dried from ferrofluid, very small coercivity of 8 Oe was observed at room temperature.

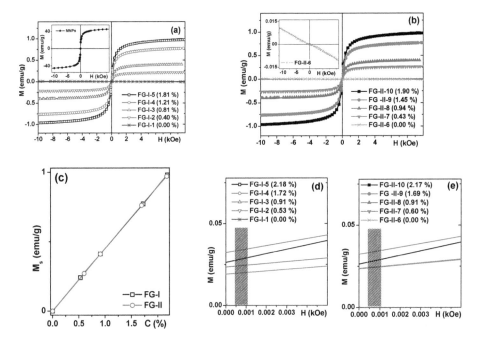

Figure 6. M(H) hysteresis loops of soft (**a**) and of dense ferrogels (**b**): in both cases concentrations of magnetic filler are obtained from the synthesis data. Inset for (**a**) shows the hysteresis loop of iron oxide MNPs from dried ferrofluid. Inset for (**b**) shows the hysteresis loop of pure gel. Concentration dependence of the saturation magnetization for soft and dense gels with different amount of magnetic filler (**c**). Parts of M(H) hysteresis loops of soft (**d**) and of dense ferrogels (**e**) approaching remanence: in both cases concentrations of magnetic filler are given for magnetic measurements data. Vertical bars indicate the field interval of 4 to 11 Oe corresponding to typical anisotropy fields of thin magnetic films used in sensor applications.

Table 2. Magnetic properties of MNPs, gels and ferrogels: H_c is coercivity; M_s = M(H = 10 kOe).

Number	FG Series	Concentration of MNPs, % from Synthesis	H_c, Oe	M_s for gel (emu/g)	Concentration of MNPs, % from M_s Data
1		0.00	0	~−0.01	0.00
2		0.40	13	0.24	0.53
3	FG-I	0.81	13	0.41	0.91
4		1.21	13	0.77	1.72
5		1.81	8	0.98	2.18
6		0.00	0	~−0.01	0.00
7		0.43	6	0.27	0.60
8	FG-II	0.94	6	0.41	0.91
9		1.45	7	0.76	1.69
10		1.90	7	0.97	2.17
11	MNPs	100	8	44.90	100.00

This can be explained by the presence of very small amounts of large particles due to the existence of particle size distribution (see Figure 3b). Even so, the behavior is quite close to the one expected for a superparamagnetic ensemble of iron oxide MNPs of this size. It is also important to mention that the contribution of pure gels was diamagnetic and very small in comparison with the magnetic signals of ferrogels. We have subtracted gel contributions when necessary from the ferrogel signals in order to obtain magnetic signals corresponding to the MNPs.

Inset Figure 6a shows the hysteresis loop of MNPs with saturation magnetization about 45 emu/g, which corresponds quite well to the average size, d_s = 24.4 nm of MNPs [28,43]. The measurements of MNPs allow us to check the iron oxide concentration in ferrogels. Table 2 shows iron oxide concentrations obtained from the experimental data related to synthesis. Although these values are reliable, since the mass fraction of particles changes when manipulating the gel due to the inevitable loss of moisture, it is very difficult to estimate the MNPs concentrations from the synthesis data with high precision. Instead one can take advantage of magnetic measurements. Knowing the magnetization of the ferrogel in H = 10 kOe, we can calculate the concentration of the magnetic filler taking into account the magnetization of MNPs in the same magnetic field. The concentrations of MNPs obtained from M_s data were reasonably close to those calculated from the synthesis data (Table 2).

Figure 6c shows the concentration dependences of the saturation magnetization for soft and dense gels with different amounts of magnetic filler—a very good linear fit is evident in both cases under consideration. Figure 6d,e show parts of M(H) hysteresis loops of soft and dense ferrogels approaching remanence: in both cases concentrations of magnetic filler were recalculated for magnetic measurement data. Vertical bars indicate the field interval of 4–11 Oe, which correspond to typical anisotropy fields of thin magnetic films used in sensor applications. Here it becomes very clear that for efficient detection the magnetic moment must be of the order of 0.025 emu/g. For a ferrogel sample with a volume of about 18 mm^3 (9 mm × 2 mm × 1 mm geometry), this means a magnetic moment of 0.0005 emu in the field near the anisotropy field.

3.4. Characterization of the Multilayered MI Element

Figure 7 shows hysteresis loops of magnetic multilayered structures in the shape of MI elongated stripes. M(H) loops were measured by MOKE in the external magnetic field applied in the plane of the multilayered structure and parallel to the long side of the stripe, i.e., in the hard magnetization direction. The shape of the M(H) curves confirms that the application of an external magnetic field during the deposition of the multilayered structures did indeed result in the formation of the induced magnetic anisotropy with easy magnetization axes oriented in the plane of the film parallel to the short side of the sensitive element. The anisotropy fields for FeNi/Cu-based multilayers were about 7 Oe.

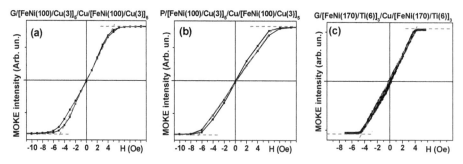

Figure 7. M(H) hysteresis loops of MI multilayered structures in the shape of rectangular elements. Numbers indicate thicknesses of the layers in nanometers. Thickness of the central Cu lead was equal to 500 nm in all cases under consideration. MI multilayers in the cases (**a,c**) were deposited onto glass substrates and onto cycloolefin COC flexible substrate in case (**b**). All units for the thickness of the layers are nanometers. The dashed red lines determine the anisotropy fields.

It is important to mention that both FeNi/Cu- based structures have quite similar magnetostatic characteristics. Previously, we conducted special studies to define the optimum parameters for FeNi/Ti-based multilayers deposition onto glass and flexible COC substrates [44]. The comparison of the M(H) loops (Figure 7) shows that FeNi/Cu- based structures with well-defined uniaxial induced magnetic anisotropy can be also successfully deposited onto both kinds of substrates. Although present day deposition techniques allow deposition of very sophisticated multilayered structures combining magnetic, conducting and insulating layers, simple structures with a smaller number of different types of layers always have a technological advantage. Actually, the deposition of [FeNi(170 nm)/Ti(6 nm)]$_3$/Cu(500 nm)/[Ti(6 nm)/FeNi(170 nm)]$_3$ multilayers requires the utilization of three targets, whereas the deposition of [FeNi(100 nm)/Cu(3 nm)]$_3$/Cu(500 nm)/[Cu(3 nm)/FeNi(100 nm)]$_5$ multilayers requires two targets during the sputtering process.

3.5. Model for Distribution of Electromagnetic Fields in Multilayered Film with Ferrogel

Let us consider a film structure [F/X]$_n$/F/C/[F/X]$_n$/F having the length l and width $w < l$. The film structure consists of a highly conductive central layer C of thickness $2d_0$ and two external magnetic multilayers consisting of soft magnetic layers F of thickness d_2 and non-magnetic separating layers X of thickness d_1. The total thickness $2t$ of the film structure is given by the following expression: $2t = 2d_0 + 2nd_1 + 2(n + 1)d_2$. It is assumed further that the material of the central layer and separating layers is the same, although the model can be extended to the case of different materials C and X. The layer of ferrogel with a thickness of d_3 is placed onto the top surface of the film structure.

The driving electric field $e = e_0\exp(-i\omega t)$ is applied to the film structure, and the external magnetic field H_e is parallel to the long side of the film. It is assumed that the film length and width are much higher than its thickness. Neglecting the edge effects, we consider that the electromagnetic fields depend only on the coordinate perpendicular to the film plane (z-coordinate).

In this approximation, the solution of Maxwell equations for the amplitudes of the longitudinal electric field and transverse magnetic field in the central and separating non-magnetic layers can be expressed as,

$$e_1^{(j)} = (c\lambda_1/4\pi\sigma_1)[A_1^{(j)}\cosh(\lambda_1 z) + B_1^{(j)}\sinh(\lambda_1 z)],$$
$$h_1^{(j)} = A_1^{(j)}\sinh(\lambda_1 z) + B_1^{(j)}\cosh(\lambda_1 z). \tag{9}$$

Here $e_1^{(j)}$ and $h_1^{(j)}$ are the amplitudes of the electric and magnetic field, $j = 1, \ldots, 2n + 1$ is the non-magnetic layer number, $A_1^{(j)}$ and $B_1^{(j)}$ are the constants, $\lambda_1 = (1 - i)/\delta_1$, $\delta_1 = c/(2\pi\omega\sigma_1)^{1/2}$ and σ_1 are the skin depth and conductivity of the non-magnetic layers, respectively, and c is the velocity of light. Note that the expressions for the field amplitudes in the central layer have an asymmetric form with respect to the central plane of the structure, $z = 0$, due to the presence of ferrogel.

The field amplitudes $e_2^{(k)}$ and $h_2^{(k)}$ in the ferromagnetic layers are given by the following expressions:

$$e_2^{(k)} = (c\lambda_2/4\pi\sigma_2)[A_2^{(k)}\cosh(\lambda_2 z) + B_2^{(k)}\sinh(\lambda_2 z)],$$
$$h_2^{(k)} = A_2^{(k)}\sinh(\lambda_2 z) + B_2^{(k)}\cosh(\lambda_2 z). \tag{10}$$

Here $k = 1, \ldots, 2n + 2$ is the ferromagnetic layer number, $A_2^{(k)}$ and $B_2^{(k)}$ are the constants, $\lambda_2 = (1 - i)/\delta_2$, $\delta_2 = c/(2\pi\omega\mu\sigma_2)^{1/2}$, σ_2 and μ are the skin depth, conductivity and transverse permeability of the soft magnetic layers, respectively.

The solution of Maxwell equations in the ferrogel can be presented in the form:

$$e_3 = [A_3\cosh(\lambda_3 z) + B_3\sinh(\lambda_3 z)]/\varepsilon^{1/2},$$
$$h_3 = A_3\sinh(\lambda_3 z) + B_3\cosh(\lambda_3 z). \tag{11}$$

Here A_3 and B_3 are the constants, $\lambda_3 = -i\omega\varepsilon^{1/2}/c$ and ε is the permittivity of ferrogel layer.

For the external regions, $z < -t$ and $z > t + d_3$, the approximate solution for the fields can be written as [44–46],

$$e_m = C_m\frac{i\omega}{c}\left[\frac{l}{2w}\log\left(\frac{R+w}{R-w}\right) - \frac{2z}{w}\arctan\left(\frac{wl}{2Rz}\right) + \frac{1}{2}\log\left(\frac{R+l}{R-l}\right)\right],$$
$$h_m = -C_m\frac{4lz}{R}\left[\frac{R^2+4z^2}{4R^2z^2+l^2w^2} - \frac{1}{R^2-w^2} - \frac{1}{R^2-l^2}\right] + C_m\frac{2}{w}\arctan\left(\frac{wl}{2Rz}\right). \tag{12}$$

Here $m = 4$ and $m = 5$, correspond to the bottom and top external region, respectively, C_m are the constants and $R = (l^2 + w^2 + 4z^2)^{1/2}$. Note that expression (11) was obtained assuming that $w \ll l$ and allows one to describe the distribution of the electromagnetic fields outside the film structure with ferrogel.

The constants in Equations (9)–(12) can be found from the continuity conditions for the amplitudes of the electric and magnetic fields at the interfaces between the ferromagnetic and non-magnetic layers. Furthermore, boundary conditions at the film structure surface and at the interface between the ferrogel and external region should be added.

Taking into account that the driving electric field is applied to the film structure only, the boundary conditions at the bottom surface of the film, $z = -t$, can be written in the following form:

$$e_2^{(1)}(-t) = e_4(-t) + e_0 \,, h_2^{(1)}(-t) = h_4(-t) \,. \tag{13}$$

Similar expressions can be found at the interface between the film and ferrogel, $z = t$,

$$e_2^{(2n+2)}(t) = e_3(t) + e_0 \,, h_2^{(2n+2)}(t) = h_3(t) \,. \tag{14}$$

The boundary conditions at the top surface of the ferrogel layer have the form:

$$e_3(t + d_3) = e_5(t + d_3) \,, h_3(t + d_3) = h_5(t + d_3) \,. \tag{15}$$

The boundary conditions allow one to find the constants $A_1^{(j)}$, $B_1^{(j)}$, $A_2^{(k)}$, $B_2^{(k)}$, A_3, B_3, C_4 and C_5 in Equations (9)–(12) and describe completely the distribution of the electromagnetic fields. When the field distribution is obtained, the impedance Z of the film with the ferrogel layer can be found as a ratio of the applied potential difference le_0 to the total current I flowing through the film:

$$Z = \frac{le_0}{I} = \frac{le_0}{w \int_{-t}^{t} \sigma(z)e(z)dz} = \frac{2\pi l}{cw} \times \frac{e_0}{h_2^{(2n+2)}(t) - h_2^{(1)}(-t)} \,. \tag{16}$$

3.6. Effect of Ferrogel Layer on Film Permeability

The MI response of the multilayered film is controlled by the transverse permeability in the ferromagnetic layers. The transverse permeability depends on many factors, such as the domain structure, anisotropy axes distribution, mode of the magnetization variation, and so on. We assume that the value of the permeability in the ferromagnetic layers is governed by the magnetization rotation only. This approximation is valid at sufficiently high frequencies, when the domain-wall motion is damped [47]. We also suppose that the ferromagnetic layers have in-plane uniaxial anisotropy, and the direction of the anisotropy axes is close to the transverse one.

The influence of ferrogel on the MI can be attributed to stray fields induced by MNPs. The stray fields change the magnetization distribution in the soft magnetic layers of the film structure, and correspondingly, affect the film permeability. In general, the spatial distribution of the stray fields from MNPs can only be found by numerical calculations.

To describe qualitatively the effect of stray fields on the MI response, let us consider the following approximate model. It is assumed that the stray fields generate an effective field H_p in the film structure. We suppose that the effective field is uniform over the film structure thickness. Note that a smooth variation of the effective field over the film thickness can be introduced in the model, with maximal value being at the top surface of the film and minimum value being at the bottom surface. Taking into account such approximate spatial distribution of the effective field results in a more complicated solution, however, it does not significantly influence the dependence of the impedances on the field and frequency.

It is assumed further that the value of H_p is proportional to the concentration of MNPs in the ferrogel. This approximation seems to be reasonable since the magnetostatic measurements demonstrate that the ferrogel saturation magnetization increases linearly with the concentration of nanoparticles [2,27]. Experimental studies have shown that the ferrogels have S-shaped hysteresis loops [2,27,48]. To simplify calculations, we present the hysteresis curve by the linear field dependence of the ferrogel magnetization at low external fields.

Thus, we suppose that the effective stray field H_p acts on the ferromagnetic layers of the film. The field H_p has the opposite direction with respect to the effective angle ϕ of the ferrogel magnetization. The dependence of the angle ϕ on the external field can be approximated as follows:

$$\sin \phi = (H_e - H_c)/H_1 , \tag{17}$$

where H_c is the coercivity and H_1 is the external field corresponding to the approach of the ferrogel saturation magnetization.

The magnetization distribution in the ferromagnetic layers can be found by minimizing the free energy. Taking into account the effective stray field H_p, the minimization procedure results in the following equation for the equilibrium magnetization angle θ:

$$H_a \sin(\theta - \psi) \cos(\theta - \psi) + H_p \sin(\theta - \phi) - H_e \cos \theta = 0 . \tag{18}$$

Here H_a is the anisotropy field in the ferromagnetic layers and ψ is the deviation angle of the anisotropy axis from the transverse direction.

The solution of the linearized Landau-Lifshitz equation leads to the following expression for the transverse permeability μ in the ferromagnetic layers:

$$\mu = 1 + \frac{\gamma 4\pi M(\gamma 4\pi M + \omega_1 - i\kappa\omega)\sin^2\theta}{(\gamma 4\pi M + \omega_1 - i\kappa\omega)(\omega_2 - i\kappa\omega) - \omega^2} , \tag{19}$$

where M is the saturation magnetization of the ferromagnetic layers, γ is the gyromagnetic constant, κ is the Gilbert damping parameter, and

$$\begin{aligned}
\omega_1 &= \gamma[H_a \cos^2(\theta - \psi) - H_p \cos(\theta - \phi) + H_e \sin\theta] , \\
\omega_2 &= \gamma[H_a \cos\{2(\theta - \psi)\} - H_p \cos(\theta - \phi) + H_e \sin\theta]
\end{aligned} \tag{20}$$

Thus, the MI response in the multilayered film with a ferrogel layer can be calculated as follows. The first step is the calculation of the equilibrium magnetization angle in the ferromagnetic layers by means of Equations (17) and (18). The second step is the determination of the transverse permeability in the ferromagnetic layers by using Equations (19) and (20). Then, the corresponding values of the skin depth in the layers can be calculated, and the distribution of the fields is determined by means of Equations (9)–(12), taking into account boundary conditions. When the field distribution is found, the impedance of the film with ferrogel can be obtained by means of Equation (16).

3.7. Model Results

According to the model proposed, the ferrogel layer makes two contributions to the MI response. The first contribution is related to the high permittivity of the ferrogel, and the second one results from the stray fields of MNPs. Let us consider, first, the dielectric contribution of the pure gel without MNPs. In this case, the stray fields from the gel are equal to zero.

Figure 8a shows the calculated dependence of $\Delta Z/Z$ on the external field H_e at the frequency of 100 MHz for different values of the gel layer thickness d_3. The MI field dependence of the film structure without gel exhibits the typical two-peak behavior (curve 1 in Figure 8a). In the presence of the gel layer, the dependence has the same shape, however, the value of $\Delta Z/Z$ increases. This fact can be ascribed to the high permittivity of the gel, which influences the MI response of the film structure.

It follows from Figure 8a that the MI ratio increases with the thickness of the gel layer. Note that a similar significant growth of the MI ratio in the presence of pure gel has been observed previously in experimental studies [2,48].

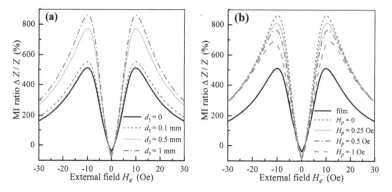

Figure 8. MI ratio $\Delta Z/Z$ as a function of the external field H_e for gel at $f = \omega/2\pi = 100$ MHz and different values of the ferrogel layer thickness d_3 (**a**). MI ratio $\Delta Z/Z$ as a function of the external field H_e for ferrogel at $f = 100$ MHz and different values of the effective stray field H_p. The parameters of the ferrogel layer are $d_3 = 1$ mm, $H_c = 4$ Oe and $H_1 = 750$ Oe (**b**). Parameters used for calculations are $l = 1$ cm, $w = 0.01$ cm, $2d_0 = 500$ nm, $d_1 = 5$ nm, $d_2 = 100$ nm, $n = 4$, $M = 750$ G, $H_a = 7.5$ Oe, $\psi = -0.05\pi$, $\sigma_1 = 5 \times 10^{17}$ s^{-1}, $\sigma_2 = 4 \times 10^{16}$ s^{-1}, $\kappa = 0.02$ and $\varepsilon = 70$.

To describe the field dependence of the MI ratio for the film with the ferrogel layer we take the following values of coercivity H_c and field H_1 in Equation (17): $H_c = 4$ Oe and $H_1 = 750$ Oe. These values of the parameters correspond to the hysteresis loops measured in [2]. Figure 8b shows the field dependence of the MI ratio $\Delta Z/Z$ for the film without gel, film with pure gel and film with ferrogel at different values of the effective stray field H_p.

The frequency dependence of the maximum value of MI ratio $\Delta Z_{max}/Z$ for the same samples is shown in Figure 9. An increase in the concentration of MNPs in the ferrogel leads to growth in the saturation magnetization of the ferrogel and to an enhancement of the effective stray field H_p. As a result, the value of the MI ratio decreases with an increase in the concentration of MNPs in the ferrogel. The calculated dependences presented in Figures 8b and 9 are in qualitative agreement with experimental results obtained in [2,48].

The model proposed allows one to explain qualitatively the main features of the experimental results concerning the MI response of multilayered films with a ferrogel layer [2,48]. It is demonstrated that high permittivity of the pure gel layer affects the MI response of the film structure, resulting in its significant enhancement. It is shown also that the contribution of the stray fields induced by MNPs in the ferrogel layer leads to a decrease in the MI ratio with an increase of the concentration of MNPs in the ferrogel. It should be noted that the model does not describe the essential shift of the maximum in the frequency dependence of $\Delta Z_{max}/Z$ to lower frequencies with an increase in the concentration of MNPs in the ferrogel [2].

Note also, that the simplified presentation of the stray fields created by the MNPs by means of the effective field H_p qualitatively describes the effect of the ferrogel layer on the MI of the multilayered film.

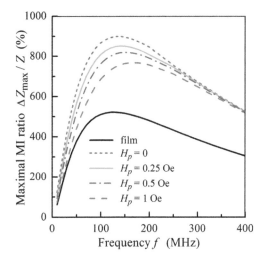

Figure 9. Frequency dependence of the maximum value of the MI ratio $\Delta Z_{max}/Z$ for ferrogel at different values of the effective stray field H_p. Other parameters used for calculations are the same as in Figure 8.

However, this approximation has some restrictions. The first restriction is related to the fact that the effective stray field is assumed to be constant over the multilayered film thickness. The spatial variation of the effective stray field can be included in the model. However, this variation does not significantly affect the MI response. The second restriction is the introduction of the effective angle ϕ of the ferrogel magnetization to describe hysteresis loops of ferrogel layer at low external fields (see Equation (17)). A more general approach consists in the approximation of the experimental magnetization curves by analytical functions.

The main disadvantage of the model is as follows. It is evident that the effective stray field H_p should be proportional to the concentration of the MNPs in the ferrogel. However, the coefficient of proportionality cannot be found in the framework of the approach proposed. To estimate the value of this coefficient, an approximate distribution of the stray fields should be found by means of a numerical solution for the magnetostatic equations.

3.8. Magnetoimpedance Measurements in Configuration of Biosensor Prototype

Let us give some examples of MI responses of multilayered sensitive elements without and with ferrogels. In a previous section, comparative analysis of the MI behavior was based on our previous experimental research results and a new model approach was developed for ferrogel detection with a MI sensitive element showing good qualitative agreement with experimental data.

As mentioned before, in our previous works we used FeNi/Ti-based magnetic multilayers deposited onto rigid glass substrates and large gel pieces of 1.0 to 0.5 g weight having cylindrical or half cylinder shapes [2,35]. The measurements were strongly conditioned by time as the mass of the gel/ferrogels changed rapidly due to water loss. In the present work, we therefore developed a special protocol for the MI measurements with gels and FG taking into account the mass loss of the piece of gel or FG of the same shape and about 18 mg mass.

Figure 10a shows an example of the mass loss for rectangular stripes of ferrogels used in MI measurements. One can see that the dependence of the mass of the sample on the time t is non-linear for a time scale of about 1 h. In a traditional regime, the complete cycle of MI measurements from magnetic saturation in a positive magnetic field to saturation in a negative magnetic field and again from magnetic saturation in a negative magnetic field to magnetic saturation in a positive magnetic field ("down" and "up" branches, accordingly [49,50]) are usually made with an averaging procedure and take about 30–60 min depending on the number of measured frequencies. This means that for small pieces of gel a much faster measuring protocol must be elaborated.

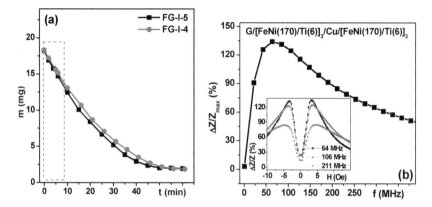

Figure 10. Mass loss for rectangular stripes of ferrogels used in MI measurements as a function of time drying in normal conditions (**a**); Frequency dependence of the maximum value of MI ratio $\Delta Z_{max}/Z$ for multilayered structure deposited onto glass substrate. Inset shows field dependence of the $\Delta Z/Z$ ratio for characteristic selected frequencies. All units for the layers' thickness are nanometers (**b**).

There are two reasons for making an attempt to detect the pieces of ferrogels of a small size. The first, is the biomedical request to detect small tumors below 1 cm in size in order to insure early stage cancer diagnostics. The second, is a purely technical issue—modeling is only possible for thin layers of gel/ferrogel due to limitations in computer capacity. We therefore elaborated a special protocol for MI measurements in order to minimize the experimental errors and to ensure reproducible procedures with minimum measurement. Instead of complete measurements of "down" and "up" branches only the first half of the "down" branch was measured: the total time of the MI measurements in a decreasing field was limited to 3 min and characterized by the linear part for a total mass loss of 10% with respect to the initial mass (see Figure 10a). We were able to ensure repeatable way of measurements with very thin gels in order to proceed with modeling of the MI responses. The next step in this research process is to design a MI device with controlled humidity inside the measuring cell.

Figure 10b shows an example of a typical frequency dependence of a FeNi/Ti-based MI multilayered sensitive element. The inset shows the field dependence of the $\Delta Z/Z$ ratio for selected characteristic frequencies (near the maximum of $\Delta Z/Z$ ratio, and slightly and significantly above it). The two peak MI responses are very consistent with uniaxial magnetic anisotropy features evaluated from the shape of the hysteresis loops (Figure 7). Both frequency and field dependence of the observed types are widely described in the literature and understood in the framework of classic electrodynamics [18,20,51]. For the next step, the MI FeNi/Ti-based multilayered sensitive element was measured in the same conditions without, and with gel/ferrogel with maximum available concentration of MNPs (Figure 11a,b).

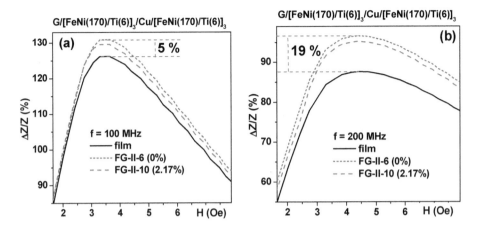

Figure 11. Field dependence of the $\Delta Z/Z$ ratio for thin film FeNi/Ti-based MI sensitive element measured either without gel/ferrogel or in the presence of gel or ferrogel with maximum available concentration of MNPs: (**a**) f = 100 MHz; (**b**) f = 200 MHz. Dashed lines show the difference between the MI element itself and the MI element covered by gel. G-glass substrate, numbers in the description of the multilayered structure are layer thicknesses. All units for the layers thicknesses are nanometers.

As before [2,35,49] the presence of the gel resulted in a sizable increase in the $\Delta Z/Z$ ratio near the anisotropy field of the MI element but the presence of a ferrogel of the same mass resulted in the displacement of the $\Delta Z/Z$ curve in the intermediate position between the position for the MI element itself and the MI element covered by pure gel. This is exactly what was predicted by the proposed electrodynamic model (compare Figures 8b and 11a,b). It is also clear that the difference between the response of the MI element itself and the MI element covered by gel/ferrogel depends on the frequency of the driving current: in the case of G/[FeNi(170 nm)/Ti(6 nm)]$_3$/Cu(500 nm)/[FeNi(170 nm)/Ti(6 nm)]$_3$. Similar behavior was also observed for FeNi/Cu-based multilayered structures deposited onto flexible substrates for both kinds of gels (see also Table 2): in all cases the presence of a thin gel/ferrogel layer on the surface of the MI multilayered element resulted in an increase in the MI response near the anisotropy field.

In order to evaluate the difference between the $\Delta Z/Z$ value of the uncovered and gel covered MI element according to the type of frequency, we measured it for different frequencies (Figure 12b, Inset). For a mathematical representation the following equation was used for each one of the fixed frequencies: $\Delta(\Delta Z/Z)_f = 100\% \times (\Delta Z/Z_{film+FG} - \Delta Z/Z_{film})/\Delta Z/Z_{film}$. One can see clearly that the $\Delta(\Delta Z/Z)_f$ parameter increases with an increase in the driving current frequency.

Despite the fact that evident qualitative agreement between the proposed model and the experimental data was obtained, for the practical purpose of biosensor application or simple use of MI for studies of gel materials, the experimental part of the measurements with gels required special efforts. Apart from an unusual time limitation due to gel dehydration, the measurements required fabricating gel samples of exactly the same shape and mass and exactly the same placement in the measuring system. For large samples previously studied in the MI-biosensor regime, both problems were less critical. Here, the major difficulties came not from the insufficient sensitivity of MI detector but rather from the difficulties of fabricating exactly the same samples and placing them into the system in the same way. Even so, we were able to obtain concentration dependences of the maximum of $\Delta Z/Z$ ratio for fixed frequencies (Figure 13) with a sensitivity of almost 2% per 1 wt % of MNPs. Here, we need to remember that all experiments were made with very diluted systems with very low concentrations of iron oxide MNPs.

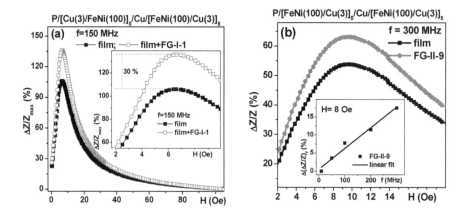

Figure 12. Field dependence of the $\Delta Z/Z$ ratio for thin film FeNi/Cu-based MI sensitive element measured either without gel/ferrogel or in the presence of gel (**a**) f = 150 MHz; (**b**) f = 300 MHz. Dashed lines (Inset (**a**)) show the difference between the MI element itself and the MI element covered by FG-I-1. P-polymer flexible substrate, numbers in the description of the multilayered structure are layer thicknesses. Inset (**b**) shows the difference in the field of 8 Oe between $\Delta Z/Z$ value for FG-II-9. All units for the layer thicknesses are nanometers.

In order to show the extraordinary capacity of the MI element, we also measured the $\Delta Z/Z$ ratio for a fixed frequency of 300 MHz for the same ferrogel rectangular element which was cut step-by-step into shorter pieces of different length, L_g (Figure 13b). Although the procedure was not ideal (peaks for 8 and 6 mm were displaced toward higher fields), we were able to detect a tiny piece of about 4 mg of FG-II-10 ferrogel placed symmetrically in the centre of the MI element. This means the detection of about 0.09 mg of MNPs. Different parameters can be used for such detection. Figure 13c shows the comparison of $\Delta Z/Z(L_g)$ and $\Delta Z/Z_{max}(L_g)$, where $\Delta Z/Z(L_g)$ was taken for the external magnetic field of 8 Oe. $\Delta Z/Z(L_g)$ shows exponential and $\Delta Z/Z_{max}(L_g)$ shows linear dependence on the gel stripe lengths.

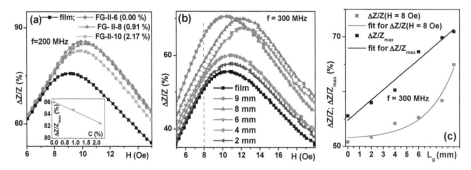

Figure 13. Field dependence of the $\Delta Z/Z$ ratio for thin film P/[FeNi(100)/Cu(3)]$_5$/Cu(500 nm)/ [FeNi(100)/Cu(3)]$_5$. MI sensitive element measured either without gel/ferrogel or in the presence of gel and ferrogels (**a**) f = 200 MHz; (**b,c**) f = 300 MHz. Inset (**a**) shows the concentration dependence of $\Delta Z/Z$ ratio. Dashed line (**b**) shows the field of 8 Oe for which exponential dependence on the lengths of the gel sample was observed (**c**).

There have been attempts to design a MI-based biosensor with different types of sensitive elements, such as amorphous ribbons, wires, and thin films [2,10–12,16,52,53]. They all have advantages and

Sensors **2018**, *18*, 872

disadvantages, but the thin film direction seems to be the most promising. For multifunctional biosensor applications we expect the creation of a magnetic biosensor which would be capable of quantitatively monitoring living tissue functions, such as muscle contraction or evaluating a number of MNP loaded cells adhered onto the surface of a ferrogel implant near damaged tissue.

We were not able to design a biosensor prototype with controlled humidity inside the measuring chamber, but this is an urgent goal for the next study in order to avoid the limitations imposed by the short time for the measurements. Additional methodology should be elaborated to control the shape of small gel samples. Here, the biological technique of sample preparation might be useful [22]. Meanwhile, we have completed the next step in the development of a new generation of magnetic biosensors by the development of ferrogels incorporating magnetic nanoparticles into chemically crosslinked hydrogels mimicking some properties of the living systems and showing that the giant magnetoimpedance effect is promising for magnetic biosensor prototype development.

4. Conclusions and Outlook

In the present work, two series of ferrogels based on polyacrylamide (PAAm) with different chemical network densities (soft and dense) were synthesized by free-radical polymerization in aqueous solution with N,N'-methylene-diacrylamide as a cross-linker and iron oxide nanoparticles as a filler. These biocompatible maghemite Fe_2O_3 MNPs were fabricated by the electrophysical technique of laser target evaporation. Their mechanical (tension-deformation behavior, Young's modulus), electrical (dependence of electrical potential of FG on the weight fraction of MNPs) and magnetic properties were comparatively analyzed.

We developed a giant magnetoimpedance sensor prototype with multilayered FeNi-based sensitive elements deposited by magnetron sputtering onto glass or polymer substrates, which was adapted for FG studies. The MI measurements in the initial state and in the presence of FG with different concentrations of MNPs allowed the characterization of stray fields of the MNPs in the FG. This is the first report on ferrogel detection with a flexible MI sensitive element. This may open the possibility of creating prototypes for deformation evaluation. We proposed an electrodynamic model to describe the MI in multilayered film with a FG and found good qualitative agreement between the model and experimental data.

Acknowledgments: This work was supported in part within the framework of the state task of the Ministry of Education and Science of Russia 3.6121.2017/8.9; RFBR grants 16-08-00609-a, 18-08-00178, and by the ACTIMAT ELKARTEK grant of the Basque Country Government. Selected studies were made at SGIKER Common Services of UPV-EHU and URFU Common Services. We thank I.V. Beketov, A.A. Chlenova, S.O. Volchkov, V.N. Lepalovskij, A.M. Murzakaev and A.A. Svalova for special support.

Author Contributions: F.A. Blyakhman, N.A. Buznikov, A.P. Safronov and G.V. Kurlyandskaya conceived and designed the experiments; A.P. Safronov, F.A. Blyakhman, T.F. Sklyar, E.V. Golubeva, A.V. Svalov, S. Yu. Sokolov, G. Yu. Melnikov, I. Orue and G.V. Kurlyandskaya performed the experiments; N.A. Buznikov proposed the model and made the calculations; A.P. Safronov, F.A. Blyakhman, N.A. Buznikov and G.V. Kurlyandskaya analyzed the data; A.P. Safronov, F.A. Blyakhman, N.A. Buznikov and G.V. Kurlyandskaya wrote the manuscript. All authors discussed the results and implications and commented on the manuscript at all stages. All authors read and approved the final version of the manuscript.

Conflicts of Interest: The authors declare no conflict of interest.

References

1. Baselt, D.R.; Lee, G.U.; Natesan, M.; Metzger, S.W.; Sheehan, P.E.; Colton, R.J. A biosensor based on magnetoresistance technology. *Biosens. Bioelectron.* **1998**, *13*, 731–739. [CrossRef]
2. Kurlyandskaya, G.V.; Fernandez, E.; Safronov, A.P.; Svalov, A.V.; Beketov, I.V.; Burgoa Beitia, A.; García-Arribas, A.; Blyakhman, F.A. Giant magnetoimpedance biosensor for ferrogel detection: Model system to evaluate properties of natural tissue. *Appl. Phys. Lett.* **2015**, *106*, 193702. [CrossRef]
3. Turner, A.P.F. Tech. Sight. Biochemistry. Biosensors—Sense and sensitivity. *Science* **2000**, *290*, 1315–1317. [CrossRef] [PubMed]

4. Kriz, C.B.; Radewik, K.; Kriz, D. Magnetic permeability measurements in bioanalysis and biosensors. *Anal. Chem.* **1996**, *68*, 1966–1970. [CrossRef] [PubMed]

5. Miller, M.M.; Prinz, G.A.; Cheng, S.-F.; Bounnak, S. Detection of a micron-sized magnetic sphere using a ring-shaped anisotropic magnetoresistance-based sensor: A model for a magnetoresistance-based biosensor. *Appl. Phys. Lett.* **2002**, *81*, 2211–2213. [CrossRef]

6. Ferreira, H.A.; Graham, D.L.; Freitas, P.P.; Cabral, J.M.S. Biodetection using magnetically labeled biomolecules and arrays of spin valve sensors. *J. Appl. Phys.* **2002**, *93*, 7281–7286. [CrossRef]

7. Ong, K.G.; Wang, J.; Singh, R.S.; Bachas, L.G.; Grimes, C.A. Monitoring of bacteria growth using a wireless, remote query resonant-circuit sensor: Application to environmental sensing. *Biosens. Bioelectron.* **2001**, *16*, 305–312. [CrossRef]

8. Rife, J.C.; Miller, M.M.; Sheehan, P.E.; Tamanaha, C.R.; Tondra, M.; Whitman, L.J. Design and performance of GMR sensors for the detection of magnetic microbeads in biosensors. *Sens. Actuators A* **2003**, *107*, 209–218. [CrossRef]

9. Megens, M.; Prins, M. Magnetic biochips: A new option for sensitive diagnostics. *J. Magn. Magn. Mater.* **2005**, *293*, 702–708. [CrossRef]

10. Kurlyandskaya, G.V.; Levit, V.I. Advanced materials for drug delivery and biosensors based on magnetic label detection. *Mater. Sci. Eng. C* **2007**, *27*, 495–503. [CrossRef]

11. Blanc-Béguin, F.; Nabily, S.; Gieraltowski, J.; Turzo, A.; Querellou, S.; Salaun, P.Y. Cytotoxicity and GMI bio-sensor detection of maghemite nanoparticles internalized into cells. *J. Magn. Magn. Mater.* **2009**, *321*, 192–197. [CrossRef]

12. Wang, T.; Guo, L.; Lei, C.; Zhou, Y. Ultrasensitive determination of carcinoembryonic antigens using a magnetoimpedance immunosensor. *RSC Adv.* **2015**, *5*, 51330–51336. [CrossRef]

13. Blyakhman, F.A.; Safronov, A.P.; Zubarev, A.Y.; Shklyar, T.F.; Makeyev, O.G.; Makarova, E.B.; Melekhin, V.V.; Larrañaga, A.; Kurlyandskaya, G.V. Polyacrylamide ferrogels with embedded maghemite nanoparticles for biomedical engineering. *Results Phys.* **2017**, *7*, 3624–3633. [CrossRef]

14. Shahinpoor, M.; Kim, K. Ionic polymer–metal composites: III. Modeling and simulation as biomimetic sensors, actuators, transducers, and artificial muscles. *Smart Mater. Struct.* **2004**, *13*, 1362–1388. [CrossRef]

15. Kennedy, S.; Roco, C.; Délérisa, A.; Spoerria, P.; Cezara, C.; Weavera, J.; Vandenburghd, H.; Mooney, D. Improved magnetic regulation of delivery profiles from ferrogels. *Biomaterials* **2018**, *161*, 179–189. [CrossRef] [PubMed]

16. Kumar, A.; Mohapatra, S.; Fal-Miyar, V.; Cerdeira, A.; Garcia, J.A.; Srikanth, H.; Gass, J.; Kurlyandskaya, G.V. Magnetoimpedance biosensor for Fe_3O_4 nanoparticle intracellular uptake evaluation. *Appl. Phys. Lett.* **2007**, *91*, 143902. [CrossRef]

17. Beach, R.; Berkowitz, A. Sensitive field-and frequency-dependent impedance spectra of amorphous FeCoSiB wire and ribbon. *J. Appl. Phys.* **1994**, *76*, 6209–6213. [CrossRef]

18. Antonov, A.S.; Gadetskii, S.N.; Granovskii, A.B.; D'yachkov, A.L.; Paramonov, V.P.; Perov, N.S.; Prokoshin, A.F.; Usov, N.A.; Lagar'kov, A.N. Giant magnetoimpedance in amorphous and nanocrystalline multilayers. *Phys. Met. Metallorgr.* **1997**, *83*, 612–618.

19. Uchiyama, T.; Mohri, K.; Honkura, Y.; Panina, L.V. Recent advances of pico-Tesla resolution magneto-impedance sensor based on amorphous wire CMOS IC MI Sensor. *IEEE Trans. Magn.* **2012**, *48*, 3833–3839. [CrossRef]

20. Landau, L.D.; Lifshitz, E.M. *Electrodynamics of Continuous Media*; Pergamon: New York, NY, USA, 1975.

21. Glaser, R. *Biophysics*; Springer: Berlin/Heidelberg, Germany; New York, NY, USA, 1999.

22. Wong, J.; Chilkoti, A.; Moy, V.T. Direct force measurements of the streptavidin–biotin interaction. *Biomol. Eng.* **1999**, *16*, 45–55. [CrossRef]

23. Pollack, G.H. *Cells, Gels and the Engines of Life*; Ebner&Sons: Seattle, WA, USA, 2001.

24. Harland, R.; Prudhomme, R. *Polyelectrolyte Gels: Properties, Preparation and Applications*; American Chemical Society: Washington, DC, USA, 1992.

25. Liu, T.; Hu, S.; Liu, T.; Liu, D.; Chen, S. Magnetic-sensitive behaviour of intelligent ferrogels for controlled release of drug. *Langmuir* **2006**, *22*, 5974–5978. [CrossRef] [PubMed]

26. Grossman, J.H.; McNeil, S.E. Nanotechnology in cancer medicine. *Phys. Today* **2012**, *65*, 38–42. [CrossRef]

27. Safronov, A.P.; Mikhnevich, E.A.; Lotfollahi, Z.; Blyakhman, F.A.; Sklyar, T.F.; Larrañaga Varga, A.; Medvedev, A.I.; Fernández Armas, S.; Kurlyandskaya, G.V. Polyacrylamide ferrogels with magnetite or strontium hexaferrite: Next step in the development of soft biomimetic matter for biosensor applications. *Sensors* **2018**, *18*, 257. [CrossRef] [PubMed]

28. Safronov, A.P.; Beketov, I.V.; Komogortsev, S.V.; Kurlyandskaya, G.V.; Medvedev, A.I.; Leiman, D.V.; Larranaga, A.; Bhagat, S.M. Spherical magnetic nanoparticles fabricated by laser target evaporation. *AIP Adv.* **2013**, *3*, 052135. [CrossRef]

29. Novoselova, I.P.; Safronov, A.P.; Samatov, O.M.; Beketov, I.V.; Medvedev, A.I.; Kurlyandskaya, G.V. Water based suspensions of iron oxide obtained by laser target evaporation for biomedical applications. *J. Magn. Magn. Mater.* **2016**, *415*, 35–38. [CrossRef]

30. Fernandez, E.; Svalov, A.V.; García-Arribas, A.; Feuchtwanger, J.; Barandiaran, J.M.; Kurlyandskaya, G.V. High performance magnetoimpedance in FeNi/Ti nanostructured multilayers with opened magnetic flux. *J. Nanosci. Nanotechnol.* **2012**, *12*, 7496–7500. [CrossRef] [PubMed]

31. Alzola, N.; Kurlyandskaya, G.V.; Larranaga, A.; Svalov, A.V. Structural peculiarities and magnetic properties of FeNi films and FeNi/Ti-based magnetic nanostructures. *IEEE Trans. Magn.* **2012**, *48*, 1605–1608. [CrossRef]

32. Svalov, A.V.; Aseguinolaza, I.R.; Garcia-Arribas, A.; Orue, I.; Barandiaran, J.M.; Alonso, J.; Fernández-Gubieda, M.L.; Kurlyandskaya, G.V. Structure and magnetic properties of thin permalloy films near the 'transcritical' state. *IEEE Trans. Magn.* **2010**, *46*, 333–336. [CrossRef]

33. Safronov, A.P.; Kamalov, I.A.; Shklyar, T.F.; Dinislamova, O.A.; Blyakhman, F.A. Activity of counterions in hydrogels based on poly(acrylic acid) and poly(methacrylic acid): Potentiometric measurements. *Polym. Sci. Ser. A* **2012**, *54*, 909–919. [CrossRef]

34. Blyakhman, F.A.; Safronov, A.P.; Zubarev, A.Y.; Shklyar, T.F.; Dinislamova, O.A.; Lopez-Lopez, M.T. Mechanoelectrical transduction in the hydrogel-based biomimetic sensors. *Sens. Actuators A Phys.* **2016**, *248*, 54–61. [CrossRef]

35. Kurlyandskaya, G.V.; Fernández, E.; Svalov, A.; Burgoa Beitia, A.; García-Arribas, A.; Larrañaga, A. Flexible thin film magnetoimpedance sensors. *J. Magn. Magn. Mater.* **2016**, *415*, 91–96. [CrossRef]

36. Pearson, W.B. *Handbook of Lattice Spacing Structures of Metals and Alloys*; Pergamon Press: London, UK, 1958.

37. Kosmulski, M. *Chemical Properties of Material Surfaces*; Marcel Dekker: New York, NY, USA; Basel, Switzerland, 2001.

38. Quesada-Perez, M.; Maroto-Centeno, J.A.; Forcada, J.; Hidalgo-Alvarez, R. Gel swelling theories: The classical formalism and recent approaches. *Soft Matter* **2011**, *7*, 10536–10547. [CrossRef]

39. Rubinstein, M.; Colby, R.H. *Polymer Physics*; Oxford University Press: New York, NY, USA, 2003.

40. Shklyar, T.F.; Safronov, A.P.; Klyuzhin, I.S.; Pollack, G.H.; Blyakhman, F.A. A Correlation between Mechanical and Electrical Properties of the Synthetic Hydrogel chosen as an Experimental Model of Cytoskeleton. *Biophysics* **2008**, *53*, 544–549. [CrossRef]

41. Shklyar, T.; Dinislamova, O.; Safronov, A.; Blyakhman, F. Effect of cytoskeletal elastic properties on the mechanoelectrical transduction in excitable cells. *J. Biomech.* **2012**, *45*, 1444–1449. [CrossRef] [PubMed]

42. Guo, H.; Kurokawa, T.; Takahata, M.; Hong, W.; Katsuyama, Y.; Uo, F.; Ahmed, J.; Nakajima, T.; Nonoyama, T.; Gong, J.P. Quantitative observation of electric potential distribution of brittle polyelectrolyte hydrogels using microelectrode technique. *Macromolecules* **2016**, *49*, 3100–3108. [CrossRef]

43. Coey, J.M.D. *Magnetism and Magnetic Materials*; Cambridge University Press: New York, NY, USA, 2010.

44. Gromov, A.; Korenivski, V.; Rao, K.V.; van Dover, R.B.; Mankiewich, P.M. A model for impedance of planar RF inductors based on magnetic films. *IEEE Trans. Magn.* **1998**, *34*, 1246–1248. [CrossRef]

45. Gromov, A.; Korenivski, V.; Haviland, D.; van Dover, R.B. Analysis of current distribution in magnetic film inductors. *J. Appl. Phys.* **1999**, *85*, 5202–5204. [CrossRef]

46. Sukstanskii, A.; Korenivski, V.; Gromov, A. Impedance of a ferromagnetic sandwich strip. *J. Appl. Phys.* **2001**, *89*, 775–782. [CrossRef]

47. Kraus, L. GMI modelling and material optimization. *Sens. Actuators A* **2003**, *106*, 187–194. [CrossRef]

48. Kurlyandskaya, G.V.; Portnov, D.S.; Beketov, I.V.; Larrañaga, A.; Safronov, A.P.; Orue, I.; Medvedev, A.I.; Chlenova, A.A.; Sanchez-Ilarduya, M.B.; Martinez-Amesti, A.; et al. Nanostructured materials for magnetic biosensing. *BBA Gen. Subj.* **2017**, *1861*, 1494–1506. [CrossRef] [PubMed]

49. Kurlyandskaya, G.V.; de Cos, D.; Volchkov, S.O. Magnetosensitive transducers for nondestructive testing operating on the basis of the giant magnetoimpedance effect: A review. *Russ. J. Nondestruct. Test.* **2009**, *45*, 377–398. [CrossRef]
50. Nishibe, Y.; Ohta, N.; Tsukada, K.; Yamadera, H.; Nomomura, Y.; Mohri, K.; Uchiyama, T. Sensing of passing vehicles using a lane marker on road with built-in thin film MI sensor and power source. *IEEE Trans. Veh. Technol.* **2004**, *53*, 1827–1834. [CrossRef]
51. Volchkov, S.O.; Cerdeira, M.A.; Gubernatorov, V.V.; Duhan, E.I.; Potapov, A.P.; Lukshina, V.A. Effects of slight plastic deformation on magnetic properties and giant magnetoimpedance of FeCoCrSiB amorphous ribbons. *Chin. Phys. Lett.* **2007**, *24*, 1357–1360. [CrossRef]
52. Beato-López, J.J.; Pérez-Landazábal, J.I.; Gómez-Polo, C. Magnetic nanoparticle detection method employing non-linear magnetoimpedance effects. *J. Appl. Phys.* **2017**, *121*, 163901. [CrossRef]
53. Chiriac, H.; Herea, D.D.; Corodeanu, S. Microwire array for giant magnetoimpedance detection of magnetic particles for biosensor prototype. *J. Magn. Magn. Mater.* **2007**, *311*, 425–428. [CrossRef]

 *sensors*

Article

Shear Elasticity of Magnetic Gels with Internal Structures

Dmitry Borin [1], Dmitri Chirikov [2] and Andrey Zubarev [2,3,*]

[1] Chair of Magnetofluiddynamics, Measuring and Automation Technology, TU Dresden, 01069 Dresden, Germany; dmitry.borin@tu-dresden.de

[2] Department of Theoretical and Mathematical Physics, Ural Federal University, Lenina Ave 51, 620083 Ekaterinburg, Russia; d.n.chirikov@urfu.ru

[3] M.N. Mikheev Institute of Metal Physics of the Ural Branch of the Russian Academy of Sciences, Sofia Kovalevskaya st., 18, 620219 Ekaterinburg, Russia

* Correspondence: A.J.Zubarev@urfu.ru; Tel.: +7-905-800-7855

Received: 27 May 2018; Accepted: 20 June 2018; Published: 27 June 2018

Abstract: We present the results of the theoretical modeling of the elastic shear properties of a magnetic gel, consisting of soft matrix and embedded, fine magnetizable particles, which are united in linear chain-like structures. We suppose that the composite is placed in a magnetic field, perpendicular to the direction of the sample shear. Our results show that the field can significantly enhance the mechanical rigidity of the soft composite. Theoretical results are in quantitative agreement with the experiments.

Keywords: magnetic gels; shear modulus; chain structures; magnetorheological effect

1. Introduction

Magnetic gels are composites of nano- and micron-sized magnetic particles in soft polymer matrixes. The combination of a rich set of physical properties of the polymer and magnetic materials is very valuable for many progressive industrial, bioengineering, and biomedical applications [1–7]. In part, for address drug delivery, for industrial and biological sensors [8–14], for the construction of soft actuators and artificial muscles [2,15], and for regenerative medicine and tissue engineering [16–41]. An overview of the works on magnetic gel synthesis and their biomedical applications can be found in [23].

One of the remarkable properties of magnetic gels is their ability to change, under the action of an external magnetic field, their microstructure, magnetic, mechanical, and other microscopic properties, size, and shape. This provides the opportunity to control, with the help of the field, mechanic behavior, transport, and electrical processes in these systems, and this possibility presents a significant advantage for biosensoric, tissue engineering, and other biological applications [8,9,20,23,39,41].

During the magnetic gels' synthesis, the particles are usually embedded in the liquid polymer, and their spatial distribution is fixed after the composite gelation. If the host polymer is cured without an external magnetic field, the particles, as a rule, are distributed more or less homogeneously and isotropically. If the composite is polymerized under the field (field of polymerization), the particles form various anisotropic structures, elongated in the field direction. The appearance of these internal structures significantly changes the sensitivity of the gels to mechanic, electrical, magnetic, and other external impacts, and changes the kinetics of the internal transport phenomena and chemical reactions, the rate of cell proliferation, and other phenomena in these systems. This opens perspectives of the tunable synthesis of the magnetically controlled sensors, and scaffolds for the growth of cell tissues with willing structure and properties, artificial muscles, and other materials for biological and industrial applications.

The simplest kind of the internal structures that are formed by the magnetic particles in liquid media are liner chains, where particles are bounded as "head to tails" by the forces of magnetic attraction. These structures appear in the systems with a low and moderate volume concentrations of magnetic particles (usually in the range of 10–15%). Some photos of these chains can be found, for example in [42,43]. In the case of the higher concentrations, the particles can form topologically more complicated branched, net-like, bulk and other structures (see, for example, [44–50]).

The effect of the magnetic field on the elastic shear modulus of the gels with a homogeneous and isotropic (gas-like) spatial distribution of magnetic particles has been theoretically studied in refs. [51,52]. It was shown that a magnetic field that is applied perpendicular to the macroscopic shear of the composite, enhances the composite elastic modulus. At the same time, the experiments [53] show that the rigidity of the composites with the chains aligned perpendicularly to the shear, is significantly more than that of the systems with a chaotic distribution of the particles. Since the micromechanic (on the level of the particles and their aggregates) situation in the composites with the heterogeneous aggregates is significantly different from that in the homogeneous systems, the microscopic analysis of macroscopic properties of the composites with internal heterogeneous structures requires the development of a special theoretical approach.

The aim of this work is to theoretically study the effect of an external magnetic field on the shear elastic modulus of magnetic gels with internal chain-like structures, that are formed by magnetizable non-Brownian spherical particles. Physically, this means that the size of the particles are supposed to range from several tens of nanometers to microns. The particles of this size are very often used for the preparation of magnetic gels for bio-medical applications.

We take into account that the chains appear at the stage of the matrix polymerization under the action of an external magnetic field. That is why all chains are parallel to this field. We suppose that the actual magnetic field **H** has the same direction as the field of polimerization, i.e., that the field **H** is parallel to the chains. We consider the situation when the macroscopic shear of the sample is perpendicular, whereas the gradient of the shear is parallel to the chains. The length of the chains is supposed to be much less than the size of the sample.

2. Physical and Mathematical Model

For maximal simplification of the mathematical part of the problem, we will suppose that the particles are identical. Like in [54,55], we will neglect fluctuations of the chain's shape and will consider them as ideally straight aggregates, aligned along the applied magnetic field **H**. This model of the chain is illustrated in Figure 1.

The typical size of the cell of the polymer matrix in ferrogels is several nanometers, and the size of the particles vary from several tens of nanometers to microns. Thus, the particles are much larger than the gel cell. That is why we will consider the host polymer as a continuous medium with respect to the particles.

We will restrict ourselves by the analysis of small deformations of the composites and will suppose the linear relations between the mechanic stress and deformations in the matrix.

We will also neglect any interactions between the chains. This approximation is based on the results of ref. [56], which shows that the effects inside the chains play a dominant role in the formation of macroscopic properties of the composites, compared with the effects of the interchain interaction.

For mathematical definiteness, we will suppose that the chain consists of an odd number of particles. This assumption is not of principle for the physical analysis.

Sensors **2018**, *18*, 2054

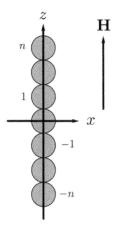

Figure 1. An illustration of the chain model and the used Cartesian coordinate system; n is the number of particles at the extremity of the chain, starting from the central one, which is number 0.

Let us denote the mean vector of a material point displacement in the composite as **u**. In the coordinate system, shown in Figure 2, the vector **u** is the mean displacement in the composite (i.e., the displacement at the infinitive distance from the chain). The vector **u** can be presented as: $u_x = \gamma x$ where γ is the mean shear of the system. We will consider small shear deformations of the composite, which means that the strong inequality $\gamma \ll 1$ is held.

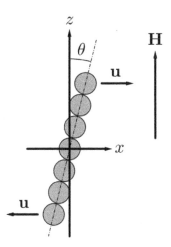

Figure 2. Illustration of the shearing of the chain.

Let the total number of the particles N in a chain be $N = 2n + 1$, where n is an integer. In the framework of the used approximations, the equations of the stationary displacement of the particles in the chain can be presented in the following form [54]:

$$3\pi G_0 d(\gamma i d - u_i) + f_i^{(m)} = 0, \qquad 0 < i \le n, \qquad -n \le i < 0,$$
$$u_0 = 0. \tag{1}$$

Here, G_0 is the shear modulus of the matrix, d is the diameter of the particles, $f_i^{(m)}$ is the force of the magnetic interaction of the i-th particle with the neighboring particles of the chain, and for the central particle $i = 0$. The equality $u_i = -u_{-i}$ for $0 < i \leq n$ follows from the symmetry of the problem.

For convenience, we introduce the dimensionless magnetic force $\widetilde{f}_i^{(m)}$ and dimensionless displacement $\widetilde{u}_i$:

$$\widetilde{f}_i^{(m)} = \frac{f_i^{(m)}}{\pi G_0 d^2}, \qquad \widetilde{u}_i = \frac{u_i}{d}, \qquad 0 \leq i \leq n. \tag{2}$$

By using these notations, one can rewrite Equation (1) as:

$$3(\gamma i - \widetilde{u}_i) + \widetilde{f}_i^{(m)} = 0, \qquad \widetilde{u}_i = \widetilde{u}_{-i}, \qquad 0 < i \leq n,$$
$$\widetilde{u}_0 = 0. \tag{3}$$

Under the assumption that $\gamma \ll 1$, the inequality $\widetilde{u}_i \ll 1$ is held.

The magnetic force $\widetilde{f}_i^{(m)}$ of the interparticle interaction can be estimated in the framework of the simplest dipole-dipole approximation. In order to calculate $\widetilde{f}_i^{(m)}$, we need to determine the magnetic moments of the particles in the chain. Strictly speaking, the value of a particle moment depends on the number i of the particle position in the chain. However, analysis [55] shows that approximation, where magnetic moments of all of the particles in the chain are supposed identical, leads to not significant deviations from the strict approach. That is why we will use the simplest approximation of identity of the particle moments in the chain.

When the composite experiences the macroscopic shear deformation, the axis of the chain deviates from the z-axis, as is illustrated in Figure 2. Because of the mutual magnetization of particles, the vector of the particle magnetic moment will also be deviated from this axis. Therefore, both components M_x and M_z of the vector **M** of the particle magnetization will take place in the deformed composite.

We estimate the magnetic force $\widetilde{f}_i^{(m)}$ by using the nearest-neighbor approximation, taking into account the magnetic dipole-dipole interaction only between the neighbor particles in the chain. By using the well-known relation for the force of the dipole-dipole interaction (see, for example, [57]), after simple but cumbersome transformations, in the linear approximation with respect to the displacements $\widetilde{u}_i$, one can get:

$$\widetilde{f}_1^{(m)} = \widetilde{f}_{1,0}^{(m)} + \widetilde{f}_{1,2}^{(m)} = \frac{\beta \widetilde{M}_z \widetilde{M}_x}{24} - \frac{\beta \widetilde{M}_z^2 \widetilde{u}_1}{12} - \left[\frac{\beta \widetilde{M}_z \widetilde{M}_x}{24} + \frac{\beta \widetilde{M}_z^2(\widetilde{u}_1 - \widetilde{u}_2)}{12}\right]$$
$$= \frac{\beta \widetilde{M}_z^2(\widetilde{u}_2 - 2\widetilde{u}_1)}{12},$$
$$\widetilde{f}_i^{(m)} = \widetilde{f}_{i,i-1}^{(m)} + \widetilde{f}_{i,i+1}^{(m)} = \frac{\beta \widetilde{M}_z \widetilde{M}_x}{24} + \frac{\beta \widetilde{M}_z^2(\widetilde{u}_{i-1} - \widetilde{u}_i)}{12} - \left[\frac{\beta \widetilde{M}_z \widetilde{M}_x}{24} + \frac{\beta \widetilde{M}_z^2(\widetilde{u}_i - \widetilde{u}_{i+1})}{12}\right] \tag{4}$$
$$= \frac{\beta \widetilde{M}_z^2(\widetilde{u}_{i+1} - 2\widetilde{u}_i + \widetilde{u}_{i-1})}{12}, \qquad 1 < i < n,$$
$$\widetilde{f}_n^{(m)} = \widetilde{f}_{n,n-1}^{(m)} = \frac{\beta \widetilde{M}_x \widetilde{M}_z}{24} + \frac{\beta \widetilde{M}_z^2(\widetilde{u}_{n-1} - \widetilde{u}_n)}{12}, \qquad \widetilde{M}_x = \frac{M_x}{M_s}, \qquad \widetilde{M}_z = \frac{M_z}{M_s}, \qquad \beta = \frac{\mu_0 M_s^2}{G_0}.$$

Here, $f_{i,i\pm1}^{(m)}$ is the force of the magnetic interaction between the i-th particle in the chain with the neighbor particle, M_s is the saturated magnetization of the particle's material, β is the parameter which defines the ratio of the energy of magnetic interaction between two magnetically saturated particles to the energy of elastic deformation of the matrix, and μ_0 is the vacuum magnetic permeability.

The main problem now is to estimate the dimensionless components $\widetilde{M}_x$ and $\widetilde{M}_z$. The strict solution of a problem of determination of magnetic moments of two closely situated magnetizable particles has not been obtained in literature because of not overcoming mathematical complexity. Here, we use the approach [58], where each particle is considered to be situated in a uniform magnetic field $H^{(e)}$, consisting of the external, with respect to the particle's field **H** (i.e., the mean field in the sample), and the field created by the other particle in the center of the first one. It is supposed that the magnetization of the particle obeys to the nonlinear Frolich-Kennelly relation [59]:

$$\widetilde{M}_k = \frac{\chi_p \widetilde{H}_k^{(i)}}{1 + \chi_p \widetilde{H}^{(i)}}, \qquad k = x, z, \qquad \widetilde{H}^{(i)} = \sqrt{\left(\widetilde{H}_x^{(i)}\right)^2 + \left(\widetilde{H}_z^{(i)}\right)^2}. \tag{5}$$

Here, χ_p is the initial magnetic susceptibility of the particle material, and $H^{(i)}$ is the magnetic field inside the particle. For the spherical particle the last field, this can be found from the general relation [60]:

$$\widetilde{H}_k^{(i)} + \frac{\widetilde{M}_k}{3} = \widetilde{H}_k^{(e)}, \qquad k = x, z. \tag{6}$$

Taking into account the magnetic interaction between the neighbor particles in the chain, in the dipole-dipole approximations we get:

$$\widetilde{H}_x^{(e)} = \frac{3\widetilde{M}_z \Psi_n^{(1)} - \widetilde{M}_x \Psi_n^{(0)}}{12N}, \qquad \widetilde{H}_z^{(e)} = \widetilde{H} + \frac{3\widetilde{M}_x \Psi_n^{(1)} + 2\widetilde{M}_z \Psi_n^{(0)}}{12N}.$$

$$\Psi_n^{(0)} = \sum_{i=1}^{2n} \frac{2n+1-i}{i^3}, \qquad \Psi_n^{(1)} = 2(a_{n-1}\widetilde{u}_{n-1} + a_n \widetilde{u}_n), \qquad a_i = \sum_{j=n+1-i}^{n+i} \frac{1}{j^4}. \tag{7}$$

Combining Equations (5)–(7) and Notations (4), after some transformations we come to the following system of nonlinear algebraic equations with respect to the dimensionless components $\widetilde{M}_x$ and $\widetilde{M}_z$ of the particle magnetization:

$$\frac{\widetilde{M}_x}{\chi_p(1 - \sqrt{\widetilde{M}_x^2 + \widetilde{M}_z^2})} + C_x \widetilde{M}_x - A\widetilde{M}_z = 0,$$

$$\frac{\widetilde{M}_z}{\chi_p(1 - \sqrt{\widetilde{M}_x^2 + \widetilde{M}_z^2})} + C_z \widetilde{M}_z - A\widetilde{M}_x - \widetilde{H} = 0, \tag{8}$$

$$C_z = \frac{1}{3}\left(1 - \frac{\Psi_n^{(0)}}{2N}\right), \qquad C_x = \frac{1}{3}\left(1 + \frac{\Psi_n^{(0)}}{4N}\right), \qquad A = \frac{\Psi_n^{(1)}}{4N},$$

This system can be solved analytically only under condition $\widetilde{M}_x \ll \widetilde{M}_z$. The last is true when the sample deformation is small (i.e., when $\gamma \ll 1$). By using the linear approximation with respect to $\widetilde{M}_x$, one gets from (8):

$$\widetilde{M}_z = \frac{D_z - \sqrt{D_z^2 - 4C_z \widetilde{H}}}{2C_z}, \qquad \widetilde{M}_x = \frac{\widetilde{M}_z(a_{n-1}\widetilde{u}_{n-1} + a_n \widetilde{u}_n)}{2NE},$$

$$D_z = \frac{1}{\chi_p} + C_z + \widetilde{H}, \qquad D_x = \frac{1}{\chi_p} + C_x + \widetilde{H}, \qquad E = D_x - C_z \widetilde{M}_z. \tag{9}$$

Parameter a_n can be found from the definition of a_i in Equation (7). Let us remind that the result (9) is obtained in the approximation [55] of the identity of magnetic moments and therefore, of is the identity of magnetization **M** of all of the particles in the chain.

Combining the relations (3), (4), and (9), we come to the system of the linear algebraic equations with respect to the dimensionless displacement $\widetilde{u}_i$:

$$2\left(18 + \beta \widetilde{M}_z^2\right)\widetilde{u}_1 - \beta \widetilde{M}_z^2 \widetilde{u}_2 = 36\gamma,$$

$$-\beta \widetilde{M}_z^2 \widetilde{u}_{i-1} + 2\left(18 + \beta \widetilde{M}_z^2\right)\widetilde{u}_i - \beta \widetilde{M}_z^2 \widetilde{u}_{i+1} = 36\gamma i, \qquad 1 < i < n, \tag{10}$$

$$-\beta \widetilde{M}_z^2(a_{n-1} + 4NE)\widetilde{u}_{n-1} + \left[4NE\left(36 + \beta \widetilde{M}_z^2\right) - \beta \widetilde{M}_z^2 a_n\right]\widetilde{u}_n = 144\gamma nNE.$$

This system can be solved either analytically, or numerically.
The total shear stress σ in the composite that is placed in magnetic field can be presented as:

$$\sigma = \sigma^{(0)} + \sigma^{(m)}. \tag{11}$$

Here, $\sigma^{(0)}$ and $\sigma^{(m)}$ are the nonmagnetic and magnetic parts of the stress that are produced by the aggregates. The nonmagnetic part of the stress appears because of the local inhomogeneous deformations of the elastic matrix, caused by the presence of the chains.

Nowadays, there is no a strict theoretical description of the elastic interaction of the chain with the environment. That is why we estimate $\sigma^{(0)}$ by using the approach [61], modeling the N-particles chain as a prolate ellipsoid of revolution with the minor and major axes equal to the particle diameter d and Nd, respectively. It is of fundamental importance that the volume of this ellipsoid is equal to the total volume of all of the particles in the chain. Therefore, the volume concentration of these ellipsoids is equal to the volume concentration of the particles in the ferrogels.

By using the results of the mechanics of suspensions of ellipsoidal particles [62], in the linear approximation with respect to the shear, we get:

$$\sigma^{(0)} = G_0\gamma + G_0\varphi\gamma \sum_N \left[\alpha_N + \frac{\zeta_N + \beta_N(1+\lambda_N)}{2} \right] F_N. \tag{12}$$

Here, φ is the volume concentration of the particles; α_N, β_N, λ_N, and ζ_N are some coefficients, which depend on the length of the chain. The explicit forms of these coefficients are given in the Appendix A; F_N is a function of distribution over the number N of the particles in the chains, normalized so that $\sum_N F_N = 1$. This function depends on many factors and features of the composite synthesis (size and concentration of the particles, viscosity and kinetics of the host polymer curing, the strength of the field of polymerization, etc.). The determination of the function F_N presents a separate problem. Theoretical study of evolution over time of this function in the magnetic suspensions with a permanent viscosity of the currier liquid has been done in [63]. This model is based on the analysis of a system of the Smoluchowski equations, which describes the kinetic of the aggregation of the chains with a various number of particles.

In the case of cured magnetic gel, this evolution must be studied by taking into account change and time of the rheological properties of the host polymer. Depending on the molecular structure of the polymer, the concentration, and the chemical properties of the curing agent, these properties of evolution can obey to different laws, which can hardly be presented in a general form. That is why here, we suppose that the function F_N is known from either independent experiments or theoretical analysis.

By using the results [64,65] for the macroscopic stress in a system of chain-like polymer macromolecules, we get the following estimate for the magnetic stress $\sigma^{(m)}$:

$$\sigma^{(m)} = -\frac{2\varphi}{v_p} \sum_N \frac{1}{N} \left[\sum_{i=1}^{n} f_{i,i-1}^{(m)} d \right] F_N. \tag{13}$$

Here, v_p is the volume of the particle.

By definition, the shear modulus of the composite G:

$$G = \frac{\sigma}{\gamma}. \tag{14}$$

Let us introduce the dimensionless stresses:

$$\tilde{\sigma} = \frac{\sigma}{G_0}, \quad \tilde{\sigma}^{(0)} = \frac{\sigma^{(0)}}{G_0}, \quad \tilde{\sigma}^{(m)} = \frac{\sigma^{(m)}}{G_0}, \quad \tilde{\sigma} = \tilde{\sigma}^{(0)} + \tilde{\sigma}^{(m)},$$
$$\tilde{\sigma}^{(0)} = \gamma + \varphi\gamma \sum_N \left[\alpha_N + \frac{\zeta_N + \beta_N(1+\lambda_N)}{2} \right] F_N,$$
$$\tilde{\sigma}^{(m)} = -12\varphi \sum_N \frac{1}{N} \left[\sum_{i=1}^{n} \tilde{f}_{i,i-1}^{(m)} \right] F_N = -12\varphi \sum_N \frac{1}{N} \sum_{i=1}^{n} \left[\frac{\beta \tilde{M}_z \tilde{M}_x}{24} + \frac{\beta \tilde{M}_z^2(\tilde{u}_{i-1}-\tilde{u}_i)}{12} \right] F_N \tag{15}$$
$$= \varphi\beta \sum_N \frac{\tilde{M}_z^2}{N} \left[\tilde{u}_n - \frac{n(a_{n-1}\tilde{u}_{n-1} + a_n\tilde{u}_n)}{4NE} \right] F_N,$$

and the dimensionless shear modulus of the composite $\widetilde{G}$:

$$\widetilde{G} = \frac{G}{G_0} = \frac{\widetilde{\sigma}}{\gamma}. \tag{16}$$

The shear modulus $\widetilde{G}^{(0)}$ of the composite without a magnetic field can be found from Equation (12):

$$\widetilde{G}^{(0)} = \frac{\widetilde{\sigma}^{(0)}}{\gamma} = 1 + \varphi \sum_N \left[\alpha_N + \frac{\zeta_N + \beta_N(1 + \lambda_N)}{2} \right] F_N. \tag{17}$$

The magnetically induced part $\Delta\widetilde{G}$ of the shear modulus can be calculated as (14):

$$\Delta\widetilde{G} = \widetilde{G} - \widetilde{G}^{(0)} = \frac{\widetilde{\sigma}^{(m)}}{\gamma} = \frac{\varphi\beta}{4E} \sum_N \frac{\widetilde{M}_z^2}{N^2} [(4NE - na_n)x_n - na_{n-1}x_{n-1}] F_N, \qquad x_i = \frac{\widetilde{u}_i}{\gamma}. \tag{18}$$

For the chain with each number N of the particles, the dimensionless displacements x_i are determined from the system of Equation (10), which can be presented in the form:

$$2\left(18 + \beta\widetilde{M}_z^2\right)x_1 - \beta\widetilde{M}_z^2 x_2 = 36,$$
$$-\beta\widetilde{M}_z^2 x_{i-1} + 2\left(18 + \beta\widetilde{M}_z^2\right)x_i - \beta\widetilde{M}_z^2 x_{i+1} = 36i, \qquad 1 < i < n, \tag{19}$$
$$-\beta\widetilde{M}_z^2(a_{n-1} + 4NE)x_{n-1} + \left[4NE\left(36 + \beta\widetilde{M}_z^2\right) - \beta\widetilde{M}_z^2 a_n\right]x_n = 144nNE.$$

Substituting the solution of Equations (19) into (18), we determine the dimensionless modulus $\Delta\widetilde{G}$.

3. Results

In this part, we compare the results of our calculations with the experiments of [44]. It should be noted that the distribution function F_N over number N of particles in the chains has not been determined in [44]. That is why we used the simplest approximation that all chains consist of an identical number N of particles. This number has been determined from the condition of the best agreement between the calculated and the measured [44] values of the "no field" modulus $G^{(0)}$. Some of the results of this comparison and the estimated magnitudes of N are given in Table 1.

Table 1. Comparison between the experimental [44] and our theoretical results for shear modulus $G^{(0)}$ (no field is applied).

Volume Concentration of the Particles	Experimental [45] Shear Modulus of the Composite without a Magnetic Field (MPa)	Theoretical Shear Modulus of the Composite without a Magnetic Field (MPa)	Estimated Number of Particles N in the Chain
10% (V/V) iron	0.26	0.27	9
20% (V/V) iron	0.74	0.81	13

With respect to the real systems, the estimated N can be considered as a characteristic number of the particles per chain.

A comparison of our calculations of the magnetically induced part ΔG of the shear modulus G with the experiments of [44] are shown in Figure 3.

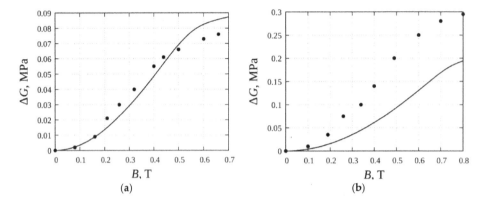

Figure 3. The magnetically induced part ΔG of the shear modulus vs. the flux density B in the composite. Lines—theory, dots—experiment [44]. The shear modulus of the matrix G_0 = 60 kPa; the initial magnetic susceptibility of the particle material χ_p = 100; the saturated magnetization of the particle material M_s = 1670 kA/m; the volume concentration of the particles φ = 0.1 (**a**) and φ = 0.2 (**b**).

For the systems with the relatively low volume concentration of the particles (Figure 3a), our results are in good agreement with experiments. Note that any unjustly fit parameters have not been used in our calculation of the modulus ΔG. For the higher concentrations (Figure 3b), the agreement is worse and rather, is only in the frame of the order of magnitude. The physical reason of the worsening of the agreement between the theory and the experiment lies in the fact that besides the linear chains, more topologically complicated branched, net-like, and bulk structures appear in magnetic suspensions with a high concentration of particles [48–54]. The spatial disposition of the particles is fixed with the host polymer gelation and determines the experimental results for the cured composite. The analysis of morphology of these structures and their effect on the macroscopic properties of the magnetic gels requires a special study. Note that, as a rule, the volume concentration of the particles in ferrogels prepared for biological applications is in the frame of several per cent, or even less than one per cent [9,39,41,66,67]. Figure 3a demonstrates that the present model leads to appropriate results for the low concentrated systems with the internal chains.

Our analysis shows that the elastic modulus of the composite significantly depends on the characteristic number N of the particles in the chains. The calculated dependencies of G on N are shown in Figure 4 for the gels, with two different magnitudes of the elastic modulus G_0 of the polymer matrix.

These results demonstrate that by varying the number N of particles in the chains, one can vary in a wide range of magnitudes, the mechanic modulus of the ferrogel. The relative increase of the modulus under the field action is high in soft gels and is less pronounced in the rigid ones. Biological ferrogels, which are used in various applications, are usually soft, with the modulus less than 10 kPa. Therefore, their mechanical properties and behavior can be effectively controlled with the help of an applied magnetic field.

The characteristic length of the chains is determined by the condition of the gel polymerization, which is in part determined by the ratio between the kinetics of the particle's aggregation and the rate of the host polymer's curing. Thus, by changing the condition of the system gelation, one can tune in a wide range of the magnitudes of the macroscopic properties of the composite material.

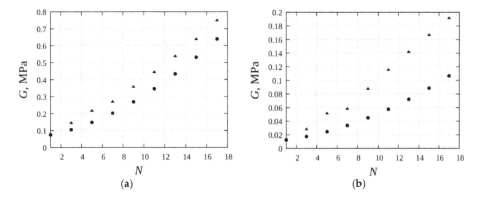

Figure 4. Shear modulus G vs. the total number of particles N. Squares—magnetic field is zero; triangles—magnetic flux $B = 0.7$ T. Parameters of the system: the initial magnetic susceptibility of the material of the particle material $\chi_p = 100$; the saturated magnetization of the particle material $M_s = 1670$ kA/m; the volume concentration of the particles $\varphi = 0.10$. Shear modulus of the elastic matrix $G_0 = 60$ kPa (**a**) and 10 kPa (**b**).

4. Discussion

We present the results of the theoretical modeling of magnetorheological effects in magnetic gels with chain-like aggregates. Unlike the previous theoretical models suggested in [44], this model does not contain any unjustly fit parameters. In the frames of applicability of the hypothesis that only linear chains appear at the stage of the composite synthesis (i.e., that only materials with low or moderate concentrations of particles are considered), the model is in good agreement with the experiments of [44] (see Figure 3a). This agreement indicates that the proposed model leads to adequate results for the composite with the concentration of particles, at least in the frames of ten per cent. For the higher concentrations, the appearance of topologically complicated structures is quite probable, and that is why these concentrations are out of the scope of this model.

Note that the volume concentration of the particles in the magnetic gels that are synthesized for biomedical applications, as a rule, is low and in the frames of several per cent (see, for example [9,39,41,66,67] and the references therein). The systems with higher concentrations are usually synthesized for various mechanical systems (dampers, actuators, etc.).

The obtained results can be considered as a theoretical background for the development of technologies of magnetically controllable biosensors, scaffolds with tunable properties, and for engineering and the regeneration of biological tissues.

In real magnetic gels, the distribution over chain size can be quite broad. At the same time, the assumption of the identity of the chains in the composite, which was used in part 2 of this work, is not necessary for the present model. Indeed, the relation 17 and relation 18 can be used to estimate the elastic modulus if the distribution function F_N is known.

Unfortunately, the law of the size of distribution has not been studied in [44]. That is why, and only because of that, that we have used the model of identical chains to compare our results with the experiments [44]. The determined number N can be considered as an estimate of the characteristic size of the chains in the real composites.

In principle, our approach allows the studying of the large shear deformations, including the rupture of the chains, and Equations (1)–(3) can be solved numerically in nonlinear approximation with respect to the particles' displacement u_i. This can be a natural continuation of the present work.

Author Contributions: A.Z. and D.C. developed the model and made the calculations; D.B. took part in the development of the model and analyzed the results.

Funding: D. Chirikov and A. Zubarev were funded by the Program of the Ministry of Education and Science of the Russian Federation, projects 02.A03.21.0006; 3.1438.2017/4.6; 3.5214.2017/6.7 as well as by the Russian Foundation of Basic Researches, project 18-08-00178. D. Borin was funded by the DFG (Deutsche Forschungsgemeinschaft) under Grant Bo 3343/2-1.

Conflicts of Interest: The authors declare no conflict of interest.

Appendix

The coefficients used in relation to (9) for the symmetric stress $\tilde{\sigma}_s$ read:

$$\alpha_N = \frac{1}{N\alpha_0'}, \ \zeta_N = \frac{4}{N(N^2+1)\beta_0'} - \frac{2}{N\alpha_0'}, \ \beta_N = \frac{2(N^2-1)}{N(N^2\alpha_0+\beta_0)}, \ \lambda_N = \frac{N^2-1}{N^2+1}.$$

Here,

$$\alpha_0 = \frac{2}{3}, \ N = 1,$$

$$\alpha_0 = -\frac{1}{N^2-1}\left[\frac{2}{N} + \frac{\ln\left(2N^2-1-2N\sqrt{N^2-1}\right)}{\sqrt{N^2-1}}\right], \ N > 1,$$

$$\beta_0 = \frac{2}{3}, \ N = 1,$$

$$\beta_0 = \frac{1}{N^2-1}\left[N - \frac{\ln\left(2N^2-1+2N\sqrt{N^2-1}\right)}{2\sqrt{N^2-1}}\right], \ N > 1,$$

$$\alpha_0' = \frac{2}{5}, \ N = 1,$$

$$\alpha_0' = \frac{1}{4(N^2-1)^2}\left[N\left(2N^2-5\right) - \frac{3\ln\left(2N^2-1-2N\sqrt{N^2-1}\right)}{2\sqrt{N^2-1}}\right], \ N > 1,$$

$$\beta_0' = \frac{2}{5}, \ N = 1,$$

$$\beta_0' = \frac{1}{(N^2-1)^2}\left[\frac{N^2+2}{N} - \frac{3\ln\left(2N^2-1+2N\sqrt{N^2-1}\right)}{2\sqrt{N^2-1}}\right], \ N > 1.$$

References

1. Bose, H.; Rabindranath, R.; Ehrlich, J. Soft magnetorheological elastomers as new actuators for valves. *J. Intell. Mater. Syst. Struct.* **2012**, *23*, 989–994. [CrossRef]
2. Filipcsei, G.; Csetneki, I.; Szilagyi, A.; Zrınyi, M. Magnetic Field-Responsive Smart Polymer Composites. *Adv. Polym. Sci.* **2007**, *206*, 137–189. [CrossRef]
3. Boczkowska, A.; Awietjan, S. Tuning active magnetorheological elastomers for damping applications. *Mater. Sci. Forum* **2010**, *636–637*, 766–771. [CrossRef]
4. Dyke, S.; Spencer, B.; Sain, M.; Carlson, J. Modeling and control of magnetorheological dampers for seismic response reduction. *Smart Mater. Struct.* **1996**, *5*, 565–575. [CrossRef]
5. Occhiuzzi, A.; Spizzuoco, M.; Serino, G. Experimental analysis of magnetorheological dampers for structural control. *Smart Mater. Struct.* **2003**, *12*, 703–711. [CrossRef]
6. Carmona, F.; Mouney, C. Temperature-dependent resistivity and conduction mechanism in carbon particle-filled polymers. *J. Mater. Sci.* **1992**, *27*, 1322–1326. [CrossRef]
7. Feller, J.; Linossier, I.; Grohens, Y. Conductive polymer composites: Comparative study of poly(ester)-short carbon fibres and poly(epoxy)-short carbon fibres mechanical and electrical properties. *Mater. Lett.* **2002**, *57*, 64–71. [CrossRef]
8. Bañobre-López, M.; Piñeiro-Redondo, Y.; de Santis, R.; Gloria, A.; Ambrosio, L.; Tampieri, A.; Tampieri, A.; Dediu, V.; Rivas, J. Poly(caprolactone) based magnetic scaffolds for bone tissue engineering. *J. Appl. Phys.* **2011**, *109*, 07B313. [CrossRef]
9. Bock, N.; Riminucci, A.; Dionigi, C.; Russo, A.; Tampieri, A.; Landi, E.; Goranov, V.A.; Marcacci, M.; Dediu, V. A novel route in bone tissue engineering: Magnetic biomimetic scaffolds. *Acta Biomater.* **2010**, *6*, 786–796. [CrossRef] [PubMed]
10. Lin, C.; Metters, A. Hydrogels in controlled release formulations: Network design and mathematical modeling. *Adv. Drug Deliv. Rev.* **2006**, *58*, 1379–1408. [CrossRef] [PubMed]
11. Langer, R. New methods of drug delivery. *Science* **1990**, *249*, 1527–1533. [CrossRef] [PubMed]

12. Mitragotri, S.; Lahann, J. Physical approaches to biomaterial design. *Nat. Mater.* **2009**, *8*, 15–21. [CrossRef] [PubMed]

13. Choi, N.W.; Cabodi, M.; Held, B.; Gleghorn, J.P.; Bonassar, L.J.; Stroock, A.D. Microfluidic scaffolds for tissue engineering. *Nat. Mater.* **2007**, *6*, 908–915. [CrossRef] [PubMed]

14. Kurlyandskaya, G.V.; Fernández, E.; Safronov, A.P.; Svalov, A.V.; Beketov, I.; Beitia, A.B.; García-Arribas, A.; Blyakhman, F.A. Giant magnetoimpedance biosensor for ferrogel detection: Model system to evaluate properties of natural tissue. *Appl. Phys. Lett.* **2015**, *106*, 193702. [CrossRef]

15. Thevenot, J.; Oliveira, H.; Sandre, O.; Lecommandoux, S. Magnetic responsive polymer composite materials. *Chem. Soc. Rev.* **2013**, *42*, 7099–7116. [CrossRef] [PubMed]

16. Hunt, N.C.; Grover, L.M. Cell encapsulation using biopolymer gels for regenerative medicine. *Biotechnol. Lett.* **2010**, *32*, 733–742. [CrossRef] [PubMed]

17. Das, B.; Mandal, M.; Upadhyay, A.; Chattopadhyay, P.; Karak, N. Bio-based hyperbranched polyurethane/Fe$_3$O$_4$ nanocomposites: Smart antibacterial biomaterials for biomedical devices and implants. *Biomed. Mater.* **2013**, *8*, 035003. [CrossRef] [PubMed]

18. De Santis, R.; Gloria, A.; Russo, T.; d'Amora, U.; Zeppetelli, S.; Dionigi, C. A basic approach toward the development of nanocomposite magnetic scaffolds for advanced bone tissue engineering. *J. Appl. Polym. Sci.* **2011**, *122*, 3599–3605. [CrossRef]

19. Gloria, A.; Russo, R.; d'Amora, U.; Zeppetelli, S.; d'Alessandro, T.; Sandri, M.; Bañobre-López, M.; Piñeiro-Redondo, Y.; Uhlarz, M.; Tampieri, A.; et al. Magnetic poly(1-caprolactone)/irondoped hydroxyapatite nanocomposite substrates for advanced bone tissue engineering. *J. R. Soc. Interface* **2013**, *10*, 20120833. [CrossRef] [PubMed]

20. Hu, S.H.; Liu, T.Y.; Tsai, C.H.; Chen, S.Y. Preparation and characterization of magnetic ferroscaffolds for tissue engineering. *J. Magn. Magn. Mater.* **2007**, *310*, 2871–2873. [CrossRef]

21. Hu, H.; Jiang, W.; Lan, F.; Zeng, X.; Ma, S.; Wu, Y.; Gu, Z. Synergic effect of magnetic nanoparticles on the electrospun aligned superparamagnetic nanofibers as a potential tissue engineering scaffold. *RSC Adv.* **2013**, *3*, 879–886. [CrossRef]

22. Lai, K.; Jiang, W.; Tang, J.Z.; Wu, Y.; He, B.; Wang, G.; Gu, Z. Superparamagnetic nano-composite scaffolds for promoting bone cell proliferation and defect reparation without a magnetic field. *RSC Adv.* **2012**, *2*, 13007–13017. [CrossRef]

23. Li, Y.; Huang, G.; Zhang, X.; Li, B.; Chen, Y.; Lu, T.; Lu, T.J.; Xu, F. Magnetic Hydrogels and Their Potential Biomedical Applications. *Adv. Funct. Mater.* **2013**, *23*, 660–672. [CrossRef]

24. Panseri, S.; Cunha, C.; Alessandro, T.; Sandri, M.; Giavaresi, G.; Marcacci, M.; Hung, C.T.; Tampieri, A. Intrinsically superparamagnetic Fe-hydroxyapatite nanoparticles positively influence osteoblast-like cell behavior. *J. Nanobiotechnol.* **2012**, *10*, 32. [CrossRef] [PubMed]

25. Skaat, H.; Ziv-Polat, O.; Shahar, A.; Last, D.; Mardor, Y.; Margel, S. Magnetic Scaffolds Enriched with Bioactive Nanoparticles for Tissue Engineering. *Adv. Healthc Mater.* **2012**, *1*, 168–171. [CrossRef] [PubMed]

26. Tampieri, A.; Landi, E.; Valentini, F.; Sandri, M.; d'Alessandro, T.; Dediu, V.; Marcacci, M. A conceptually new type of bio-hybrid scaffold for bone regeneration. *Nanotechnology* **2011**, *22*, 015104. [CrossRef] [PubMed]

27. Tampieri, A.; d'Alessandro, T.; Sandri, M.; Sprio, S.; Landi, E.; Bertinetti, L.; Panseri, S.; Pepponi, G.; Goettlicher, J.; Bañobre-López, M.; et al. Intrinsic magnetism and hyperthermia in bioactive Fe-doped hydroxyapatite. *Acta Biomater.* **2012**, *8*, 843–851. [CrossRef] [PubMed]

28. Zeng, X.B.; Hu, H.; Xie, L.Q.; Lan, F.; Jiang, W.; Wu, Y.; Gu, Z.W. Magnetic responsive hydroxyapatite composite scaffolds construction for bone defect reparation. *Int. J. Nanomed.* **2012**, *7*, 3365–3378. [CrossRef] [PubMed]

29. Zeng, X.B.; Hu, H.; Xie, L.Q.; Lan, F.; Wu, Y.; Gu, Z.W. Preparation and Properties of Supermagnetic Calcium Phosphate Composite Scaffold. *J. Inorg. Mater.* **2013**, *28*, 79–84. [CrossRef]

30. Zhu, Y.; Shang, F.; Li, B.; Dong, Y.; Liu, Y.; Lohe, M.R.; Hanagatad, N.; Kaskel, S. Magnetic mesoporous bioactive glass scaffolds: Preparation, physicochemistry and biological properties. *J. Mater. Chem. B* **2013**, *1*, 1279–1288. [CrossRef]

31. Ziv-Polat, O.; Skaat, H.; Shahar, A.; Margel, S. Novel magnetic fibrin hydrogel scaffolds containing thrombin and growth factors conjugated iron oxide nanoparticles for tissue engineering. *Int. J. Nanomed.* **2012**, *7*, 1259–1274. [CrossRef] [PubMed]

32. Singh, R.K.; Patel, K.D.; Lee, J.H.; Lee, E.J.; Kim, J.H.; Kim, T.H.; Kim, H.W. Potential of magnetic nanofiber scaffolds with mechanical and biological properties applicable for bone regeneration. *PLoS ONE* **2014**, *9*, e91584. [CrossRef] [PubMed]

33. Lopez-Lopez, M.T.; Scionti, G.; Oliveira, A.C.; Duran, J.D.G.; Campos, A.; Alaminos, M.; Rodriges, I.A. Generation and Characterization of Novel Magnetic Field-Responsive Biomaterials. *PLoS ONE* **2015**, *10*, e0133878. [CrossRef] [PubMed]

34. Nicodemus, G.D.; Bryant, S.J. Cell encapsulation in biodegradable hydrogels for tissue engineering applications. *Tissue Eng. Part B* **2008**, *14*, 149–165. [CrossRef] [PubMed]

35. Ladet, S.; David, L.; Domard, A. Multi-membrane hydrogels. *Nature* **2008**, *452*, 76–79. [CrossRef] [PubMed]

36. Caló, E.; Khutoryanskiy, V.V. Biomedical applications of hydrogels: A review of patents and commercial products. *Eur. Polym. J.* **2015**, *65*, 252–267. [CrossRef]

37. Nair, L.S. (Ed.) *Injectable Hydrogels for Regenerative Engineering*; Imperial College Press: London, UK, 2016.

38. Yun, H.M.; Ahn, S.J.; Park, K.R.; Kim, M.J.; Kim, J.J.; Jinc, G.Z.; Kim, H.W.; Kim, E.C. Magnetic nanocomposite scaffolds combined with static magnetic field in the stimulation of osteoblastic differentiation and bone formation. *Biomaterials* **2016**, *85*, 88–98. [CrossRef] [PubMed]

39. Rodriguez-Arco, L.; Rodriguez, I.A.; Carriel, V.; Bonhome-Espinosa, A.B.; Campos, F.; Kuzhir, P.; Duran, J.D.G.; Lopez-Lopez, M.T. Biocompatible magnetic core-shell nanocomposites for engineered magnetic tissues. *Nanoscale* **2016**, *8*, 8138–8150. [CrossRef] [PubMed]

40. Lopez-Lopez, M.T.; Rodriguez, I.A.; Rodriguez-Arco, L.; Carriel, V.; Bonhome-Espinosa, A.B.; Campos, F.; Zubarev, A.; Duran, J.D.G. Synthesis, characterization and in vivo evaluation of biocompatible ferrogels. *J. Magn. Magn. Mater.* **2017**, *431*, 110–114. [CrossRef]

41. Bonhome-Espinosa, A.B.; Campos, F.; Rodriguez, I.A.; Carriel, V.; Marins, J.A.; Zubarev, A.; Duran, J.D.G.; Lopez-Lopez, M.T. Effect of particle concentration on the microstructural and macromechanical properties of biocompatible magnetic hydrogels. *Soft Matter* **2017**, *13*, 2928–2941. [CrossRef] [PubMed]

42. Coquelle, E.; Bossis, G. Mullins effect in elastomers filled with particles aligned by a magnetic field. *Int. J. Solids Struct.* **2006**, *43*, 7659–7672. [CrossRef]

43. Saxena, P.; Pelteret, J.P.; Steinmann, P. Modelling of iron-filled magneto-active polymers with a dispersed chain-like microstructure. *Eur. J. Mech. A Solids* **2015**, *50*, 132–151. [CrossRef]

44. Jolly, M.R.; Carlson, J.D.; Muñoz, B.C.; Bullions, T.A. The Magnetoviscoelastic Response of Elastomer Composites Consisting of Ferrous Particles Embedded in a Polymer Matrix. *J. Intell. Mater. Syst. Struct.* **1996**, *6*, 613–622. [CrossRef]

45. Danas, K.; Kankanala, S.V.; Triantafyllidis, N. Experiments and modeling of iron-particle-filled magnetorheological elastomers. *J. Mech. Phys. Solids* **2012**, *60*, 120–138. [CrossRef]

46. Butter, K.; Bomans, P.H.H.; Frederik, P.M.; Vroege, G.J.; Philipse, A.P. Direct observation of dipolar chains in iron ferrofluids by cryogenic electron microscopy. *Nat. Mater.* **2003**, *2*, 88–91. [CrossRef] [PubMed]

47. Klokkenburg, M.; Erne, B.H.; Meeldijk, J.D.; Wiedenmann, A.; Petukhov, A.V.; Dullens, R.P.A.; Philipse, A.P. In Situ Imaging of Field-Induced Hexagonal Columns in Magnetite Ferrofluids. *Phys. Rev. Lett.* **2006**, *97*, 185702. [CrossRef] [PubMed]

48. Safronov, A.P.; Terziyan, T.V.; Istomina, A.S.; Beketov, I.V. Swelling and contraction of ferrogels based on polyacrylamide in a magnetic field. *Polym. Sci. Ser. A* **2012**, *54*, 26–33. [CrossRef]

49. Benkoski, J.J.; Deacon, R.M.; Land, H.B.; Baird, L.M.; Breidenich, J.L.; Srinivasan, R.; Clatterbaugh, G.V.; Keng, P.Y.; Pyun, J. Dipolar assembly of ferromagnetic nanoparticles into magnetically driven artificial cilia. *Soft Matter* **2010**, *6*, 602–609. [CrossRef]

50. Cheng, G.; Romero, D.; Fraser, G.T.; Hight Walker, A.R. Magnetic-field-induced assemblies of cobalt nanoparticles. *Langmuir* **2005**, *21*, 12055–12059. [CrossRef] [PubMed]

51. Lopez-Lopez, M.T.; Iskakova, L.Y.; Zubarev, A.Y. To the theory of shear elastic properties of magnetic gels. *Physica A* **2017**, *486*, 908–914. [CrossRef]

52. Lopez-Lopez, M.T.; Borin, D.Y.; Zubarev, A.Y. Shear elasticity of isotropic magnetic gels. *Phys. Rev. E* **2017**, *96*, 022605. [CrossRef] [PubMed]

53. Böse, H.; Röder, R. Magnetorheological Elastomers with High Variability of Their Mechanical Properties. *J. Phys. Conf. Ser.* **2009**, *149*, 012090. [CrossRef]

54. Zubarev, A.Y.; Iskakova, L.Y.; Lopez-Lopez, M.T. Towards a theory of mechanical properties of ferrogels. Effect of chain-like aggregates. *Physica A* **2016**, *455*, 98–103. [CrossRef]

55. Zubarev, A.Y.; Chirikov, D.N.; Borin, D.Y.; Stepanov, G.V. Hysteresis of the magnetic properties of soft magnetic gels. *Soft Matter* **2016**, *12*, 6473–6480. [CrossRef] [PubMed]

56. Coquelle, E.; Bossis, G.; Szabo, D.; Giulieri, F. Micromechanical analysis of an elastomer filled with particles organized in chain-like structure. *J. Mater. Sci.* **2006**, *41*, 5941–5953. [CrossRef]

57. Landau, L.D.; Lifshitz, E.M. *The Classical Theory of Field*; Butterworth-Heinemann: Oxford, UK, 1975; ISBN 978-0-7506-2768-9.

58. Biller, A.M.; Stolbov, O.V.; Raikher, Y.L. Mesoscopic magnetomechanical hysteresis in a magnetorheological elastomer. *Phys. Rev. E* **2015**, *92*, 023202. [CrossRef] [PubMed]

59. Bozorth, R. *Ferromagnetism*; Wiley: New York, NY, USA, 1993; ISBN1 13 978-0780310322, ISBN2 10 0780310322.

60. Landau, L.D.; Lifshitz, E.M. *Electrodynamics of Continuous Media*; Pergamon Press: New York, NY, USA, 1960; ISBN 978-0-08-009105-1.

61. Chirikov, D.N.; Fedotov, S.P.; Iskakova, L.Y.; Zubarev, A.Y. Viscoelastic properties of ferrofluids. *Phys. Rev. E* **2010**, *82*, 051495. [CrossRef] [PubMed]

62. Pokrovskii, V.N. *Statistical Mechanics of Dilute Suspensions*; Nauka: Moscow, Russia, 1978. (In Russian)

63. Bossis, G.; Lancon, P.; Meunier, A.; Iskakova, L.; Kostenko, V.; Zubarev, A. Kinetics of internal structures growth in magnetic suspensions. *Physica A* **2013**, *392*, 1567–1576. [CrossRef]

64. Doi, M.; Edwards, S.F. *The Theory of Polymer Dynamics*; University Press: Oxford, UK; New York, NY, USA, 1986; ISBN 0 19 852033 6.

65. Larson, R.G. *The Structure and Rheology of Complex Fluids*; Oxford University Press: New York, NY, USA, 1999; ISBN 0-19-512197-X.

66. Van Berkum, S.; Dee, J.T.; Philipse, A.P.; Erne, B.H. Frequency-Dependent Magnetic Susceptibility of Magnetite and Cobalt Ferrite Nanoparticles Embedded in PAA Hydrogel. *Int. J. Mol. Sci.* **2013**, *14*, 10162–10177. [CrossRef] [PubMed]

67. Weeber, R.; Hermes, M.; Schmidt, A.M.; Holm, C. Polymer architecture of magnetic gels: A review. *J. Phys. Condens. Matter* **2018**, *30*, 063002. [CrossRef] [PubMed]

Article

Accurate Determination of the Q Quality Factor in Magnetoelastic Resonant Platforms for Advanced Biological Detection

Ana Catarina Lopes [1], Ariane Sagasti [1,*], Andoni Lasheras [2], Virginia Muto [3], Jon Gutiérrez [1,4], Dimitris Kouzoudis [5] and José Manuel Barandiarán [1]

1   BCMaterials, Bld. Martina Casiano, 3rd Floor, UPV/EHU Science Park, Barrio Sarriena s/n, 48940 Leioa, Spain; catarina.lopes@bcmaterials.net (A.C.L.); jon.gutierrez@ehu.eus (J.G.); manu@bcmaterials.net (J.M.B.)
2   Departamento de Matemática Aplicada, Universidad del País Vasco UPV/EHU, Torres Quevedo 1, C.P., 48013 Bilbao, Spain; andoni.lasheras@ehu.eus
3   Departamento de Matemática Aplicada y Estadística e Investigación Operativa, Universidad del País Vasco UPV/EHU, P.O. Box 644, 48080 Bilbao, Spain; virginia.muto@ehu.eus
4   Departamento de Electricidad y Electrónica, Universidad del País Vasco UPV/EHU, P.O. Box 644, 48080 Bilbao, Spain
5   Department of Chemical Engineering, University of Patras, 26504 Patras, Greece; kouzoudi@upatras.gr
*   Correspondence: ariane.sagasti@bcmaterials.net; Tel.: +34-946-128-811

Received: 8 February 2018; Accepted: 14 March 2018; Published: 16 March 2018

Abstract: The main parameters of magnetoelastic resonators in the detection of chemical (i.e., salts, gases, etc.) or biological (i.e., bacteria, phages, etc.) agents are the sensitivity S (or external agent change magnitude per Hz change in the resonance frequency) and the quality factor Q of the resonance. We present an extensive study on the experimental determination of the Q factor in such magnetoelastic resonant platforms, using three different strategies: (a) analyzing the real and imaginary components of the susceptibility at resonance; (b) numerical fitting of the modulus of the susceptibility; (c) using an exact mathematical expression for the real part of the susceptibility. Q values obtained by the three methods are analyzed and discussed, aiming to establish the most adequate one to accurately determine the quality factor of the magnetoelastic resonance.

Keywords: magnetic biosensors; quality factor; magnetoelastic resonance

1. Introduction

Magnetoelastic resonators used as sensing devices present advantages like allowing remote "query and answer" [1,2] as well as low cost and low power consumption [3]. Due to these reasons, chemical and many other parameters can be detected: aqueous chemicals including pH [1], salt, and glucose concentrations [3], as well as inorganic salt depositions [4], gas humidity [5], gases such as carbon dioxide [6], or toxic volatile organic compounds (VOCs) such as benzene or hexane, among others [7]. In recent years, they have become a hot topic as novel wireless biosensors for bacteria, potentially lethal for humans, such as *Salmonella* [8,9], *Bacillus anthracis* [9], or *Escherichia coli* [10]. Such detection will be achieved if the surface of the magnetoelastic resonator is coated with an appropriate smart functionalized film that interacts with the target of interest.

The detection process in such sensors is based on the shift of the magnetoelastic resonance (MER) frequency under the action of an external agent, easily seen when measuring the magnetic susceptibility versus the frequency of the applied magnetic field. In the case of biological agents, the adhesion of different bacteria to the resonators causes an increase in the total mass, which leads to a decrease in the MER (see Figure 1).

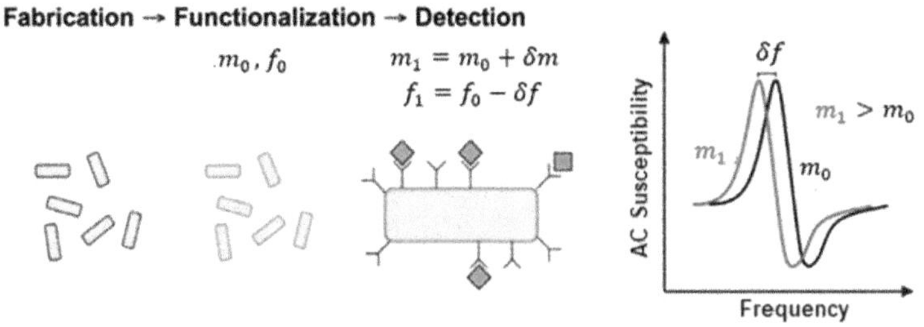

Fabrication → Functionalization → Detection

Figure 1. Principle of detection of biological targets using magnetoelastic resonators. The adhesion of the bacteria to the materials leads to an increase in the total mass of the system, which is detected as a shift (always decreasing) in the measured magnetoelastic resonance frequency.

Among the magnetoelastic materials to be used as biological sensors, Fe-based amorphous ferromagnetic alloys in the form of a ribbon are among the most suitable, mainly due to their high magnetoelastic coupling coefficient (k), high saturation magnetostriction (λ_S), and high saturation magnetization (M_S) [11].

The good performance of a magnetoelastic sensing device is mainly determined by two parameters: sensitivity and quality factor. The sensitivity is related to the lowest detectable frequency change. This depends on the experimental system, but also on the sharpness of the resonance, which in turns depends on the Q factor. In some cases, the determination of Q can be more sensitive than that of the resonance frequency for detecting small changes in mass. With m_0 and f_0 the unloaded mass and corresponding MER frequency of a magnetoelastic resonator, respectively, its sensitivity to a change in mass due to an external target is given by the relationship:

$$S = -\frac{\delta f}{\delta m} = \frac{f_0}{2m_0},$$ (1)

where δf represents the resonance frequency shift caused by the presence of an external agent that causes a change of mass δm, mass change of the MER film. Thus, a high sensitivity value means a large δf shift for a given mass change. The linearity expressed by Equation (1) is valid for small mass changes compared to the initial MER film mass. Nevertheless, it is just an approximation of a more general expression [12] and is still subject to revision and discussion by the authors [13,14].

Concerning the quality factor Q, it has been already experimentally observed that damping strongly affects both resonant frequency and magnetoelastic resonance curve shape (see, for example, [15–17]). A high Q value means a sharp resonance frequency and, consequently, a well-defined f_r resonance frequency. From the measured susceptibility curve around the magnetoelastic resonance, the Q quality factor can be estimated as the resonance curve full bandwidth Δf signal (or full width at half maximum power) relative to its susceptibility maximum frequency f_r, that is:

$$Q_0 = \frac{f_r}{\Delta f},$$ (2)

which is a dimensionless number [18,19]. This classical empirical first approximation can lead to errors as high as 20% in the correct Q value determination, as previously noted by Kaczkowski [20]. Therefore, in biological detection based on the magnetoelastic resonance frequency shift, accurate determination of the Q quality factor beyond the empirical expression Equation (2) turns out to be a key parameter.

In the present work, we present an extensive study of the determination of the Q factor in magnetoelastic resonant platforms. To do this, strips ($L = 4$ cm) of Fe-rich $Fe_{64}Co_{17}Si_{6.6}B_{12.4}$ composition homemade metallic glass have been used. Determination of the Q quality factor value has been performed in three different ways: (a) analyzing the full susceptibility curve around the resonance (real and imaginary components); (b) numerical fitting of the magnitude (modulus) of the susceptibility; and (c) using an exact mathematical expression for the Q value arising from analysis of the real part of the susceptibility curve at the resonance.

2. Experimental

2.1. Material: Magnetic and Magnetostrictive Characterization

In the present study, Fe-based $Fe_{64}Co_{17}Si_{6.6}B_{12.4}$ composition homemade metallic glass ribbons were used. They were prepared by the single roller quenching method in the form of a long ribbon. Equal length strips ($L = 4$ cm) were cut to perform all the magnetic and magnetoelastic characterizations. Room-temperature hysteresis loops were measured by a classical induction method, obtaining a saturation magnetization (given as internal magnetic induction in Tesla) of $\mu_0 M_S \approx 1.6$ T and a magnetic susceptibility $\chi \approx 15,000$. A magnetostriction value of $\lambda_S \approx 22$ ppm was measured using strain gauges connected to an electronic Wheatstone bridge. Figure 2 shows the obtained hysteresis loop and magnetostriction curves.

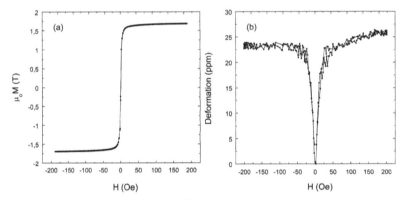

Figure 2. $Fe_{64}Co_{17}Si_{6.6}B_{12.4}$ composition metallic glass magnetic characterization: (**a**) hysteresis loop and (**b**) magnetostriction curve.

2.2. Magnetoelastic Characterization

The metallic glasses of the present study show excellent coupling between magnetic and elastic properties, that is, the applied mechanical stress and the magnetic field generating equivalent effects in the magnetization and deformation of the materials. A direct consequence of the magnetoelastic coupling is the dependence of the elastic constants of magnetostrictive materials on the external magnetic field H, in particular the dependence of the longitudinal Young's modulus on H, known as the ΔE effect ($\Delta E = 1 - E(H)/E_S$, with E_S being the Young's modulus measured at magnetic saturation (a detailed mathematical formula can be found in [21]).

This ΔE effect is easy to measure experimentally through the change in mechanical resonance as a function of the field. The resonance can be excited by an alternating field and detected by the changes in magnetic susceptibility. For this purpose, we used a home-mounted, computer-controlled magnetoelastic resonance detection apparatus [22,23] that automatically changes the DC external applied magnetic field H_{dc}, also known as bias field, and sweeps the frequency of the AC magnetic field H_{ac} in order to drive the sample to magnetoelastic resonance at a given bias. This is the so-called resonance–antiresonance detection method. We use an HP 3589A Spectrum Analyzer in order to quickly measure the magnitude (χ) of the susceptibility curve at the magnetoelastic resonance and store the resonant frequency f_r at the maximum and antiresonance frequency f_a at the minimum signals, together with the signal amplitude at the resonance. Besides the susceptibility χ, we also measure its real (χ') and imaginary (χ'') components separately with the help of a Signal Recovery 7280 Lock-in Amplifier. In these measurements, we extract the frequencies f'_M and f'_m, at the maximum and the minimum signal of the χ', respectively, and to the frequency f''_M at maximum χ''.

All these measured frequencies and in particular the resonance (f_r) vary with the bias field H_{dc}, and so does the Young's modulus, determined as $E(H) = \left[2Lf_r^2(H)\right]^2 \rho$ [24], where L and ρ are the length and density, respectively, of the sample. Other useful magnetoelastic parameters that can be determined from these measurements are the magnetoelastic coupling coefficient $(k = \sqrt{(\pi^2/8)\left(1 - (f_r/f_a)^2\right)})$ [25] and the quality factor of the resonance Q. All such quantities are a function of the applied external magnetic field.

Figure 3 shows the typical external applied magnetic field dependence of Young's modulus $E(H)$ and magnetoelastic coupling coefficient $k(H)$ for our magnetostrictive material. It can be seen that there is a minimum in the $E(H)$ curve that happens at a value of the applied external magnetic corresponding to H_k or effective anisotropy field of the sample. In the same field, the maximum of $k(H)$ and minimum of $Q(H)$ occur. This is an expected behavior since in fact the k value is high when the difference between f_r and f_a is also high, that is, the resonance curve is broad, and so its corresponding Q value (quality of the resonance curve) is poor. While the maximum of the coupling value k guarantees the best sensitivity S of a magnetoelastic resonator working as a biological or chemical sensor [13], the simultaneous occurrence of the worst Q value jeopardizes the accurate determination of the magnetoelastic resonance frequency. Bearing this in mind, we will study the Q factor under external bias field conditions for poor ($k = 0.065$ at 16 *Oe*), medium ($k = 0.176$ at 0.6 *Oe*), and good ($k = 0.282$ at 2 *Oe*) magnetoelastic coupling, aiming to establish the most adequate method (analysis of the real and imaginary part of the susceptibility, numerical fitting of modulus of the susceptibility, and analytical calculations) to determine the quality factor of a magnetoelastic resonance.

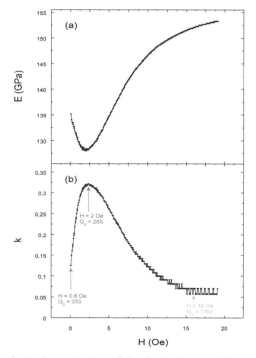

Figure 3. Magnetoelastic characterization of $Fe_{64}Co_{17}Si_{6.6}B_{12.4}$ metallic glass: magnetic field dependence of **(a)** Young's modulus $E(H)$; **(b)** magnetoelastic coupling coefficient $k(H)$.

3. Results: Determination of the Q Quality Factor Value

3.1. Full Susceptibility Curve Analysis at Resonance

Figure 4 shows the magnetic susceptibility modulus (χ) curves for our $Fe_{64}Co_{17}Si_{6.6}B_{12.4}$ ($L = 4$ cm) composition metallic glass, measured around the magnetoelastic resonance frequency for all the applied magnetic field cases under study. In these measurements, the frequency step between consecutive points was 10 Hz.

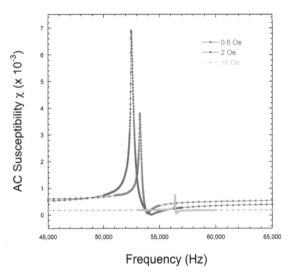

Figure 4. Susceptibility modulus (χ) measured for the $Fe_{64}Co_{17}Si_{6.6}B_{12.4}$ ($L = 4$ cm) composition metallic glass, for all the bias magnetic field cases.

From these susceptibility modulus (χ) curves, we can directly extract the resonance (f_r) and antiresonance (f_a) frequencies and so estimate the magnetoelastic coupling coefficient k and the quality factor Q, from Equation (2), as Table 1 summarizes. As mentioned before, however, the values of Q_0 are quite inaccurate.

Table 1. Resonance and antiresonance frequencies for $Fe_{64}Co_{17}Si_{6.6}B_{12.4}$ and calculated k and Q_0 values determined directly from the experimental data (this last one obtained using Equation (2)).

H (Oe)	f_r (Hz)	f_a (Hz)	Δf	k	Q_0
0.6	53,260	53,940	160	0.176	333
2	52,496	54,280	184	0.282	285
16	56,380	56,476	32	0.065	1762

However, a careful measurement of those magnetoelastic resonance curves, using a Lock-in Amplifier, allows us to record the susceptibility real and imaginary parts (χ' and χ'', respectively) as Figure 5 shows.

With the measured frequencies corresponding to maximum and minimum values of χ' (f'_M and f'_m, respectively, see Figure 5a), we can use the first approximated expression often used to give an accurate value of this Q quality factor [20]:

$$Q_1 \approx \frac{f_r}{f'_m - f'_M} . \tag{3}$$

Table 2 summarizes the experimentally obtained frequency data for maximum and minimum values of χ' (f'_M and f'_m, respectively) and calculated Q_1 values from Equation (3), as well as the relative difference between Q_0 and Q_1.

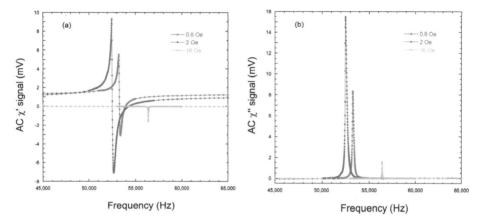

Figure 5. (a) Real, χ', and (b) imaginary, χ'', parts of the magnetic susceptibility for the $Fe_{64}Co_{17}Si_{6.6}B_{12.4}$ ($L = 4$ cm) metallic glass ribbon, for the three bias magnetic field cases under study.

Table 2. Frequencies for the maximum and minimum of χ' (f'_M and f'_m, respectively), calculated Q_1 values (using Equation (3)) and relative error between Q_0 and Q_1.

H (Oe)	f'_M (Hz)	f'_m (Hz)	Q_1	Relative Difference (%) (Q_0, Q_1)
0.6	53,212	53,396	290	13
2	52,456	52,672	243	15
16	56,372	56,412	1410	20

It must be noted that to separately obtain the real and imaginary parts of the susceptibility is a quite difficult and time-consuming experimental task. Therefore, a method to obtain Q based only on the magnitude or modulus of the susceptibility is highly desirable, though it demands more complex numerical treatment afterwards.

3.2. Numerical Fitting of the Magnitude of the Susceptibility Curve

In 1978 Savage et al. [25] derived the following expression for the susceptibility around the magnetoelastic resonance in a free-standing cylinder-shaped sample:

$$\chi(\omega) = \chi_0 \left[1 - \frac{8k^2}{\pi^2} \sum_n \frac{1}{n^2} \times \frac{1}{1 - \frac{\omega_n^2}{\omega^2} + iQ^{-1}\frac{\omega_n}{\omega}} \right], \tag{4}$$

where k is the magnetoelastic coupling coefficient, $\omega_n = 2\pi f_n$ is the frequency of the nth harmonic of the excited fundamental mode ($n = 1$), Q^{-1} is a phenomenological damping coefficient, and χ_0 is the magnetic susceptibility measured at a frequency far below the resonance [26]. Equation (4) also applies to rectangular section ribbons with a proper choice of the k factor. Figure 6 shows an example of the calculated magnetic susceptibility, up to the fifth harmonic, by using Equation (4).

A different approach to estimate Q of a magnetoelastic resonance curve is the numerical fitting of the modulus or magnitude of the experimental susceptibility around its first resonant mode ($n = 1$) by using Equation (4). Thus, we proceed to perform numerical fittings using Mathematica© software (v.11.0), following two different strategies: (a) by using the measured

f_r, f_a, χ_0 values as fixed parameters and (b) by leaving these parameters to vary around the experimentally obtained ones. In both cases, our goal is to search for the optimum Q value that minimizes the L^2 *norm* between the fit and experimental data. We define such a *norm* (also called *residual*) as: $\mathcal{R} = \frac{1}{N} \sum_{i=1,N} \left(\frac{\chi_{exp,i} - \chi_{fit,i}}{\chi_{max}} \right)^2$, where $\chi_{max} = \max\left(\chi_{max,exp}, \chi_{max,fit} \right)$ and N is the number of experimental points. In our measurements, $N = 397$, 691 and 639 for $H = 0.6$, 2 and 16 *Oe*, respectively. With such a definition, $0 \leq \mathcal{R} \leq 1$, and a value of R close to 0 means very good fitting.

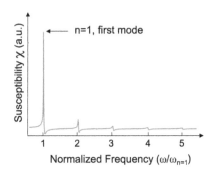

Figure 6. Calculated magnetic susceptibility vs. frequency behavior for a magnetoelastic ribbon, up to the fifth harmonic, using Equation (4).

3.2.1. Numerical Fitting of the Susceptibility Curve Using Fixed Parameters

Figure 7 shows the measured magnetoelastic resonance curve at $H = 2\ Oe$ and the fitted one when procedure a) is used. The only free parameter was the quality factor Q with an initial value of $Q_0 = 285$. The sweep range for Q was 50–380, and the optimum fitting was found for $Q_{fit1} = 178$ (see Figure 7 inset). For all the fits performed, the frequency step between consecutive points will be 1 Hz. The fit of Figure 7 is the best one obtained by following the a procedure. It looks satisfactory, but can still be improved.

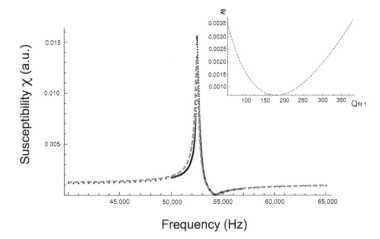

Figure 7. Measured resonance curve at $H = 2\ Oe$ (black dots) and fitted one (magenta marks). The inset shows the *residual* $\mathcal{R}$ change versus Q values, all obtained in calculations using procedure a.

Table 3 summarizes the results obtained for the three different magnetic field cases. While all obtained Q_{fit1} values are lower than the previous Q_0 ones, it is noticeable that the worst fit corresponds to the applied magnetic field, where magnetoelastic coupling is maximum but quality factor is minimum.

Table 3. Q values from the fitting of the χ susceptibility modulus, using fixed experimental parameters (procedure a).

$H\ (Oe)$	$f_r\ (Hz)$	$f_a\ (Hz)$	Q_{fit1}	$\mathcal{R}$
0.6	53,260	53,940	208	0.0041
2	52,496	54,280	178	0.0039
16	56,380	56,476	1067	0.00031

3.2.2. Numerical Fitting of the Susceptibility by Leaving All Parameters Free

Figure 8 shows the measured magnetoelastic resonance curve at $H = 2\ Oe$ and the fitted one when this second procedure is used: all parameters were left free in a range around the starting guess given by the experimental values appearing in Table 1. In this case we swept the Q value in the range 50–440, finding the best fit for $Q_{fit2} = 229$, for the case of applied magnetic field $H = 2\ Oe$ (see Figure 8 inset). Now the fit has greatly improved, as the *norm* values are much lower, especially for the curve with the highest Q value curve (*at* $H = 16\ Oe$) (see Table 4).

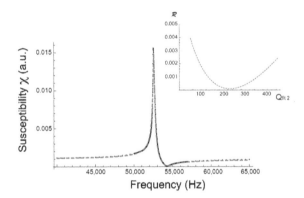

Figure 8. Measured resonance curve at $H = 2\ Oe$ (black dots) and fitted one (magenta marks). The inset shows the *residual* $\mathcal{R}$ change versus Q values, all obtained in calculations when using procedure b.

Table 4 summarizes the results obtained for the three applied magnetic field cases. Again, as with the first simulation procedure, all obtained Q_{fit2} values are lower than the previous Q_0 ones and the worst fit corresponds to the applied magnetic field where magnetoelastic coupling is maximal but the quality factor is minimal.

Table 4. Obtained parameters (resonance and antiresonance frequencies and Q) from the fitting leaving all parameters free (procedure b).

$H\ (Oe)$	$f_r\ (Hz)$	$f_a\ (Hz)$	Q_{fit2}	$\mathcal{R}$
0.6	53,301	53,927	283	0.00027
2	52,566	54,296	229	0.00060
16	56,390	56,466	1321	0.000011

3.3. An Exact Expression for the Q Factor

Starting with Equation (4) and taking into account the shape of the $\chi'(\omega)$ curve, one of us (J.G.) has derive an exact analytical expression for the Q factor value calculation. Since magnetic susceptibility χ can be described by Equation (4) as a complex number, $\chi = \chi' + i\chi''$, we can separate the real and imaginary parts of this complex expression (taking into account only the first harmonic, $n = 1$) as follows:

$$\frac{\chi'}{\chi_0} = 1 - \frac{8k^2}{\pi^2} \frac{\omega^2(\omega^2 - \omega_1{}^2)}{(\omega^2 - \omega_1{}^2)^2 + \left(\frac{\omega\omega_1}{Q}\right)^2}, \tag{5}$$

and

$$\frac{\chi''}{\chi_0} = \frac{8k^2}{\pi^2} \frac{\frac{\omega^3\omega_1}{Q}}{(\omega^2 - \omega_1{}^2)^2 + \left(\frac{\omega\omega_1}{Q}\right)^2}. \tag{6}$$

The real part shows resonance at its maximum: (ω'_M) and antiresonance at its minimum: (ω'_m) (see Figure 5a). It is well known that at the local maxima and minima, the derivative vanishes so we can apply this criteria for the frequencies $\omega = \omega'_M$ and $= \omega'_m$:

$$\frac{d}{d\omega}\left(\frac{\chi'}{\chi^\sigma}\right) = 0 \Rightarrow \frac{d}{d\omega} \frac{\omega^2(\omega^2 - \omega_1{}^2)}{(\omega^2 - \omega_1{}^2)^2 + \left(\frac{\omega\omega_1}{Q}\right)^2} = 0. \tag{7}$$

After a short calculation, and taking into account that $Q > 0$ and single-valued (which means $\omega_M < \omega_r < \omega_m$), the Q factor value can be written in the following two ways:

$$Q = \frac{\omega_m{}^2}{\omega_m{}^2 - \omega_1{}^2} \text{ and } Q = \frac{\omega_M{}^2}{\omega_1{}^2 - \omega_M{}^2}. \tag{8}$$

Solving these two expressions, first for $\omega_M = \omega_r$ and afterwards for Q, we finally get an exact analytical expression for the Q factor value:

$$Q_{cal} = \frac{\omega_m^2 + \omega_M^2}{\omega_m^2 - \omega_M^2} = \frac{1 + u^2}{1 - u^2}, \tag{9}$$

where $u = \omega'_M/\omega'_m = f'_M/f'_m$. Table 5 summarizes the experimentally measured values for those frequencies and the subsequently obtained Q_{cal} values.

Table 5. $Fe_{64}Co_{17}Si_{6.6}B_{12.4}$ experimentally obtained data for resonance and antiresonance frequencies of the real part of the magnetic susceptibility, and calculated Q values using Equation (9).

H (Oe)	f_M (Hz)	f_m (Hz)	Q_{cal}
0.6	53,212	53,396	290
2	52,456	52,672	243
16	56,372	56,412	1410

4. Discussion

All the Q values obtained by the different calculation procedures explained in this study are summarized in Table 6, while Table 7 shows the estimated errors (in %, calculated as *Error* (%) = $\left|\frac{Q_1 - Q}{Q_1}\right| \cdot 100$) among those obtained Q values. The first thing to notice is that in all cases (or calculation procedures) the higher the estimated Q value, the higher the error in its determination, and this always corresponds to the highest applied field. This is a direct consequence of the sharpness of the measured resonance curves, as well as of its low amplitude (magnetic susceptibility), as can be seen in Figure 5a.

As predicted previously by other authors [20], this estimated error (if the classical Q_0 definition is used) can be as high as 20%. Surprisingly, the obtained *norm* of the fitting for this high applied field case is the lowest, this is, the numerical fits are the best for this case.

Table 6. Q values for all the applied magnetic field cases, obtained by the different procedures shown in this study.

$H\,(Oe)$	$Q_0 = \frac{f_r}{\Delta f}$	$Q_1 \approx \frac{f_r}{f_m - f_M}$	Q_{fit1}	Q_{fit2}	$Q_{cal} = \frac{f_m^2 + f_M^2}{f_m^2 - f_M^2}$
0.6	333	290	208	283	290
2	285	243	178	229	243
16	1762	1410	1067	1321	1410

Table 7. Comparison and estimated errors of the Q values obtained with the different calculation procedures.

$H\,(Oe)$	Error (%) Respect to Q_1 or Q_{cal}		
	Q_0	Q_{fit1}	Q_{fit2}
0.6	13	28	2.4
2	15	27	5.7
16	20	24	6.3

We also found that the values obtained for Q_1 (the value given by Kaczkowski [20]) and Q_{cal} (exact analytical expression) are almost equal. While Kaczkowski's expression was an approximation obtained graphically from the impedance circle of an electrical circuit, our exact formula comes from the analysis of the real part of the magnetic susceptibility around the magnetoelastic resonance.

So, if we only have the possibility of measuring or working with the magnetic susceptibility modulus, numerical fitting of the measured magnetoelastic resonance curve is needed. From the results of the numerical fittings used (as can be observed in Figures 7 and 8 and Table 7), it is clear that the second procedure (leaving all parameters f_r, f_a and χ_0 free) leads to a much better result than the first one, as deduced from the obtained lowest *norm* values for procedure b. That is, we can affirm that the Q_{fit2} value obtained by using fitting procedure b can be considered the best approximation of the true Q quality factor of the magnetoelastic resonance curve.

On the other hand, if we compare the Q values obtained from numerical fits with the approximated Q_1 value given by Kaczkowski or with the Q_{cal} value, the estimated error when using fitting procedure a is always higher than 20%, while for fitting (procedure b) the range is only approximately 2–6%. Thus, Q_1 or Q_{cal} can be taken as approximated initial values when performing a numerical fit in order to get the most accurate Q value of a susceptibility magnetoelastic resonance curve.

It is also noticeable that Q values obtained with fitting (procedure b) are systematically lower than Q_1 or Q_{cal}, but this is a fact that should be expected: Equation (9) gives us the exact quality factor Q of the real part of the magnetic susceptibility around the frequency at which a magnetoelastic resonance happens, with this real part being a sharper curve than the corresponding measured susceptibility modulus. As is already well known, the sharper the curve, the higher the quality factor value. Finally, from the obtained error values, we can affirm that when using a magnetoelastic resonant platform for biological or chemical detection purposes, it is convenient to apply a bias field in the range $0 < H_{bias} < H_k$, searching for a compromise between moderate magnetoelastic coupling and low enough error in Q value determination.

5. Conclusions

We have presented an extensive study of the determination of the Q factor of a magnetoelastic resonance curve. This type of resonance is of great interest in order to fabricate devices for biological or chemical detection purposes. The use of the numerical fitting of the magnetic susceptibility modulus around that magnetoelastic resonance turns out to be a useful tool to give accurate Q quality factor values. These differ by up to 20% compared with the Q values determined by following the classical definition. Comparison with approximated Q value given by Kaczkowski and by the exact analytical solution obtained from the real part of the measured susceptibility shows, as expected, that in these two last cases the quality factor value obtained is always slightly higher than that estimated from the numerical fitting. This is a direct consequence of the fact that, while the numerical fit is performed over the magnetic susceptibility modulus, Kaczkowski's and the exact expression for have been obtained from the real part of that susceptibility curve, which is always sharper than the susceptibility modulus one.

Future work should aim to obtain an analytical expression for the Q quality factor directly obtained from the magnetic susceptibility modulus measured around the magnetoelastic resonance.

Acknowledgments: Ana Catarina Lopes thanks MSCA-IF-2015 (Marie Skłodowska Curie Actions) of the European Union's Horizon 2020 Programme for the received funds under grant agreement no. [701852]. Ariane Sagasti wishes to thank BCMaterials Centre for financial support. Jon Gutierrez, Andoni Lasheras, and José Manuel Barandiarán would like to acknowledge the financial support from the Basque Government under the ACTIMAT project (Etortek 2018 program) and Research Groups IT711-13 project. Dimitris Kouzoudis is thankful for financial support under Erasmus + Mobility Agreement between the University of Patras and the University of the Basque Country (UPV/EHU). Technical and human support provided by SGIker (UPV/EHU, MICINN, GV/EJ, ESF) is gratefully acknowledged.

Author Contributions: A.C. Lopes, A. Lasheras, J. Gutiérrez and J.M. Barandiarán conceived and designed the work and needed measurements; A.C. Lopes, A. Sagasti and A. Lasheras performed the experiments; A.C. Lopes and V. Muto performed the numerical calculations; A. Sagasti, J. Gutiérrez, D. Kouzoudis and J.M. Barandiarán analysed the data; A. Sagasti, A. Lasheras, J. Gutiérrez, D. Kouzoudis and J.M. Barandiarán wrote the manuscript. All authors discussed the results and implications, and commented on the manuscript at all stages. All authors read and approved the final manuscript.

Conflicts of Interest: The authors declare no conflict of interest.

References

1. Grimes, C.A.; Seitz, W.R.; Horn, J.; Doherty, S.A.; Rooney, M.T. A remotely interrogatable magnetochemical pH sensor. *IEEE Trans. Magn.* **1997**, *33*, 3412–3414. [CrossRef]

2. Stoyanov, P.G.; Doherty, S.A.; Grimes, C.A.; Seitz, W.R. A remotely interrogatable sensor for chemical monitoring. *IEEE Trans. Magn.* **1998**, *34*, 1315–1317. [CrossRef] [PubMed]

3. Grimes, C.A.; Mungle, C.S.; Zeng, K.; Jain, M.K.; Dreschel, W.R.; Paulose, M.; Ong, G.K. Wireless magnetoelastic resonance sensors: A critical review. *Sensors* **2002**, *2*, 294–313. [CrossRef]

4. Bouropoulos, N.; Kouzoudis, D.; Grimes, C.A. The real-time, in situ monitoring of calcium oxalate and brushite precipitation using magnetoelastic sensors. *Sens. Actuators B* **2005**, *109*, 227–232. [CrossRef]

5. Grimes, C.A.; Kouzoudis, D.; Dickey, E.C.; Kiang, D.; Anderson, M.A.; Shahidain, R.; Lindsey, M.; Green, L. Magnetoelastic sensors in combination with nanometer-scale honeycombed thin film ceramic TiO_2 for remote query measurement of humidity. *J. Appl. Phys.* **2000**, *87*, 5341–5343. [CrossRef] [PubMed]

6. Cai, Q.Y.; Cammers-Goodwin, A.; Grimes, C.A. A wireless, remote query magnetoelastic CO_2 sensor. *J. Environ. Monit.* **2000**, *2*, 556–560. [CrossRef] [PubMed]

7. Baimpos, T.; Gora, L.; Nikolakis, V.; Kouzoudis, D. Selective detection of hazardous VOCs using zeolite/Metglas composite sensors. *Sens. Actuators A* **2012**, *186*, 21–31. [CrossRef]

8. Lakshmanan, R.S.; Guntupalli, R.; Hu, J.; Kim, D.J.; Petrenko, V.A.; Barbaree, J.M.; Chin, B.A. Phage immobilized magnetoelastic sensor for the detection of *Salmonella typhimurium*. *J. Microbiol. Methods* **2007**, *71*, 55–60. [CrossRef] [PubMed]

9. Huang, S.; Yang, H.; Lakshmanan, R.S.; Johnson, M.L.; Wan, J.; Chen, I.H.; Wikle, H.C., III; Petrenko, V.A.; Barbaree, J.M.; Chin, B.A. Sequential detection of *Salmonella typhimurium* and *Bacillus anthracis* spores using magnetoelastic biosensors. *Biosens. Bioelectron.* **2009**, *24*, 1730–1736. [CrossRef] [PubMed]

10. Ruan, C.; Zeng, K.; Varghese, O.K.; Grimes, C.A. Magnetoelastic immunosensors: Amplified mass immunosorbent assay for detection of *Escherichia coli* O157:H7. *Anal. Chem.* **2003**, *75*, 6494–6498. [CrossRef] [PubMed]

11. Luborsky, F.E. Chapter 6: Amorphous ferromagnets in Ferromagnetic Materials. In *Handbook of Ferromagnetic Materials*; Wohlfart, E.P., Ed.; Elsevier: Amsterdam, The Netherlands, 1980; Volume 1, ISBN 0-444-85311-1.
12. Stoyanov, P.G.; Grimes, C.A. A remote query magnetostrictive viscosity sensor. *Sens. Actuators A* **2000**, *80*, 8–14. [CrossRef]
13. Sagasti, A.; Gutiérrez, J.; Sebastián, M.S.; Barandiarán, J.M. Magnetoelastic resonators for highly specific chemical and biological detection: A critical study. *IEEE Trans. Magn.* **2017**, *53*, 4000604. [CrossRef]
14. Sagasti, A. Functionalized Magnetoelastic Resonant Platforms for Chemical and Biological Detection Purposes. Ph.D. Thesis, University of the Basque Country (UPV/EHU), Leioa, Spain, 2018, unpublished.
15. Grimes, C.A.; Kouzoudis, D.; Ong, K.G.; Crump, R. Thin-film magnetoelastic microsensors for remote query biomedical monitoring. *Biomed. Microdevices* **1999**, *2*, 51–60. [CrossRef]
16. Bravo-Ímaz, I.; García-Arribas, A.; Gorritxategi, E.; Arnaiz, A.; Barandiarán, J.M. Magnetoelastic viscosity sensor for on-line status assessment of lubricant oils. *IEEE Trans. Magn.* **2013**, *49*, 113–116. [CrossRef]
17. Sagasti, A.; Bouropoulos, N.; Kouzoudis, D.; Panagiotopoulos, A.; Topoglidis, E.; Gutierrez, J. Nanostructured ZnO in Metglas/ZnO/Hemoglobin modified electrode to detect the oxidation of the hemoglobin simultaneously by cyclic voltammetry and magnetoelastic resonance. *Materials* **2017**, *10*, 849. [CrossRef] [PubMed]
18. Peterson, P.J.; Anlage, S.M. Measurement of resonant frequency and quality factor of microwave resonators: Comparison of methods. *J. Appl. Phys.* **1998**, *84*, 3392–3402. [CrossRef]
19. Cory, D.; Hutchinson, I.; Chaniotakis, M. *Introduction to Electronics, Signals and Measurements*; Springer: Berlin, Germany, 2006; Chapter 17.
20. Kaczkowski, Z. Piezomagnetic dynamics as a new parameter of magnetostrictive materials and transducers. *Bull. Pol. Acad. Sci.* **1997**, *45*, 19–42.
21. Du Tremolet de Laichesserie, E. *Magnetostriction: Theory and Application of Magnetoelasticity*; CRC Press: Boca Raton, FL, USA, 1993; ISBN 0849369347.
22. Gutiérrez, J.; Barandiarán, J.M.; Nielsen, O.V. Magnetoelastic properties of some Fe-rich Fe-Co-Si-B metallic glasses. *Phys. Status Solidi A* **1989**, *111*, 279–283. [CrossRef]
23. Gutiérrez, J. Propiedades Magnéticas y Magnetoelásticas de Nuevas Aleaciones Amorfas de Interés Tecnológico. Ph.D. Thesis, University of the Basque Country (UPV/EHU), Leioa, Spain, 1992.
24. Landau, L.D.; Lifshitz, E.M. *Theory of Elasticity*; Oxford Pergamon Press: Oxford, UK, 1975; p. 116.
25. Savage, H.; Abbundi, R. Perpendicular susceptibility, magnetomechanical coupling and shear modulus in $Tb_{0.27}Dy_{0.73}Fe_2$. *IEEE Trans. Magn.* **1978**, *14*, 545–547. [CrossRef]
26. Hernando, A.; Madurga, V.; Barandiarán, J.M.; Liniers, M. Anomalous eddy currents in magnetostrictive amorphous ferromagnets: A large contribution from magnetoelastic effects. *J. Magn. Magn. Mater.* **1982**, *28*, 109–116. [CrossRef]

MDPI

St. Alban-Anlage 66

4052 Basel

Switzerland

Tel. +41 61 683 77 34

Fax +41 61 302 89 18

www.mdpi.com

Sensors Editorial Office

E-mail: sensors@mdpi.com

www.mdpi.com/journal/sensors

CPSIA information can be obtained
at www.ICGtesting.com
Printed in the USA
LVHW070230120620
657588LV00009B/492